INTRACELLULAR CHOLESTEROL TRAFFICKING

INTRACELLULAR CHOLESTEROL TRAFFICKING

edited by

T. Y. Chang
Dartmouth Medical School

and

Dale A. Freeman
University of Oklahoma Health Sciences Center

KLUWER ACADEMIC PUBLISHERS
Boston / Dordrecht / London

Distributors for North, Central and South America:
Kluwer Academic Publishers
101 Philip Drive
Assinippi Park
Norwell, Massachusetts 02061 USA
Telephone (781) 871-6600
Fax (781) 871-6528
E-Mail <kluwer@wkap.com>

Distributors for all other countries:
Kluwer Academic Publishers Group
Distribution Centre
Post Office Box 322
3300 AH Dordrecht, THE NETHERLANDS
Telephone 31 78 6392 392
Fax 31 78 6546 474
E-Mail <services@wkap.nl>

 Electronic Services <http://www.wkap.nl>

Library of Congress Cataloging-in-Publication Data

A C.I.P. Catalogue record for this book is available
from the Library of Congress.

Printed on acid-free paper.

Printed in the United States of America

CONTENTS

vi

Sterol Carrier Protein-2 and Cholesterol Transport Proteins

Caveolae and Caveolin

INVITED CONTRIBUTORS

E. Joan Blanchette-Mackie
National Institute of Diabetes and Digestive and
Kidney Diseases
National Institutes of Health

Catherine C.Y. Chang
Department of Biochemistry
Dartmouth Medical School

T.Y. Chang
Department of Biochemistry
Dartmouth Medical School

Robert Farese, Jr.
Gladstone Institute of Cardiovascular Disease and
Department of Medicine
University of California, San Francisco

F. Jeffry Field
Department of Internal Medicine
University of Iowa School of Medicine

Christopher J. Fielding
Cardiovascular Research Institute
University of California, San Francisco

Phoebe E. Fielding
Cardiovascular Research Institute
University of California, San Francisco

Dale A. Freeman
Department of Medicine
University of Oklahoma Health Sciences Center

William J. Johnson
Department of Biochemistry
Allegheny University of the Health Sciences

Yvonne Lange
Department of Pathology
Rush-Presbyterian - St. Lukes Medical Center

viii

Laura Liscum
Department of Physiology
Tufts University School of Medicine

Edward B. Neufeld
Lipid Cell Biology Section
National Institute of Diabetes and Digestive and
Kidney Diseases
National Institutes of Health

Peter G. Pentchev
National Institute of Neurological Disorders and Stroke
National Institutes of Health

Terence J. Scallen
Department of Biochemistry
University of New Mexico Health Sciences Center

Friedhelm Schroeder
Department of Physiology and Pharmacology
Texas A & M University Veterinary Medical College

Udo Seedorf
Institute for Arteriosclerosis Research
Westfalian Wilhems - University

Eric J. Smart
Departments of Physiology and Internal Medicine
University of Kentucky Medical Center

Jerome F. Strauss
Department of Obstetrics and Gynecology
University of Pennsylvania Medical Center

Stephen L. Sturley
Institute of Human Nutrition
Columbia University College of Physicians and Surgeons

Ira Tabas
Department of Molecular Medicine
Columbia University College of Physicians and Surgeons

PREFACE

INTRODUCTION AND RATIONALE FOR INTRACELLULAR CHOLESTEROL TRAFFICKING

This volume is an elaboration of an earlier small meeting held in St. Louis, Missouri. In April 1997, many of the authors met for a two-day meeting devoted entirely to intracellular cholesterol trafficking. The rationale for this meeting was that investigators interested in this topic worked in a variety of fields, and rarely, if ever, all met together. Everybody knew each other's papers but mostly worked in isolation from one another. Understanding of cholesterol trafficking also appeared to have reached the point where it would start to rapidly expand beyond these few laboratories. Understanding of cholesterol trafficking was moving from a largely descriptive science into the molecular age. It seemed a good time to get together and see how much we agreed upon up to this point.

More authors contributed to this volume than attended the St. Louis meeting. That meeting was generously funded by grants from Bristol-Myers Squibb, Merck and Company and Parke-Davis, however, the total funding available limited the size of the meeting. For the book, we are not so limited and have tried to be as inclusive as possible and pretty much invited everyone who is presently active in this area. We were quite fortunate to successfully recruit the authors we sought for each of these chapters. The authors and their contributions can be organized by particular interests and particular areas of expertise.

Although all authors are interested in cholesterol transport, the largest group became interested in this topic by way of studying a human and mouse genetic disease, Niemann-Pick Type C disease. Authors primarily identified with these studies include Blanchette-Mackie, Chang, Liscum, Neufeld and Penchev. Blanchette-Mackie also is the world's most knowledgeable lipid morphologist. The next largest group were persons interested in steroid hormone synthesizing cells and included Freeman, Scallen, Stocco and Strauss. Tabas and Field study other specialized cells, the macrophage and intestinal epithelial cell, respectively. Phillips and Johnson have studied cholesterol efflux from cells and from organelles. Still another group became involved with cholesterol transport through studies of caveolin and include the Fieldings and Smart. Several authors have looked for cholesterol transport mutations including Sturley in his yeast studies, Chang, Liscum, Neufeld and Penchev with humans, mouse and CHO cells. Farese and Seedorf have made and characterized respectively ACAT deficient and SCP_2/SCPx deficient mice. A couple of authors have given us candidate transport proteins. Scallen has long studied SCP_2, in recent times often in collaboration with Strauss. Stocco identified te StAR protein, a proten that translocates cholesterol across mitochondrial membranes. His laboratory and Strauss' alone and together continue to study this fascinating protein. One author, Lange, has studied in a generic sense transport in many cells. Another author, Schroeder, is the resident physical chemist in the lipid transport area. Finally, another group of authors was interested in the acyl-CoA cholesterol acyl transferase enzyme (ACAT) in cells and in whole animals. These authors include Chang, Farese and Tabas. As is obvious from this short overview, many authors at one time or another have worked together and most authors' interests overlap.

One characteristic of this area became quite apparent even at the St. Louis meeting. Individual investigators in this area are much like the blind men describing the elephant. They understand transport through the portion of the elephant they have experienced. At the end of this volume, T.Y. Chang will try to create something like a consensus of the field as it exists today. This was attempted by both editors in St. Louis but should be much more successful now that he can write without the vigorous protests of the other investigators.

Dale A. Freeman

THE STEROL-SPECIFIC REGULATION OF ACAT-1 AND SREBPs IN MAMMALIAN CELLS AND IN LIVER

Ta-Yuan Chang, Catherine C.Y. Chang, and Oneil Lee

Department of Biochemistry, Dartmouth Medical School, Hanover, N.H. 03755; e-mail: Ta.Yuan.Chang@Dartmouth.Edu

KEY WORDS: ACAT, SREBP, SCAP, allosteric regulation, cholesterol sensors

Abstract

Within a single cell, low-density lipoprotein receptor (LDLR), 3-hydroxy-3-methylglutaryl CoA (HMG-CoA) reductase, and Acyl-coenzyme A: cholesterol acyltransferase (ACAT) are three of the major cellular components that control intracellular cholesterol homeostasis. The main mode of sterol-specific regulation of ACAT is at the post-translational level, while a major mode of regulation of LDLR, HMG-CoA reductase, and various other cholesterologenic enzymes is at the transcriptional level; transcription control involves a pair of sterol-regulated protein factors called sterol regulatory element binding proteins (SREBPs). In this chapter, we summarize the molecular studies that suggest how sterol may act to regulate ACAT and SREBPs. We also summarize the current knowledge on the suggested physiological roles of ACAT and SREBPs in livers of intact animals.

CONTENTS

INTRODUCTION

Acyl-coenzyme A: cholesterol acyltransferase (ACAT) is an intracellular enzyme that catalyzes the formation of cholesteryl esters from long chain fatty acyl-coenzyme A and cholesterol. The enzyme activity is found in a variety of tissues. In liver, the enzyme is involved in producing cholesteryl esters that constitute as part of the core lipid in very low density lipoprotein (VLDL); in small intestines, it is involved in converting dietary and biliary cholesterol into cholesteryl esters that constitute as part of the core lipid in chylomicrons. (For recent reviews on this subject, see (1, 2)). VLDL and chylomicrons are the major lipoprotein carriers for triacylglycerol transport in the blood. VLDL catabolizes to form LDL, which is the major lipoprotein carrier for cholesterol transport in the blood. ACAT also plays an important role in macrophages; a dynamic cholesterol-cholesteryl ester cycle involving ACAT and neutral cholesteryl esterase exists in macrophages (3) and in other cell types (4). Under pathophysiological conditions, accumulation of cholesteryl esters as cytoplasmic lipid droplets in macrophages and in smooth muscle cells comprise the early lesions of atherosclerotic plaques.

In the 1970s, the involvement of ACAT in cellular cholesterol homeostasis was demonstrated by the classical studies of Brown, Goldstein, and their co-workers, who elucidated the low-density lipoprotein receptor (LDLR) mediated pathway for regulation of cholesterol metabolism in mammalian cells (5). Their studies showed that LDLR, 3-hydroxy-3-methylglutaryl Co-A (HMG-CoA) reductase, a rate-controlling enzyme in endogenous cholesterol biosynthesis, and ACAT are three of the major cellular components that control the intracellular cholesterol homeostasis. During the last few years, Brown, Goldstein, and co-workers have shown that LDLR and various cholesterologenic enzymes are regulated at the transcriptional level by a pair of transcription factors called sterol regulatory element binding proteins (SREBPs). The SREBP-1 gene was independently discovered by its ability to activate fatty acid biosynthesis and adipocyte differentiation in a rat pre-adipose cell line (6).

A detailed review on various biochemical and cell biological aspects of ACAT is available (7). In addition, a review focusing on molecular aspects of ACAT reaction (8), and a review focusing on isoforms of ACAT gene in mammals, as well as structure-function studies on ACAT-1 (9), are also available. For studies on SREBPs, a detailed review by Brown and Goldstein (10) and a review focusing on the molecular aspects of the SREBP processing steps (11) are available. In this Chapter, we focus on summarizing the new information on molecular mechanisms of sterol-dependent regulation of ACAT in single cells, and discuss the current knowledge on the physiological roles of ACAT in the liver. We also summarize the new information on the mode of sterol dependent regulation of SREBPs in single cell studies, and discuss the physiological roles of SREBPs in mice.

ACAT

ACAT-1 and ACAT-2

The ACAT cDNA was first cloned and functionally expressed in 1993 (12); its homologues in various species have also been cloned (reviewed in (7)). This gene is now designated as ACAT-1. Recently, ACAT-1 gene knock-out mice have been generated (13). The homozygous knock-out mice showed almost negligible amounts of cholesteryl esters in adrenal glands and in peritoneal macrophages, indicating that in mice, ACAT-1 plays a major role in these tissues/cells. In contrast, liver ACAT activity was not reduced in these mice, suggesting that a second ACAT gene may exist, and the structure of the liver ACAT may be different from the one in the adrenal glands and peritoneal macrophages. Additional results showed that ACAT-1 protein is present in liver and small intestines of mice at very low level (14), suggesting that the ACAT-1 gene product may not play a significant role in the liver and in intestines of mice. In human, the functional significance of ACAT-1 protein in various human cell lines and various human tissues has been assessed by using a biochemical approach (15, O Lee, CCY Chang, W Lee, TY Chang, *J. Lipid Res.* 39, august issue) : using the specific polyclonal anti-ACAT-1 antibodies, immunoblot analysis showed that the antibodies cross-reacted with a single 50 kDa protein band in homogenates prepared from various human cells and tissues. Anti-ACAT-1 antibodies quantitatively immunoprecipitated the ACAT-1 protein, and effectively depleted more than 90% of detergent-solubilized ACAT activities from six different human cell lines, including the human hepatoma HepG2 cells, demonstrating that the 50 kDa protein is the major ACAT catalytic component in these cells (15). In the study involving tissue culture cells (15), the detergent used for ACAT enzyme solubilization was deoxycholate, which is a strong and anionic detergent that tends to cause protein denaturation and enzyme inactivation. To assess the functional significance of ACAT-1 protein in various native human tissues, a modified immunodepletion procedure was used, using the milder and zwitterionic detergent CHAPS for solubilizing ACAT activity (O Lee, CCY Chang, W Lee, TY Chang, *J. Lipid Res.* 39, august issue). The results showed that ACAT activity was immunodepleted by approximately 90% in liver (83% in hepatocytes), 98% in adrenal gland, 91% in macrophages, 80% in kidney, and 19% in small intestines, suggesting that ACAT-1 protein plays a major catalytic role in all of the human tissue/cell homogenates examined except in the small intestines. An important difference between the results obtained in human cells and tissues and the results obtained in the ACAT-1 gene knockout mice is that while the ACAT-1 protein may not play a major role in mouse liver (13), (14), it does play a major role in the adult human liver. These results suggest that the relative importance of ACAT-1 and ACAT-2 in the liver may be different in different species. Additional results show that human intestinal ACAT activity is largely resistant to immunodepletion, and is much more sensitive to inhibition by the ACAT inhibitor Dup 128 than human liver ACAT activity (O Lee, CCY Chang, W Lee, TY Chang, *J. Lipid Res.* 39, august issue). These results suggest that most of the human intestinal ACAT activity may be due to the presence of a different ACAT gene. A second ACAT gene has been identified by Farese and colleagues, and by Rudel, Sturley, and colleagues (16), (17). Based on results currently available, it is tempting to speculate that the physiological roles of ACAT-1 and ACAT-2 proteins may be different in different species. This Chapter focuses on recent studies on ACAT-1.

Mode of Sterol-Specific Regulation of ACAT-1

Using ACAT-1 cDNA and anti-ACAT-1 antibodies as tools, molecular evidence has been provided demonstrating that in human fibroblasts cells, HepG2 cells, and Chinese hamster ovary (CHO) cells, the main mode of regulation by LDL bound cholesterol or by oxysterol does not involve the alteration in ACAT-1 protein content (15). Additional experiments showed that cholesterol loading did not affect the ACAT-1 mRNA levels in HepG2 cells (18) or in rabbit liver cells (19). These results show that in various mammalian cells, unlike LDLR and various cholesterologenic enzymes, ACAT-1 is not regulated by sterol at the transcription level.

In various tissue culture cells and in the liver, ACAT protein is mainly located in the ER (15), (19); reviewed in (7). It has been shown that the ACAT activity in the reconstituted vesicle responds in a cooperative (sigmoidal) manner to the cholesterol concentration in the vesicle (reviewed in 7). In addition, using an *in vitro* system, Cheng et al. (20) showed that cholesterol itself can serve as an ACAT activator, in addition to its role as an ACAT substrate, suggesting the possibility that ACAT may be an allosteric enzyme. Based on these results, it has been proposed that the main mode of regulation of ACAT by cholesterol is by allosteric regulation; i.e., the enzyme activity may respond in a cooperative manner to the cholesterol pool in the vicinity of ACAT within the ER membrane (7).

The ACAT enzyme is an integral membrane protein. Its enzyme activity can not be solubilized without the use of detergent. Due to its scarcity, little progress has been made towards purifying the enzyme to homogeneity (reviewed in 7). Recently, using recombinant DNA technology, a CHO cell line that stably expresses the human ACAT-1 protein bearing a hexa-histidine tag at its N-terminus has been established. Using this cell line as the starting material, through the steps of subcellular fractionation, detergent solubilization by the zwitterionic detergent CHAPS, and two different affinity column chromatography (an ACAT-1 monoclonal antibody affinity column and an immobilized metal affinity column), the recombinant ACAT-1 protein has been purified several thousand fold from crude cell extracts. The final step yielded a protein with an apparent molecular weight of 54 kDa; it is enzymatically active and is homogeneous as judged by SDS-PAGE. The ACAT enzyme activity of the purified protein responds to cholesterol concentration present in the vesicle in a highly cooperative (sigmoidal) manner. This result shows that only a single protein is required for ACAT enzyme catalysis; it also strengthens the hypothesis that ACAT is an allosteric enzyme regulated by cholesterol (21).

ER is also the organelle from which newly synthesized cholesterol originates. Most (if not all) of the newly synthesized cholesterol is destined to move to the plasma membrane (PM) (22, 23) ; for a recent review see (24). To minimize interference by ACAT of the sterol translocation process between the ER and the PM, significant esterification of newly synthesized cholesterol by ACAT should not occur, unless the cholesterol concentration in the ER exceeds a certain critical threshold. Conversely, in cells with high intracellular cholesterol trafficking activity, endogenously synthesized cholesterol can constitute a significant portion of the substrate pool for ACAT. The latter situation may exist in cells with high intracellular cholesterol trafficking activity, such as in macrophages. In macrophage-like cells grown under high cholesterol trafficking condition, a dynamic cholesterol-cholesteryl ester cycle exists; most of the cholesteryl ester constituting this cycle is derived from cholesterol endogenously synthesized (4). The allosteric regulation model discussed here provides a plausible mechanism to explain how ACAT could utilize endogenously synthesized cholesterol in efficient manner. A similar situation may exist in hepatocytes, where intracellular cholesterol trafficking activity is high: the allosteric properties of ACAT would allow the enzyme to efficiently utilize sterol endogenously synthesized, to produce cholesteryl esters that serve as part of the core lipid for VLDL synthesis and assembly.

Hepatic ACAT

Liver is the major organ for synthesizing and processing body cholesterol on a daily basis (25). To export cholesterol to the blood stream, liver synthesizes VLDL. VLDL contains triacylglycerol and cholesteryl ester as its core lipid. It has been suggested by various investigators that cholesterol and/or cholesteryl esters in the VLDL may play a certain regulatory role in stimulating the synthesis and secretion of VLDL. This subject has been extensively reviewed elsewhere (26-29). Therefore, only a few representative papers are cited here: in a series of studies, Hamberg and colleagues showed that cholesterol and/or its metabolite(s) is required for secretion of VLDL, in perfused rat liver and in intact rats (30-33). In cultured human hepatocytes, adding cholesterol in the culture medium stimulated the secretion of apo-B (34). In HepG2 cells, Craig et al. (35) demonstrated that ß-VLDL or chylomicron remnants caused an increase in cellular cholesteryl esters and stimulated secretion of newly synthesized apo-B; Sniderman and colleagues (36) showed that in oleate supplemented HepG2 cells, blocking endogenous cholesterol synthesis or blocking ACAT with ACAT inhibitors decreased cholesteryl ester content and decreased apo-B secretion. They also showed that the increase in apo-B secretion correlated better with the cellular cholesteryl ester content than with the rate of cholesteryl ester synthesis (37). The link between cholesteryl ester content and apo-B secretion in HepG2 cells was not always demonstrable (38,39). HepG2 cells only secrete low levels of apo-B, due to a defect in coupling the mobilization of stored triacylglycerol for apo-B secretion (40), (41); the same situation may also exist in primary human hepatocytes (42). It is possible that in cells that exhibit low VLDL synthesis rate, the addition of a large cholesterol source, such as ß-VLDL or chylomicron remnants, or a large amount of the water soluble cholesterol analog 25-hydroxycholesterol, is needed to cause a large increase in cholesteryl ester synthesis rate, which may be needed in order to stimulate VLDL synthesis and secretion in these cells (39).

The concept that there may be a link between cholesteryl ester content and apo-B secretion is further supported by the following studies: in cultured rabbit hepatocytes, Kita and co-workers (43) showed that cholesterol loading increases VLDL production; conversely, blocking endogenous cholesterol synthesis decreases VLDL production; blocking ACAT with ACAT inhibitors blocked VLDL synthesis and secretion; under these conditions, cellular contents of free cholesterol, triacylglycerol and phospholipid did not change. In perfused livers isolated from African green monkeys fed with high cholesterol diet, blocking ACAT with three structurally distinct ACAT inhibitors consistently reduced the secretion of triglyceride-rich lipoproteins in the perfusates (44). In intact animal studies, Krause et al. (45) showed that feeding ACAT inhibitors to casein fed rabbits decreases LDL cholesterol level in the plasma; Huff et al. (46) showed that in miniature pigs fed with a cholesterol-free diet, feeding an ACAT inhibitor decreased apo-B secretion; Wrenn et al. (46) showed that feeding ACAT inhibitors to African green monkeys reduced apo-B containing lipoproteins in their plasma without affecting cholesterol absorption. At present, it is not clear exactly what role cholesteryl esters may play during the VLDL synthesis and assembly process. An early cotranslational lipidation step by cholesteryl esters during the apo-B synthesis process has been postulated (28),29).

It is well-known that treating various species with HMG-CoA reductase inhibitors increases their hepatic LDLR levels. In addition, recent studies have demonstrated that modulation of hepatic apo-B synthesis and secretion appears to be an important mechanism for the reductase inhibitors to lower the plasma concentrations of cholesterol (reviewed in (27)). Conde et al. (47) treated reductase inhibitor to guinea pigs that are producing VLDL at elevated levels, and showed that the reductase inhibitor reduced microsomal ACAT activity, presumably by diminishing its substrate supply, thus causing a reduction in the VLDL synthesis and assembly.

These results support the hypothesis that cholesterol synthesized endogenously efficiently serves as the ACAT substrate; the cholesteryl esters thus produced play a certain regulatory role in stimulating the assembly and secretion of the VLDL particle.

SREBPs

Mode of Sterol-Specific Regulation of SREBPs

Transcriptional control has been implicated as a major mode of regulation by sterol for controlling the LDLR and various cholesterologenic enzymes. A pair of sterol regulatory element binding proteins (SREBP-1 and SREBP-2) (48-50) serve as sterol-specific transcription factors. The mature SREBPs activate transcription by binding to the 10-base pair sterol regulatory element (SRE-1; reviewed in (51)) within the promoters of the genes encoding the LDL receptor and various other cholesterologenic enzymes. The SREBPs are probably also involved in activating transcription of genes encoding the two enzymes involved in the saturated fatty acid synthesis, acetyl CoA carboxylase, and fatty acid synthase (52),53); and the enzyme responsible for unsaturated fatty acid synthesis, stearoyl-CoA desaturase-1, as well as enzymes involved in glycerolipid biosynthesis (54). It is possible that the main function of SREBPs may be to control membrane lipid synthesis (55). Two different genes produce at least three different SREBP proteins: SREBP-1a and SREBP-1c as the two isoform proteins produced from the same gene, and SREBP-2 from a separate gene. SREBP-1a is much more active in stimulating transcription than SREBP1c. In tissue culture cells, the SREBP-1a and SREBP-2 are the two predominant transcripts. The functions of these three proteins are similar but not identical, they seem to act independently (reviewed in (10)). The precursor forms of SREBPs exist as integral membrane proteins in the ER and the nuclear membrane. The N-terminal segment of the SREBP precursor contains the basic helix-loop-helix leucine zipper domain necessary for recognizing the SRE in order to activate transcription. This segment is held from entering the nucleus by an anchor segment formed by two transmembrane sequences and a short loop segment in the ER lumen. When cells are deprived of cholesterol, the SREBP precursor undergoes two sequential proteolytic cleavages (56). The first cleavage occurs within the lumen of the ER, producing an intermediate that contains the SRE-recognition segment but remains attached to the ER membrane through the first transmembrane domain. This cleavage is sensitive to sterols; adding sterols to the medium inactivates this step. The putative enzyme responsible for this cleavage step has been designated as S1P; its identity is unknown at present. A new gene called SREBP activating protein (SCAP) has been shown to be required in this process (57, 58). The SCAP protein, also located in the ER, is believed to activate the first proteolytic cleavage of SREBPs in a sterol-sensitive manner. SCAP is a membrane protein with multiple membrane-spanning segments, five of which show high homology with the membrane spanning domains of HMG-CoA reductase (57). The SCAP protein may contain an element, postulated as "the sterol sensing domain", that senses the cholesterol level in the ER. An aspartic acid-asparagine mutation within the "sterol-sensing domain" converts the SCAP protein to be insensitive to sterol dependent suppression of its normal function (57). The membrane topography of SCAP in the ER is not known at present, but its C-terminus is known to be at the cytoplasmic side of the ER (59). The C-terminus of the SCAP protein contains repeating units of the WD domain, and interacts with the C-terminus of the precursor SREBPs, which is also known to be located at the cytoplasmic side of the ER (60), to form protein complexes (59, 61). As shown by co-immunoprecipitation experiments, this interaction does not seem to be influenced by the cellular cholesterol content in the cells (59, 61); however, the interaction between SREBP and SCAP is necessary for the appropriate action of the S1P (61). A second cleavage of SREBP probably occurs within the first transmembrane domain of the SREBP intermediate, causing the release of the mature SREBP from the membrane (56). This cleavage is caused by a different protease, designated as S2P (62). The S2P is insensitive to sterols. The S1P and S2P cleavages occur in a sequential manner; without the S1P, the S2P does not cleave the precursor SREBPs; without the S2P, the SREBP intermediates accumulate in the membrane and cannot be released to the cytosolic form (56). The S2P enzyme has been identified as a novel metalloprotease that contains multiple transmembrane regions, but it does not contain the putative sterol sensing domain present in HMG-CoA reductase and in SCAP (62).

Fig.1 depicts the current knowledge about the two-step proteolytic cleavage of membrane bound SREBPs that involves S1P, SCAP, and S2P. The identity of the S1P, and the role of sterol in affecting the interplay between these protein components awaits further investigation.

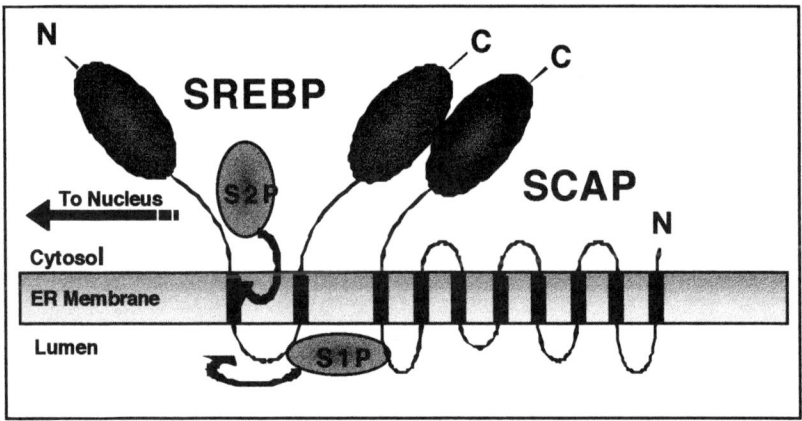

Fig.1. **The two-step sequential proteolysis of the precursor SREBPs to produce the mature transcription factor.** The C-terminus of the precursor SREBP interacts with the C-terminus of the activating protein SCAP in the ER membrane. This interaction is necessary for the proper action of the putative protease S1P. The SCAP protein may contain the sterol-sensing domain. The S1P may act within the lumenal side of the ER. Sterols present in growth medium of cells inactivate the action of S1P. After the first cleavage by S1P, a second protease, the S2P, cleaves the SREBP intermediate within its first transmembrane region, producing the mature, soluble factor that acts in the nucleus to activate transcription rates of various sterol-sensitive genes. The action of S2P is not sensitive to sterols.

Hepatic SREBPs

Studies in intact hamsters show that in liver, SREBP-2 is regulated by sterols as it is in cultured cells, but SREBP-1 is regulated differently (reviewed in (10)). Transgenic mice and gene knockout mice have been produced to further define the functions of the SREBPs in the liver. Mice over-expressing the active SREBP-1a protein caused marked elevations of messages for LDLR, as well as messages for enzymes in cholesterol and fatty acid biosynthesis. As a consequence, these mice over-accumulated cholesterol and triacylglycerols in the liver (63). These results demonstrate the consequences of unregulated expression of SREBP in the liver. The SREBP-1 gene knockout mice were next produced (64): without a functional SREBP-1 gene, most of the mice (-/- mice) died around embryonic day 11. The embryonic lethality in the -/- mouse suggests that the SREBP-1 gene plays a certain important but non-essential function in embryonic development. Fortunately, there were enough mice that survived for analyses. The surviving -/- mice appeared physically normal at birth and throughout the adulthood. Enough mice survived to permit biochemical analyses. These mice showed a moderate elevation in the SREBP-2 mRNA and a moderate increase (2 to 3 fold) in the mature SREBP-2 protein in the liver nuclei; this elevation may be due to a compensatory response for lacking the SREBP-1 gene product (s) in the liver. Sterol synthesis rate was increased approximately 3-fold, and mRNAs for enzymes in the cholesterol synthesis pathway in livers of these mice were significantly elevated (more than several fold). In contrast, the hepatic LDLR level, and mRNAs for enzymes in the saturated and unsaturated fatty acid synthesis were not significantly altered. The fatty acid synthesis rate was in fact decreased. The mRNAs for apoAI, apoB, apoE, and for microsomal triglyceride transfer protein (necessary for hepatic lipoprotein assembly) were unchanged. The available data are consistent with the idea that in the liver, the SREBPs play a dominant role in regulating cholesterol synthesis, and an auxiliary role in regulating the LDLR level and in saturated and unsaturated fatty acid synthesis (10). The surprising result was that so far, none of the manipulations of SREBPs in mouse liver has led to profound changes in plasma cholesterol levels or plasma triacylglycerol levels. The definite answer regarding the roles of SREBP-1 and SREBP-2 in the intact animals awaits the production of mice in which the SREBP-1 and SREBP-2 genes are both inactivated.

CHOLESTEROL FLUX IN AND OUT OF THE ER MAY REGULATE ACAT AND THE SREBP-SCAP COMPLEX

While LDLR is located in the plasma membrane, the HMG-CoA reductase, ACAT, SREBPs, and SCAP, are all mainly located in the ER. It has been suggested by many investigators that each of the cholesterol sensing proteins in the ER may be regulated directly or indirectly by certain specific cholesterol pools in the vicinity of these proteins (see (7) for detailed references). Early studies (reviewed in 51 showed that the LDL-bound cholesterol needs to be released from the lysosomes before it could serve as the regulatory signal. Most of the LDL-bound cholesterol released from the lysosome rapidly emerges at the plasma membrane (reviewed in 65; 7)).

10

Mutations at the NPC-1 gene locus cause this step to be sluggish (66, 67). After cholesterol arrives at the plasma membrane, the NPC-1 gene product may also be involved in mediating the translocation of plasma membrane cholesterol to the ER for esterification by ACAT (68-70); reviewed in (7).

Treating intact mammalian cells with sphingomyelinase is known to cause a translocation of plasma membrane cholesterol to the ER for esterification by ACAT (71; reviewed in 7). Scheek et al. (72) showed that treating plasma membranes of CHO cells with sphingomyelinase causes diminishment of the mature SREBP-2 in the nuclei. These results are consistent with the idea that translocation of plasma membrane cholesterol to the ER causes a sterol specific suppression of the SREBP maturation process; whether this process involves the participation of the NPC-1 gene product is currently unknown.

A working model, assuming that the intracellular cholesterol movement in and out of the ER serves as the main determinant in controlling the activities of ACAT as well as various other cholesterol-sensing proteins in the ER, including HMG-CoA reductase, the SREBP-SCAP complex, has been summarized as a diagram shown in Fig. 2. The cholesterol fluxes towards the ER are depicted as dark lines, cholesterol fluxes towards the PM are depicted as light lines. The early version of this model was discussed in detail in a previous review (7).

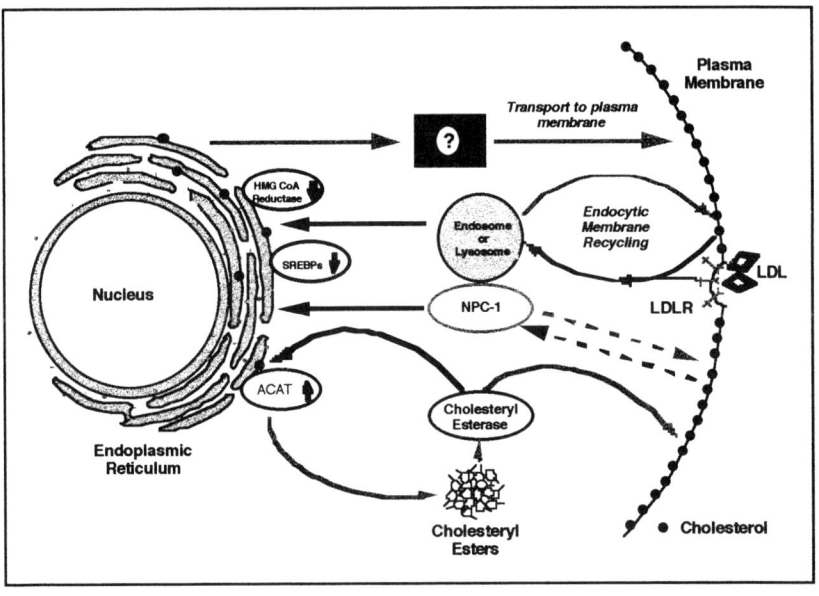

Fig.2. **A working model linking regulation of ACAT, SREBP-SCAP complex, and HMG-CoA reductase with intracellular cholesterol movements**. The cholesterol fluxes towards the ER are depicted as lines towards the left, cholesterol fluxes towards the PM are depicted as lines towards the right.

Most of the LDL-bound cholesterol released from the lysosome rapidly emerges at the plasma membrane. Mutations at the NPC-1 gene locus cause this step to be sluggish. After cholesterol arrives at the plasma membrane, the NPC-1 gene product may also be involved in mediating the translocation of plasma membrane cholesterol to the ER for esterification by ACAT (see text). This model assumes that cholesterol flux in and out of the ER may serve to regulate ACAT and other cholesterol sensors in the ER. Under high cholesterol trafficking condition, a dynamic cholesterol-cholesteryl ester cycle exists. Under this condition, most of the cholesterol ester constituting this cycle is derived from cholesterol endogenously synthesized (4). An early version of this model was described in (7).

The NPC-1 gene in human and in mouse have been identified (73, 74); sequence analyses revealed that the predicted NPC-1 protein contains multiple membrane spanning regions, part of these regions shares extensive homology with the putative "sterol sensing domain" that is present in HMG-CoA reductase and in SCAP (reviewed in 75, 76).

CONCLUSIONS

A simple model, hypothesizing that ACAT may be an allosteric enzyme that responds to a specific cholesterol pool in the ER, has been proposed to be the main mechanism for the sterol-dependent regulation of ACAT. In contrast, the mechanism for sterol-dependent regulation of SREBPs is much more complex, involving interactions between the proteins SCAP, SREBP, and the putative sterol sensitive protease S1P in the ER. Cholesterol flux in and out of certain specialized region(s) of the ER may serve as the key regulatory signal(s) to regulate ACAT and SREBPs. The NPC-1 gene product(s) may be intimately involved in delivering LDL-derived cholesterol to the ER.

To test the validity of the sterol-dependent ACAT allosteric regulation hypothesis, molecular studies on the ACAT protein is necessary. It is essential that the ACAT protein be purified to homogeneity with retention of biological activity. This task has recently been accomplished in the authors' laboratory. It can be expected that detailed structure-activity analysis of ACAT will emerge in the next few years. The molecular studies on SREBPs have been progressing at a vigorous pace. It is anticipated that soon the identity of the putative sterol sensitive protease (S1P) will be revealed, and the molecular basis of interactions between S1P with sterol, SCAP, and SREBPs will soon be understood.

We also discuss the possible roles of ACAT in hepatic VLDL assembly and secretion. The major core lipid in the VLDL is triacylglycerol, not cholesteryl esters. However, cholesteryl esters and triacylglycerol may play different roles in VLDL synthesis and secretion. Two different lipidation steps, one by cholesteryl esters and the other by triacylglycerol during the cotranslational apo-B synthesis process have been postulated (28, 29). The molecular nature of these steps, if exist, needs to be identified.

12

The physiological roles of SREBPs begin to emerge. From the results available thus far, it seems that SREBPs play a dominant role in regulating cholesterol synthesis and an auxiliary role in fatty acid and phospholipid synthesis. In contrast to the studies derived from tissue culture studies, the SREBPs do not seem to control the hepatic LDLR level in a very sensitive manner . These findings are consistent with the early studies of Dietchy and co-workers, who showed that in the animal liver, the hepatic LDLR level is much less sensitive to sterol-specific regulation than enzymes in the cholesterol biosynthetic pathway (77). The exact roles of the SREBP proteins in intact animals will be fully understood when mice containing different SREBP mutations are available for analyses.

REFERENCES

1. Wilson MD and Rudel LL. 1994. *J. Lipid Res*. 35: 943-955
2. Krause BR and Bocan TMA. 1995. ACAT inhibitors: Physiologic mechanisms for hypolipidemic and anti-atherosclerotic activities in experimental animals. In *Inflammation: Mediators and pathways*, ed. R. R. Ruffolo Jr. and M. A. Holliinger, 173-198. Boca Raton: CRC Press, Inc. 173-198 pp.
3. Brown MS, Ho YK and Goldstein JL. 1980. *J. Biol. Chem*. 255: 9344-9352
4. Klansek JJ, Warner GJ, Johnson WJ and Glick JM. 1996. *J. Biol. Chem*. 271: 4923-4929
5. Brown MS and Goldstein JL. 1986. *Science* 232: 34-47
6. Tontonoz P, Kim JB, Graves RA and Spiegelman BM. 1993. *Mol. Cell Biol*. 13: 4753-4759
7. Chang TY, Chang CCY and Cheng D. 1997. *Ann. Rev. Biochem*. 66: 613-638
8. Sturley S. 1997. *Curr .Opin .Lipidol* 8: 167-173
9. Farese RV, Jr. 1998. *Curr. Opin. Lipidol*. 9: 119-124
10. Brown MS and Goldstein JL. 1997. *Cell* 89: 331-340
11. Osborne TF. 1997. *Curr. Biol*. 7: R172-R174
12. Chang CCY, Huh HY, Cadigan KM and Chang TY. 1993. *J. Biol. Chem*. 268: 20747-20755
13. Meiner VL, Cases S, Myers HM, Sande ER, Bellosta S, et al. 1996a. *Proc. Natl. Acad. Sci. U.S.A*. 93: 14041-14046
14. Meiner VL, Tam, C, Gunn, MD, Dong, LM, Weisgraber, KH, Novak, S, Myers, HM, Erickson, SK, and Farese, Jr., RV 1997. *J. Lipid Res*. 38: 1928-1933
15. Chang CCY, Chen J, Thomas M, Cheng D, Del Priore VA, et al. 1995. *J. Biol. Chem*. 270: 29532-29540
16. Cases S, Zheng, YW, Novak, S, Meiner, V,Myers, H, Erickson, SK, and Farese, Jr., RV. 1997. *Circulation* 96(Supplement-I): 230-231 (Abstract)
17. Sturley SL, Oelker, PM, and Behari, A. 1997. *Circulation* 96 (Supplement-I): 411(Abstract)
18. Matsuda H, Hakamada H, Miyazaki A, Sakai M, Chang CCY, et al. 1996. *Biochim. Phiphys. Acta*. 1301: 76-84
19. Khelef N, Buton, X, Beatini, N, Wang, H, Meiner, V Chang, TY, Farese, Jr., RV, Maxfield, FR, and Tabas, I 1998. *J. Biol. Chem*. 273: 11218-11224
20. Cheng D, Chang CCY, Qu XM and Chang TY. 1995. *J. Biol. Chem*. 270: 685-695
21. Chang CCY, Lee, G, Chang, E, Cruz, J, and Chang, TY1998. *FASEB J*. 12: A1387 (Abstract)
22. DeGrella RF and Simoni RD. 1982. *J. Biol. Chem*. 257: 14256-14262
23. Lange Y and Matthies HJG. 1984. *J. Biol. Chem*. 259: 14624-14630
24. Liscum L and Underwood KW. 1995. *J. Biol. Chem*. 270: 15443-15446

25. Dietschy JM, Turley, S.D., and Spady, D.K. 1993. *J. Lipid Res.* 34: 1637-1659
26. Suckling KE and Stange EF. 1985. *J. Lipid Res.* 26: 647-671
27. Thompson GR, Naoumova RP and Watts GF. 1996. *J. Lipid Res.* 37: 439-447
28. Pease RJ and Leiper JM. 1996. *Curr. Opin. Lipidol.* 7: 132-138
29. Huff MW and Burnett JR. 1997. *Curr. Opin. Lipidol.* 8: 138-145
30. Goh EH and Heimberg M. 1977. *J. Biol. Chem.* 252: 2822-2826
31. Khan B, Wilcox, HG, Heimberg, M 1989. *Biochem. J.* 258: 807-816
32. Khan B, Fungwe, TV, Wilcox, HG, and Heimberg, M 1990. *Biochim. Biophys. Acta* 1044: 297-304
33. Fungwe TV, Cagen, L, Wilcox, HG, and Heimberg, M 1992. *J. Lipid Res.* 33: 179-191
34. Kosykh VA, Preobrazhensky SN, Fuki IV, Zaikina OE, Tsibulsky VP, et al. 1985. *Biochim. Biophys Acta* 836: 385-389
35. Craig WY, Nutik R and Cooper AD. 1988. *J. Biol. Chem.* 263: 13880-13890
36. Cianflone KM, Yasruel, Z, Rodriguez, MA, Vas, D, and Sniderman, AD 1990. *J. Lipid Res.* 31: 2045-2055
37. Kohen-Avramoglu R, Cianflone, K, and Sniderman, AD 1995. *J.Lipid Res.* 36: 2513-2528
38. Wu X, Nobuhiro S, Lui E and Ginsberg HN. 1994. *J. Biol. Chem.* 269: 12375-12382
39. Musanti R, Giorgini L, Lovisolo P, Pirillo A, Chiari A and Ghiselli G. 1996. *J. Lipid Res.* 37: 1-14
40. Gibbons GF, Khurana A, Odwell A and Seelaender M. 1994. *J. Lipid Res.* 35: 1801-1808
41. Wu X, Shang A, Jiang H and Ginsberg HN. 1996. *J. Lipid Res.* 37: 1198-1206
42. Lin Y, Smit MJ, Havinga R, Verkade HJ, Vonk RJ and Kuipers F. 1994. *Biochim. Biophys. Acta* 1256: 88-96
43. Tanaka M, Jingami H, Otani H, Cho M, Ueda Y, et al. 1993. *J. Biol. Chem.* 268: 12713-12718
44. Carr TP, Hamilton RLJ and Rudel LL. 1995. *J. Lipid Res.* 36: 25-36
45. Krause BR, Kieft K, Auerbah B, Stanfield R, Bousley R and Bisgaier C. 1994. *Arterioscler Thromb.* 14: 598-604
46. Wrenn J, S.M., Parks JS, Immermann FW and Rudel LL. 1995. *J. Lipid Res.* 36: 1199-1210
47. Conde K, Vergara-Jimenez M, Krause BR, Newton RS and Fernandez ML. 1996. *J. Lipid Res.* 37: 2372-2382
48. Wang X, Briggs MR, Hua X, Yokoyama C, Goldstein JL and Brown MS. 1993. *J. Biol. Chem.* 268: 14497-14504
49. Yokoyama C, Wang X, Briggs MR, Admon A, Wu J, et al. 1993. *Cell* 75: 187-197
50. Hua X, Yokoyama C, Wu J, Briggs MR, Brown MS, et al. 1993. *Proc. Natl. Acad. Sci. U.S.A.* 90: 11603-11607
51. Goldstein JL and Brown MS. 1990. *Nature* 343: 425-430
52. Lopez JM, Bennett MK, Sanchez HB, Rosenfeld JM and Osborne TF. 1996. *Proc. Natl. Acad. Sci. U.S.A.* 93: 1049-1053
53. Bennett MK, Lopez JM, Sanchez HB and Osborne TF. 1995. *J. Biol. Chem.* 270: 25578-25583
54. Ericsson J, Jackson SM, Kim JB, Spiegelman BM and Edwards PA. 1997. *J. Biol. Chem.* 272: 7298-7305
55. Chang TY. 1997. *J. Clinical Invest.* 100: 1905-1906
56. Sakai J, Duncan EA, Rawson RB, Hua X, Brown MS and Goldstein JL. 1996. *Cell* 85: 1037-1046
57. Hua X, Nohturfft A, Goldstein JL and Brown MS. 1996b. *Cell* 87: 415-426
58. Nohturfft A, Hua X, Brown MS and Goldstein JL. 1996. *Proc. Natl. Acad. Sci. U.S.A.* 93: 13709-13714

14

59. Sakai J, Nohturfft, A., Cheng, D., Ho, Y.K., Brown, M.S., and Goldstein, JL 1997. *J. Biol. Chem.* 272: 20213-20221
60. Hua X, Sakai J, Brown MS and Goldstein JL. 1995. *J. Biol. Chem.* 270: 29422-29427
61. Sakai J, Nohturfft A, Goldstein JL and Brown MS. 1998. *J. Biol. Chem.* 273: 5785-5793
62. Rawson RB, Zelenski, N.G., Nijhawan, D., Ye, J., Sakai, J., Hasan, MT, Chang, TY, Brown, MS, and Goldstein, JL. 1997. *Molecular Cell* 1: 47-57
63. Shimano H, Horton JD, Hammer RE, Shimomura I, Brown MS and Goldstein JL. 1996. *J. Clin. Invest.* 98: 1575-1584
64. Shimano H, Shimomura I, Hammer RE, Herz J, Goldstein JL, et al. 1997. *J. Clin. invest.* 100: 2115-2124
65. Tabas I. 1995. *Curr. Opin. Lipidol.* 6: 260-268
66. Pentchev PG, Comly, ME, Kruth, HS, Tokoro, T, Butler, J, Sokol, J, Filling-Katz, M, Quirk, JM, Marshall, DC, Patel, S, Vanier, MT, and Brady, RO 1987. *FASEB J.* 1: 40-45
67. Liscum L and Faust JR. 1987. *J. Biol. Chem.* 262: 17002-17008
68. Pentchev PG, Comley ME, Kruth HS, Vanier MT, Wenger DA, et al. 1985. *Proc. Natl. Acad. Sci. U.S.A.* 82: 8247-8252
69. Byers DM, Morgan MW, Cook HW, St. C. Palmer FB and Spence MW. 1992. *Biochim. Biophys. Acta* 1138: 20-26
70. Spillane DM, Reagan JW, Kennedy NJ, Schneider DL and Chang TY. 1995. *Biochim. Biophys. Acta* 1254: 283-294
71. Slotte JP and E.L. B. 1988. *Biochem. J.* 250: 653-658
72. Scheek S, Brown MS and Goldstein JL. 1997. *Proc. Natl. Acad. Sci. U.S.A.* 94: 11179-11183
73. Carsea ED, Morris JA, Coleman KG, Loftus SK, Zhang D, et al. 1997. *Science* 277: 228-231
74. Loftus SK, Morris JA, Carsea ED, Gu JZ, Cummings C, et al. 1997. *Science* 277: 232-235
75. Liscum L and Klansek JJ. 1998. *Curr. Opin. Lipidol.* 9: 131-135
76. Osborne TF and Rosenfeld JM. 1998. *Curr, Opin. Lipidol.* 9: 137-140
77. Spady DK, Woollett LA and Dietschy JM. 1993. *Ann. Rev. Nutr.* 13: 355-381

INTRACELLULAR CHOLESTEROL MOVEMENT AND HOMEOSTASIS

Yvonne Lange

Department of Pathology, Rush-Presbyterian-St Luke's Medical Center, Chicago, Illinois 60612; e-mail: ylange@rush.edu

ABSTRACT

The distribution of cholesterol in eukaryotic cells is highly non-uniform. The major fraction is localized in the plasma membrane, however, the effectors of cholesterol metabolism are mostly in the endoplasmic reticulum. Homeostasis could be regulated by a sensor of plasma membrane cholesterol which adjusts the pool in the endoplasmic reticulum.

CONTENTS

INTRODUCTION

Cells obtain cholesterol through biosynthesis, from the ingestion of serum lipoproteins, and by the hydrolysis of cholesterol esters stored in cytoplasmic droplets (1). Despite impressive advances in the understanding of the biology of cholesterol, certain basic mechanisms underlying its homeostasis remain to be elucidated. For example, how does a cell know how much cholesterol it has, how much it needs and how to initiate and regulate appropriate adjustments?

In this chapter we shall discuss the distribution and dynamics of intracellular cholesterol. This will be followed by the description of a unifying hypothesis for cholesterol homeostasis.

I. ESTIMATION OF CELLULAR CHOLESTEROL POOLS

It was recognized several years ago that cholesterol is enriched in the plasma membranes of mammalian cells (2,3). However, a precise quantitation of the relative abundance of cholesterol at the cell surface was hindered by the limitations of subcellular fractionation; this technique invariably led to contamination of all membrane fractions with plasma membrane fragments (4). Because of the high cholesterol content of the plasma membrane, such contamination resulted in inflated estimates for the amount of cholesterol in cytoplasmic membranes.

Several alternative approaches to the measurement of membrane cholesterol have now been developed. These utilize filipin, cholesterol oxidase, and cyclodextrin. Each method has advantages and drawbacks, as described below.

i. Filipin

The polyene antibiotic, filipin, has been widely applied as a cytological probe of membrane cholesterol. It has been used to advantage in demonstrating the presence of cholesterol in intracellular membranes such as Golgi and lysosomes (5). As described below, it has also found a use in assessing the penetration of cholesterol oxidase into cells (6,7). On the other hand, its application does not permit quantitation.

ii. Cholesterol oxidase

Ideally, this probe offers the possibility of differentiating between cell surface and intracellular cholesterol and eliminating the former so as to reveal the latter. However, under physiological conditions, plasma membrane cholesterol is not a good substrate for the enzyme. Two methods of increasing reactivity have been described. In the first, suspended, glutaraldehyde-fixed cells are warmed in low ionic strength buffer containing the enzyme (8). In the second, cells are fixed with glutaraldehyde in the culture flask and treated with sphingomyelinase and cholesterol oxidase in physiological buffer (6,7). Both these treatments foster rapid and complete oxidation of plasma membrane cholesterol.

Various control experiments were performed to determine whether intracellular cholesterol is attacked in these reactions. Newly-synthesized cholesterol emerging from the ER was not oxidized in the fixed cells (9-11). With time, the biosynthetic cholesterol became increasingly oxidized, suggesting its transfer to the plasma membrane. Because intracellular membranes bearing nascent cholesterol were later shown not to be a substrate for cholesterol oxidase even in cell homogenates (12), the question of enzyme penetration is not settled by these data.

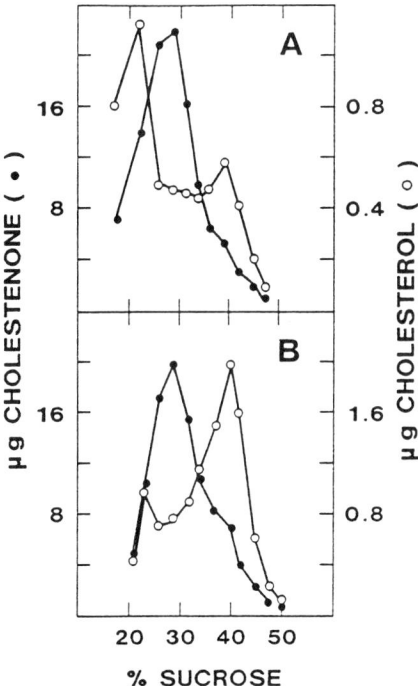

Fig. 1 **Buoyant density profile of cholesterol oxidase-accessible and inaccessible cholesterol**. A homogenate prepared from cholesterol oxidase-treated fibroblasts was treated with digitonin (or ethanol for the control) and analyzed by equilibrium sucrose density gradient centrifugation. The distribution of cholestenone (●) and cholesterol (O) in the control (panel A) and digitonin-treated (panel B) samples are shown. (From ref. 12, with permission).

Further investigation verified that the small amount of unoxidizable cholesterol in intact cells was not associated with the plasma membrane (12). Fibroblasts were treated with cholesterol oxidase by the first of the above methods (*i.e.*, at low ionic strength) and homogenized. Half of the preparation was treated with digitonin, which shifts cholesterol-rich membranes to a higher density (13). The two portions were analyzed in parallel by equilibrium sucrose gradient centrifugation. The profiles of oxidized and unoxidized cholesterol were distinctly different in the minus digitonin control; that is, the former had the density of the plasma membrane whereas the latter was in two peaks, neither of which coincided with the plasma membrane (Fig. 1A). After digitonin treatment, the distribution of cholestenone remained unchanged while a portion of the unoxidized cholesterol was shifted to higher density (Fig. 1B). These data suggest that the unoxidizable fraction of cell cholesterol (i) was not in the plasma membrane; and (ii) was at least in part associated with a cholesterol-enriched membrane. It is likely that such membranes are endosomal or Golgi apparatus.

Fluorescence microscopy of filipin-stained cells has been used to examine the distribution of cell cholesterol and the effect of cholesterol oxidase treatment. Since cholestenone, the oxidation product, does not bind filipin, membranes which have been treated with cholesterol oxidase are no longer fluorescent. Intracellular cholesterol in cells treated with sphingomyelinase plus cholesterol oxidase retained its punctate fluorescence distribution whereas the plasma membrane was no longer visible (6,7). This finding suggests that plasma membrane cholesterol was completely oxidized whereas the enzyme was excluded from the cytosol. In contrast, in the case of cells treated with cholesterol oxidase at low ionic strength, the intracellular cholesterol-rich organelles were no longer visible in filipin-stained cells (7).

The data showing loss of filipin staining in cells treated with cholesterol oxidase at low ionic strength are compelling. However, the underlying significance is puzzling. Is cholesterol in the lumen of these organelles also oxidized? Do the lysosomes open to the cell exterior at low ionic strength? Can a molecule the size of cholesterol oxidase (Mr = 35,000) rapidly penetrate not only the plasma membrane, but also the presumably cross-linked cytosolic matrix of the fixed cell and the lysosomal membranes?

iii. Cyclodextrin

β-Cyclodextrins are water-soluble compounds that can sequester cholesterol in their inner cavities. They promote the rapid removal of a large amount of cholesterol from cells (14,15). Not surprizingly, the extraction of cholesterol eventually leads to membrane instability and cell leakage (14). However, no change in fibroblast permeability was detected after limited extractions with short treatments at low concentration of cyclodextrin (14). Moreover, cytochemical studies using filipin suggested that the agent removed plasma membrane cholesterol without affecting intracellular pools (15).

II. CELL CHOLESTEROL DISTRIBUTION

Cholesterol in the plasma membrane

The fraction of cell cholesterol in the plasma membrane of rat liver cells and BHK cells estimated from the lipid compositions of subcellular fractions together with morphometric measurements was less than 40% and less than 24% respectively, (16) (Table 1). The two methods of cholesterol oxidase treatment described above both reported much higher values, from 85% - 95% (8,9,17). Finally, fractionation of fibroblast homogenates by equilibrium sucrose gradient centrifugation and two-phase aqueous partition also showed that approximately 90% of cell cholesterol was in the plasma membrane (4).

TABLE I

Measurements of plasma membrane cholesterol

Cells	Technique	PM cholesterol	Ref.
		% of total	
Rat liver	Morphometry	< 40	16
BHK	"	< 24	16
HFF	C.O. low μ	94	8
CHO	" "	92	8
Hepatocyte	" "	80	8
MA-10	C.O. low μ	66	17
HFF	" "	89	17
J774 macrophage	C.O. low μ	60-80	9
HFF	C.O. low μ	85	6
HFF	C.O. Smase	85	6
CHO	C.O. low μ	93	7
	C.O. Smase	81	7
HFF	Gradient fractionation	90	4

HFF, human fibroblasts; C.O., cholesterol oxidase; Smase, sphingomyelinase. MA-10, a clonal strain of Leydig tumor cells

Cholesterol in the Golgi apparatus

Golgi cholesterol has not been quantified accurately but has been demonstrated qualitatively in different ways. For example, it was seen as microscopically-visible complexes in cells treated with filipin and digitonin (5). Furthermore, C_6-NBD-ceramide selectively stains the Golgi in a cholesterol-dependent fashion (18). Digitonin, which increases the buoyant density of cholesterol-rich membranes, shifts *trans* Golgi markers (19). At present, the role of Golgi in cholesterol homeostasis is obscure. For example, brefeldin A, an agent which disrupts the Golgi, does not block the movement of nascent cholesterol to the plasma membrane, suggesting that the sterol by-passes this organelle and does not travel together with proteins (20). On the other hand, Golgi cholesterol appears to increase when cholesterol depleted cells are supplied with LDL (18). Furthermore, a recent study using cyclodextrin to probe cellular cholesterol suggested that the

Golgi apparatus could play a role in the transfer of lysosomal cholesterol to the plasma membrane (15).

Cholesterol in the endoplasmic reticulum

We recently developed a novel approach to measuring the size of the ER cholesterol pool (21). The method is based on the premise that the small pool of cholesterol associated with the ER at the moment when cells are homogenized can be determined from its esterification *in vitro*. Fibroblasts were labeled with exogenous [^{3}H]cholesterol, incubated to allow the label to equilibrate with the ER, washed, and homogenized. The homogenates were reacted to completion with oleoyl CoA and the label in the ester fraction determined. Alternatively, homogenates of unlabeled cells were reacted with [^{14}C]oleoylCoA. In both cases, 0.1-2.0% of total cellular cholesterol was esterified when the reaction was allowed to exhaust the cholesterol substrate. A variety of control experiments supported the inference that the pool of cholesterol esterified *in vitro* was just that associated with the ER immediately prior to homogenization. For example, the amount of cholesterol in the ER varied in parallel with the rate of esterification in the intact cells. Thus, the *in vivo* esterification of cholesterol in fibroblasts increased 4-fold when cells maintained in serum-free medium were switched to serum-containing medium for 15 h. The amount of cholesterol attributable to the ER of cells treated in parallel also increased by a factor of approximately four (Fig. 2).

Oxysterols are synthesized in mitochondria and act as potent effectors of sterol homeostasis (22). 25-hydroxycholesterol inhibits cholesterol biosynthesis, and stimulates both steroidogenesis and plasma membrane cholesterol esterification (23,24). The effect on esterification could be due to the known direct activation of ACAT (25). However, several lines of evidence suggest that oxysterols stimulate ACAT activity by raising the level of cholesterol in the ER (21). Indeed, treatment of fibroblasts with 25-hydroxycholesterol causes a linear increase in ER cholesterol mass without a significant delay (Fig. 3). This finding also substantiates the premise that the ER pool turns over rapidly (21). The effect of 25-hydroxycholesterol on steroidogenesis could similarly reflect its effect on the amount of substrate cholesterol in the mitochondria (23).

Other treatments known to alter cholesterol esterification in vivo (e.g., exposure to sphingomyelinase and cholesterol oxidase) increased and decreased the pool measured by this assay *in vitro* in the predicted direction (21).

Cholesterol in the lysosomes

Lysosomal membranes have high levels of phosphatidylcholine and sphingomyelin but very little phosphatidlyserine (26). We recently determined that lysosomes contain ~6% of total cell unesterified cholesterol in normal fibroblasts (27). This value, obtained by subcellular fractionation, included both membrane and lumenal cholesterol.

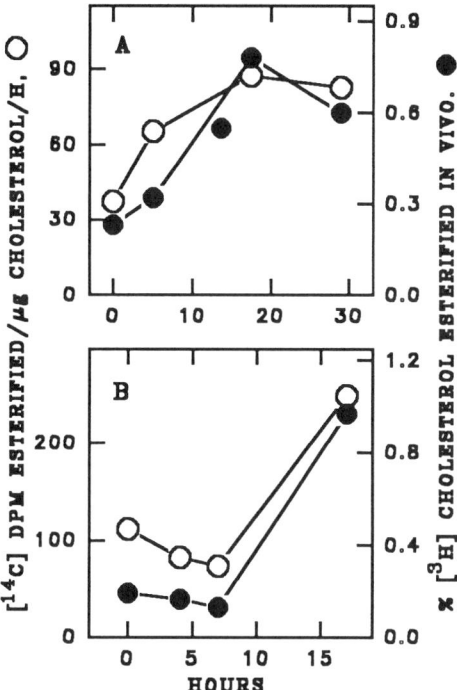

Fig. 2 **Correlation of the rate of cholesterol esterification** *in vivo* **with cholesterol esterification** *in vitro*.   In two experiments fibroblasts were preincubated for 25 h in medium containing 5% lipoprotein-deficient serum.   The cells were processed immediately or incubated in medium containing serum for the times indicated. Two assays were performed: a) conversion of plasma membrane [³H]cholesterol into the ester form *in vivo* (expressed as the fractional rate of esterification in % per hour, right ordinate); b) the mass of cholesterol esterified *in vitro* (calculated from dpm cholesterol ester synthesized/ μg cell cholesterol, left ordinate). (From ref. 21 with permission)

III. CHOLESTEROL MOVEMENT

The movement of nascent sterols to the cell surface

This transfer does not seem to involve a multi-step pathway but rather the movement through (turnover of) an intermediate pool. The half-time for delivery is on the order of 10-20 min (28,29). The process appears to depend on metabolic energy (28) and may by-pass the Golgi apparatus (20). In addition to cholesterol, sterol intermediates such as lanosterol, desmosterol and zymosterol are also rapidly transferred from the ER to the plasma membrane (29).

Inward movement of cell-surface sterols

While nascent cholesterol and that released from lysosomes are used with high efficiency for the biosynthesis of cholesteryl esters, steroids and serum lipoproteins, it is nevertheless the plasma membrane pool which provides the major source of the cholesterol used for these purposes (30,31). Presumably this is because the circulation of plasma membrane cholesterol rapidly flushes out the cytoplasmic pools. We estimate that one plasma membrane equivalent of cholesterol moves each hour through the ER pool (30). Since the ER pool is ~1% of that in the plasma membrane (21), it would appear to turn over at a rate of about once per minute. Exogenous [³H]zymosterol also moves efficiently from the plasma membrane to the ER, where it is converted to [³H]cholesterol. This product is promptly returned to the cell surface where it is accessible to cholesterol oxidase (29). Such data suggest a rapid bi-directional movement of sterols between the cell surface and the ER. This circulation could provide the link between the plasma membrane pool and the cytoplasmic elements which regulate it.

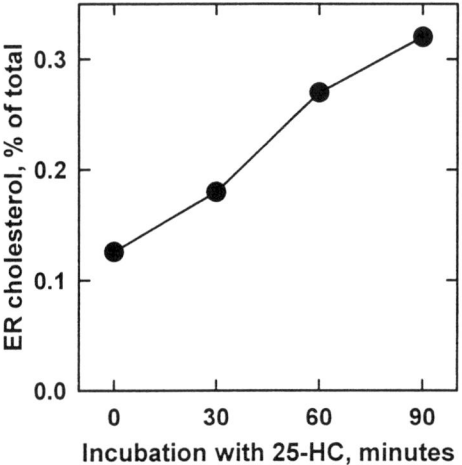

Fig. 3 **The effect of 25-hydroxycholesterol on ER cholesterol**. Replicate flasks of human fibroblasts were preincubated for 24 h in medium containing 5% lipoprotein-deficient serum. At time zero, 20 μg/ml 25-hydroxycholesterol was added and the cultures incubated for the times indicated. The cells were then assayed for ER cholesterol, expressed as % total cell cholesterol, as described (21).

Movement of LDL cholesterol out of lysosomes

Cholesteryl esters in the LDL particles taken up by receptor-mediated endocytosis are hydrolyzed in the lysosomes (1). The cholesterol formed exits the organelle rapidly and is transferred to the plasma membrane with a half-time on the order of an hour (32,33). A small fraction of this cholesterol is re-esterified by ACAT in the ER. However, the intracellular pathway followed by the sterol is controversial. While it is generally accepted that a significant fraction of the cholesterol emerging from lysosomes moves to the plasma membrane prior to esterification (7,15,33), there is evidence both for and against a direct route from lysosomes to ER. Studies of fibroblasts using cyclodextrin as an extracellular cholesterol acceptor provided indirect evidence for such a pathway (15). More recently, cholesterol oxidase treatment, coupled with the use of agents which inhibit various modes of intracellular cholesterol transfer, led to the suggestion that direct transfer from lysosomes to ER occurred in CHO cells (7). On the other hand, studies of esterification kinetics in fibroblasts provided no evidence for this direct route (33).

Movement of plasma membrane cholesterol to lysosomes

We have documented that plasma membrane cholesterol is transferred rapidly to lysosomes. For example, [^{3}H]cholesterol introduced into the plasma membrane of fibroblasts moved to the lysosomes at a rate corresponding to approximately 5% of total cell cholesterol per hour (27). This rate is similar to that reported for other plasma membrane constituents (34). However, it would seem that perhaps 95% of endocytosed cholesterol is returned to the plasma membrane, presumably from an intermediate endosomal compartment.

Thus it appears that cholesterol flows in both directions between the plasma membrane and three of the major cholesterol processing organelles within the cell: the ER, the lysosomes and the mitochondria.

Agents which affect intracellular cholesterol movement

Cells respond to exogenous oxysterols as they do to excess cholesterol. That is, sterol biosynthesis and LDL receptors are down-regulated and esterification of endogenous cholesterol is stimulated (22). Because 25-hydroxycholesterol increases esterification of [^{3}H]cholesterol inserted into the plasma membrane, it has been postulated that it stimulates transfer of this cholesterol to the ER (24). In contrast, a second set of compounds inhibits the transfer of plasma membrane cholesterol to the ER. These agents include nigericin, monensin, lysophosphatidylcholine, chloroquine, 3-β-[2-(diethylamino)ethoxy] androst-5-en-17-one (U18666A), imipramine, colchicine, progesterone and various other steroids (24). There is evidence that the two classes of compounds are antagonistic. The first and second class of compounds have been named Class 1 and Class 2 agents, respectively (24,35).

Class 2 agents affect other aspects of intracellular cholesterol movement as well. In steroidogenic cells, the compounds down-regulate steroidogenesis, presumably by inhibiting transfer of cholesterol from the plasma membrane to the mitochondria (36). Whether the transfer of cholesterol from lysosomes to plasma membrane also is affected is controversial. Some studies reported inhibition of this step (7,37); however, the absence of such an effect has also been observed (Fig. 4, ref. 33).

Long-term consequences of treatment with class 2 compounds

A large number of disparate compounds are known to induce cellular cholesterol accumulation and the formation of lamellar bodies in lysosomes (38). It was shown

24

recently that treatment of cultured cells with progesterone (37) and U18666A (7) caused accumulation of lysosomal cholesterol. The accretion of cholesterol by fibroblasts treated with U18666A or imipramine is linear in time and leads to an approximately three fold increase in total cell unesterified cholesterol over 60 hours. A large fraction of this increment is found in the lysosomes; however, plasma membrane cholesterol also appears to increase. Similar increases in cell cholesterol occur with the other class 2 agents tested to date (27).

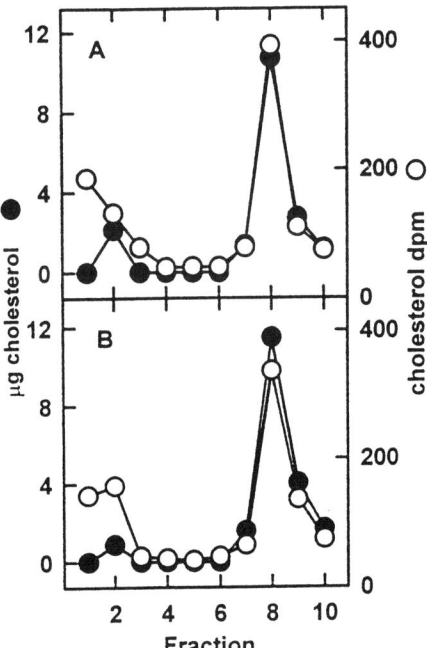

Fig. 4 **Effect of progesterone on the transfer of cholesterol from lysosomes to the plasma membrane**. Replicate flasks of fibroblasts were preincubated for 19 h in medium containing 5% lipoprotein-deficient serum. [³H]cholesteryl ester-labeled LDL was then added to the medium together with 12 µM progesterone or ethanol as solvent control. The cells were incubated for 3 h at 37 °C and then homogenized and analyzed on Percoll gradients for the distributions of cholesterol mass (●) and [³H]cholesterol (○). Fraction 1 was the bottom of the gradient. Panels A and B give the results for the control and progesterone-treated cells respectively. (Taken from ref. 33, with permission.)

Recent studies suggested that the accumulation of lysosomal cholesterol affects the kinetics of transfer to the plasma membrane of [³H]cholesterol ingested in LDL (27). Cells were treated for approximately 16 h with the class 2 agent U18666A, incubated for 1 hour with LDL labeled with [³H]cholesteryl ester and then chased for 1 hour. Transfer to the plasma membrane of [³H]cholesterol released in the lysosomes was measured as in Figure 4. In contrast to the findings illustrated there, less labeled cholesterol was transferred in the treated cells than in the controls. However, when the dilution of the specific activity of the [³H]cholesterol by the significant mass of lysosome cholesterol was taken into account, the actual flux of cholesterol had not changed. This finding could explain the discrepant results of measurements of rates of lysosome cholesterol transfer to the plasma membrane in cells treated with class 2 agents (see above and refs. 7,33,37)

IV. A MODEL FOR CHOLESTEROL HOMEOSTASIS

The rapid flux of cholesterol between the plasma membrane and cell compartments and the dynamic nature of organelle cholesterol bear on possible mechanisms of cholesterol homeostasis. Here is one hypothesis with two essential features (24,35). First, the major pool of cell cholesterol is in the plasma membrane. Second, the ER contains very little cholesterol and is the site of the major regulatory enzymes of cholesterol metabolism. The level of plasma membrane cholesterol could be signalled to the ER by a bidirectional stream of cholesterol. This circulation could be regulated by a sensor which responds to the level of cholesterol in the plasma membrane. Cholesterol fluxes into and out of the ER establish the size of its cholesterol pool which in turn regulates the activity of the key enzymes therein.

Support for this hypothesis comes from the observation that blocking cholesterol transfer from the plasma membrane to the ER by class 2 agents elicits the same response as depletion of cell cholesterol: suppression of esterification (24), stimulation of the genes for the transcription of LDL receptors and 3-hydroxy-3-methylglutaryl coenzyme A reductase (39), and increased activity of the latter (24).

CONCLUSION

Recent findings on cellular cholesterol distribution and trafficking have suggested a unifying model for cholesterol homeostasis. Further experiments will be needed to substantiate this model and to elucidate the nature of the putative cell sensor of cholesterol.

REFERENCES

1. Brown, MS, Goldstein, JL. Receptor-mediated control of cholesterol metabolism. Science 1976;191:150-154
2. Colbeau, A, Nachbaur, J, Vignais, PM. Enzymic characterization and lipid composition of rat liver subcellular membranes. Biochim Biophys Acta 1971;249:462-492
3. Steck, TL, and Wallach, DFH. The isolation of plasma membranes. In *Methods in Cancer Research,* H. Busch, ed. New York, NY: Academic Press, 1970;5:93-153
4. Lange, Y, Swaisgood, MH, Ramos, BV, Steck, TL. Plasma membranes contain half the phospholipid and 90% of the cholesterol and sphingomyelin in cultured human fibroblasts. J Biol Chem 1989;264:3786-3793
5. Blanchette-Mackie, EJ, Dwyer, NK, Amende, LM, Kruth, HS, Butler, JD, Sokol, J, Comly, ME, Vanier, MT, August, JT, Brady, RO, Pentchev, PG. Type-C Niemann-Pick disease: low density lipoprotein uptake is associated with premature cholesterol accumulation in the Golgi complex and excessive storage in lysosomes. Proc Nat'l Acad Sci 1988;85:8022-8026

6. Porn, MI, Slotte, JP. Localization of cholesterol in sphingomyelinase-treated fibroblasts. Biochem J 1995;308:269-274

7. Underwood, KW, Jacobs, NL, Howley, A, Liscum, L. Evidence for a cholesterol transport pathway from lysosomes to endoplasmic reticulum that is independent of the plasma membrane. J Biol Chem 1998;273:4266-4274

8. Lange, Y, Ramos, BV. Analysis of the distribution of cholesterol in the intact cell. J Biol Chem 1983;258:15130-15134

9. Tabas, I, Rosoff, WJ, Boykow, GC. Acyl Coenzyme A: cholesterol acyl transferase in macrophages utilizes a cellular pool of cholesterol oxidase-accessible cholesterol as substrate. J Biol Chem 1988;263:1266-1272

10. Lange, Y, Matthies, HJG. Transfer of cholesterol from its site of synthesis to the plasma membrane. J Biol Chem 1984;259:14624-14630

11. Field, JE, Born, E, Murthy, S, Mathur, SN. Transport of cholesterol from the endoplasmic reticulum to the plasma membrane is constitutive in CaCo-2 cells and differs from the transport of plasma membrane cholesterol to the endoplasmic reticulum. J Lipid Res 1998;39:333-343

12. Lange, Y. Disposition of intracellular cholesterol in human fibroblasts. J Lipid Res 1991;32:329-339

13. Lange, Y, Steck, TL. Cholesterol-rich intracellular membranes: a precursor to the plasma membrane. J Biol Chem 1985;260:15592-15597

14. Kilsdonk, EPC, Yancey, PG, Stoudt, GW, Bangerter, FW, Johnson, WJ, Phillips, MC, Rothblat, GH. Cellular cholesterol efflux mediated by cyclodextrins. J Biol Chem 1995;270:17250-17256

15. Neufeld, EB, Cooney, AM, Pitha, J., Dawidowicz, EA, Dwyer, NK, Pentchev, PG, Blanchett-Mackie, EJ. Intracellular trafficking of cholesterol monitored with a cyclodextrin. J Biol Chem 1996;271:21604-21613

16. van Meer, G. Plasma membrane cholesterol pools. Trends Biol Sci 1987;12:375-376

17. Freeman, DA. Cyclic AMP mediated modification of cholesterol traffic in Leydig tumor cells. J Biol Chem 1987;262:13061-13068

18. Martin, OC, Comly, ME, Blanchette-Mackie, EJ, Pentchev, PG, Pagano, RE. Cholesterol deprivation affects the fluorescence properties of a ceramide analog at the Golgi apparatus of living cells. Proc. Nat'l Acad Sci USA 1993;90:2661-2665

19. Lange, Y, Muraski, MF. Topographic heterogeneity in cholesterol biosynthesis. J Biol Chem 1988;263:9366-9373

20. Urbani, L, Simoni, RD. Cholesterol and vesicular stomatitis virus G protein take separate routes form the endoplasmic reticulum to the plasma membrane. J Biol Chem 1990;265:1919-1923

21. Lange, Y, Steck, TL. Quantitation of the pool of cholesterol associated with acyl-CoA:cholesterol acyltransferase in human fibroblasts. J Biol Chem 1997; 272:13103-13108

22. Goldstein, JL, Brown, MS. Regulation of the mevalonate pathway. Nature 1990;343:425-430

23. Rosenblum, MF, Huttler, CR, Strauss, JF. Control of sterol metabolism in cultured rat granulosa cells. Endocrinology 1981; 109:1518-1527

24. Lange, Y, Steck, TL. Cholesterol homeostasis. Modulation by amphiphiles. J Biol Chem 1994;269:29371-29374

25. Chang, TY, Chang, CCY, Cheng, D. Acyl-coenzyme A: cholesterol acyltransferase. Ann Rev Biochem 1997;66:613-638

26. van Meer, G. Lipid traffic in animal cells. Ann Rev Cell Biol 1989;5:247-275

27. Lange, Y, Ye, J, Steck, TL. Circulation of cholesterol between lysosomes and the plasma membrane. J Biol Chem 1998;273:18915-18922

28. De Grella, RF, Simoni, RD. Intracellular transport of cholesterol to the plasma membrane. J Biol Chem 1982;257:14256-14262

29. Lange, Y, Echevarria, F, Steck, TL. Movement of zymosterol, a precursor of cholesterol, among three membranes in human fibroblasts. J Biol Chem 1991;266:21439-21443

30. Lange, Y, Strebel, F, Steck, TL. Role of the plasma membrane in cholesterol esterification in rat hepatoma cells. J Biol Chem 1993;268:13838-13843

31. Nagy, L, Freeman, DA. Cholesterol movement between the plasma membrane and the cholesteryl ester droplets of cultured Leydig tumour cells. Biochem J 1990;271:809-814

32. Braesamle, DL, Attie, AD. Rapid intracellular cholesterol transport of LDL-derived cholesterol to the plasma membrane in cultured fibroblasts. J Lip Res 1990;31:103-112

33. Lange, Y, Ye, J, Chin, J. The fate of cholesterol exiting lysosomes. J Biol Chem 1997;272:17018-17022

34. Draye, J-P, Courtoy, PJ, Quintart, J, and Baudhuin, PA Quantitative model of traffic between plasma membrane and secondary lysosomes: evaluation of inflow, lateral diffusion and degradation. J. Cell Biol 1988;107:2109-2115

35. Lange, Y, Steck, TL. The role of intracellular cholesterol transport in cholesterol homeostasis. Trends Cell Biol 1996;6: 205-208

36. Nagy, L, Freeman, DA. Effect of cholesterol transport inhibitors on steroidogenesis and plasma membrane cholesterol transport in cultured MA-10 Leydig tumor cells. Endocrinology 1990;126:2267-2276

37. Butler, JD, Blanchette-Mackie, J, Goldin, E, O'Neill, RR, Carstea, G, Roff, CF, Patterson, MC, Patel, S, Comly, ME, Cooney, A, Vanier, M, Brady, RO, Pentchev, PG. Progesterone blocks cholesterol translocation from lysosomes. J Biol Chem 1992;267:23797-23805

38. Hruban, Z. Pulmonary and generalized lysosomal storage induced by amphiphilic drugs. Environ Health Perspect 1984; 55:53-76

39. Lange, Y, Duan, H, Mazzone, T. Cholesterol homeostasis is modulated by amphiphiles at transcriptional and post-transcriptional loci. J Lipid Res 1996;37:534-539

INTRACELLULAR STEROL ESTERIFICATION: TWO ACYL COA:CHOLESTEROL ACYLTRANSFERASES IN MAMMALS

Robert V. Farese, Jr., Sylvaine Cases, and Sabine Novak

Gladstone Institute of Cardiovascular Disease, Cardiovascular Research Institute, and Department of Medicine, University of California, San Francisco, California 94110; e-mail: bfarese@gladstone.ucsf.edu

KEY WORDS: sterol, cholesterol, cholesterol ester, acyl CoA:cholesterol acyltransferase, inhibitors, knockout mice, lipoprotein

---

ABSTRACT

The formation of sterol esters from free sterols and fatty acyl CoAs is a fundamental pathway in lipid metabolism of eukaryotic cells. In vertebrates, the sterol esterification reaction is catalyzed by acyl CoA:cholesterol acytransferase (ACAT; EC 2.3.1.26), an enzyme located primarily in the endoplasmic reticulum. In addition to its role in cellular cholesterol homeostasis, ACAT has been hypothesized to participate in a number of processes involving mammalian cholesterol metabolism. Recent studies have provided evidence that more than one ACAT enzyme exists in mammals. Here we review the recent research that has led to the identification and characterization of two mammalian ACAT enzymes, ACAT-1 and ACAT-2.

CONTENTS

ACAT INHIBITORS

CONCLUSION

INTRODUCTION

Sterols play a crucial role in the biology of eukaryotes, where they serve to modulate the fluidity and permeability of cell membranes (1). Cholesterol, the chief sterol found in vertebrates, is also the precursor of steroid hormones and bile acids and has been shown recently to covalently modify proteins (2). An important pathway in sterol metabolism involves the formation of sterol esters from free sterols and fatty acyl CoAs. Together with the sterol regulatory element binding protein (SREBP)–mediated pathway (reviewed in ref. 3), sterol esterification participates in regulating the concentration of cellular free sterols.

The formation of cholesterol esters is catalyzed by acyl CoA:cholesterol acyltransferase (ACAT, EC 2.3.1.26), an integral membrane protein localized to the endoplasmic reticulum (for reviews, see refs. 4-7) (Figure 1). In addition to participating in cellular sterol homeostasis, the formation of cholesterol esters by ACAT has been hypothesized to participate in physiologic processes such as cholesterol absorption in the intestine (8), lipoprotein synthesis and assembly in the liver or the intestine (9), and foam cell formation in macrophages of atherosclerotic lesions (10).

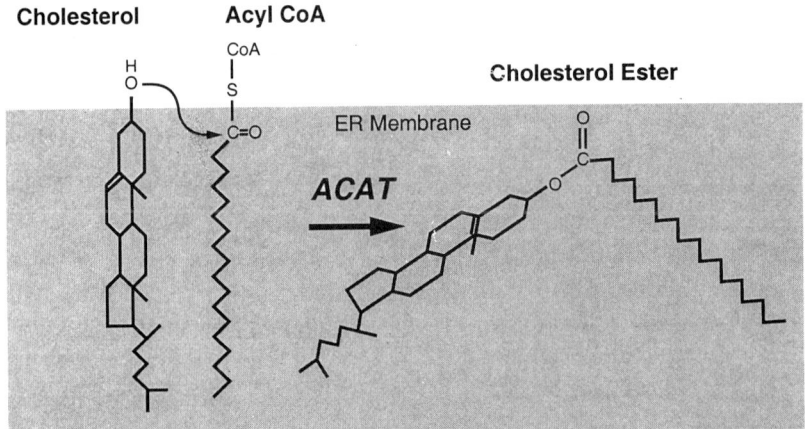

Fig. 1. **ACAT Reaction**. Localized primarily at the ER membrane, ACAT catalyzes the formation of cholesterol esters from free cholesterol and fatty acyl CoA substrates.

Recent studies have provided evidence that sterol esterification is mediated by more than one enzyme in eukaryotes. The esterification of sterols in yeast, in which ergosterol is the primary sterol found in membranes, is carried out by two sterol acyltransferases, ACAT-related enzyme 1 (ARE1) and ARE2 (11), also known as sterol acyltransferase 1 (SAT1) and SAT2 (12). ARE2 accounts for the bulk of sterol esterification activity, and the inactivation of both *ARE1* and *ARE2* genes is required to abolish sterol esterification activity (11, 12). Tissue expression (13) and gene disruption (14) studies of the originally identified mammalian ACAT enzyme (15) (henceforth designated ACAT-1) led to the hypothesis that cholesterol esterification in mammals is also mediated by more than one sterol esterification enzyme. A second mammalian ACAT, designated ACAT-2, has recently been identified (S. Cases et al., manuscript submitted). Here we review the recent research that has led to the identification and characterization of the two ACAT enzymes in mammals.

CHOLESTEROL ESTERS: BIOCHEMISTRY AND PHYSIOLOGY

Considerable knowledge has accumulated with regard to the biochemistry of cholesterol esters and ACAT in mammals (for reviews, see refs. 4-7, 16). Cholesterol esters were first identified in the plasma in 1895 (17) and as a major component of atheromatous aortas in 1910 (18). The formation of cholesterol esters in the plasma, which is carried out by lecithin:cholesterol acyltransferase (LCAT) (reviewed in ref. 19), was first described in 1935 (20). In the 1940s and 1950s, several studies identified a cellular cholesterol esterification activity in rat liver preparations (reviewed in ref. 21), which was later shown to proceed through a reaction involving cholesterol and a fatty acyl CoA ester (22). This pathway was localized to the particulate subcellular fraction (21). In the past three decades, extensive characterization of this biochemical activity has been accomplished.

The intracellular formation of cholesterol esters by ACAT occurs in most mammalian tissues and has been detected in the liver, intestine, adrenal gland, ovary, aorta, endometrium, gall bladder, placenta, mammary gland, brain, kidney, heart and skeletal muscle (reviewed in refs. 6, 7). ACAT activity has also been demonstrated in macrophages (23). Fatty acyl CoA substrate specificity for the ACAT reaction has been studied for ACAT activity in rat liver (21). Under the conditions of the assay, the relative rates of cholesterol ester formation were oleoyl > palmitoyl > stearoyl > linoleoyl. The sterol specificity for the ACAT reaction has also been extensively investigated for rat liver ACAT (reviewed in refs. 4, 7). The enzyme activity exhibits a high degree of specificity for cholesterol, and compounds that differ from cholesterol in side chain length or saturation have reduced or absent esterification. Plant sterols are poor substrates.

The formation of cholesterol esters has been implicated as being important in several physiologic processes that affect plasma cholesterol levels (reviewed in ref. 5). In the small intestine, cholesterol ester formation has been hypothesized to play a role in cholesterol absorption. Cholesterol is taken up by enterocytes from the intestinal lumen in the unesterified form, but most of the absorbed cholesterol is subsequently secreted as cholesterol esters in chylomicron particles. It is believed that cholesterol ester formation in the enterocyte helps to maintain a free cholesterol gradient to promote the absorptive process and provides cholesterol esters for incorporation into chylomicrons. Studies using ACAT inhibitors in animals have largely supported this hypothesis (reviewed in ref. 24). In the liver, cholesterol ester formation has been hypothesized to play a role in very low density lipoprotein (VLDL) assembly and secretion by providing neutral lipid (in addition to triacylglycerol) for incorporation into nascent lipoproteins. Studies employing ACAT inhibitors in primate liver perfusion experiments demonstrated that ACAT inhibition decreased lipoprotein secretion (25). However, many studies of lipoprotein synthesis and secretion in HepG2 cells, a human hepatoma cell line, have not supported a rate-limiting role for cholesterol ester formation in lipoprotein assembly (reviewed in ref. 9). The requirement for cholesterol esters in the assembly of apolipoprotein (apo-)B–containing lipoproteins is currently unsettled. Of note, the suncus, a shrew-like animal, has defective apo-B secretion from the liver (26) that is associated with marked reductions in ACAT activity in this tissue (27).

Cholesterol ester formation in macrophages is prominent in the development of atherosclerotic lesions. Macrophages that have entered the subendothelial space of early lesions take up large quantities of cholesterol-rich lipoproteins via receptor-mediated endocytosis. The hydrolysis of these lipoproteins in the endocytic pathway generates large amounts of free cholesterol, much of which is subsequently transported to cellular sites containing ACAT. Here the cholesterol is esterified and then stored as cytosolic cholesterol ester droplets, giving macrophages their foam cell appearance (28). Cholesterol esterification in macrophages likely plays an important role in protecting the cells from accumulating toxic levels of free sterol. Supporting this possibility, the inhibition of ACAT in cultured macrophages under conditions of cholesterol loading results in cell death (29, 30).

ACAT-1

After years of unsuccessful attempts by investigators to purify ACAT from cells, a cDNA for an ACAT gene was isolated from human macrophages by using an expression cloning strategy in an ACAT-deficient Chinese hamster ovary cell line (AC29) (15). Confirmation that this cDNA encoded a catalytic component of ACAT was established by expressing the cDNA in insect cells, which normally lack cholesterol esterification activity, and demonstrating that cell membranes expressing the putative ACAT enzyme had high levels of ACAT activity (31). The protein encoded by this cDNA is now called ACAT-1. Mammalian orthologs for ACAT-1 have been cloned from mouse (32, 33), rabbit (34), and hamster (35). In addition to the yeast AREs, homologous genes for ACAT-1 have been identified in *Drosophila melanogaster* (accession #AA141382) and *Caenorhabditis elegans* (accession #Z68131) through genome sequencing projects.

The mouse ACAT-1 cDNA encodes a protein of 540 amino acids that migrates at ~46–50 kDa on SDS-PAGE (13, 36). The protein contains multiple hydrophobic domains and was originally proposed to have at least two transmembrane domains (15); a different transmembrane prediction program (http://ulrec3.unil.ch/software/TMPRED_form.html) favors a model with seven or eight transmembrane domains. Although ACAT activity is largely found in microsomes, the predicted protein lacks a signal peptide and known endoplasmic reticulum retention signals (15). The protein has two potential N-linked glycosylation sites and a possible tyrosine phosphorylation motif. A serine residue in hamster ACAT-1 has been demonstrated to be essential for catalytic activity (35); this serine and nearby residues are conserved in ACAT-1 homologs (S. Cases et al., manuscript submitted). In general, the ACAT proteins are highly similar in their carboxyl termini (S. Cases et al., manuscript submitted).

In agreement with previous biochemical studies localizing ACAT activity to microsomes (21), the ACAT protein has been localized by immunocytochemistry studies mainly to the endoplasmic reticulum (ER) in several cell types. In human melanoma cells, human fibroblasts, and AC29 cells expressing human ACAT-1, the protein exhibits an ER distribution primarily (36). In mouse peritoneal macrophages, the major portion of ACAT-1 reactivity was found in the ER, but a small portion localized to an as yet undefined perinuclear compartment (37). Under certain conditions, a portion of ACAT-1 may also be expressed on the cell surface (33, 37).

From experiments expressing ACAT-1 in heterologous systems with low levels of endogenous sterols, data are accumulating with regard to substrate specificity for ACAT-1. Cholesterol and 25-hydroxycholesterol were found to be substrates for human ACAT-1 expressed in insect cells (31). Cholesterol, but not ergosterol or 7-dehydrocholesterol, were substrates for human ACAT-1 expressed in yeast lacking *ARE1* and *ARE2* (38). More recent studies have shown that a number of oxysterols, in addition to cholesterol, may act as substrates for mouse ACAT-1 expressed in insect cells (S. Cases et al., manuscript submitted). In fact, in this expression system, oxysterol substrates appear to be preferred to cholesterol. Some of the oxysterols that are ACAT substrates are also ligands for the nuclear receptor LXRα (39), raising the possibility that ACAT activity could modulate oxysterol signaling pathways. Human ACAT-1 expressed in yeast utilized a variety of fatty acyl CoAs as substrates, including arachidonyl CoA (38). Of note, both the adrenal gland and macrophages, which express high levels of ACAT-1 (see below), are characterized by a relatively large proportion of cholesteryl arachidonate in their cholesterol esters (16, 40).

Using sensitive techniques such as ribonuclease protection assays, ACAT-1 mRNA can be detected in most mammalian tissues (15, 32, 34), with the highest expression levels observed in the preputial gland, ovary, aorta, adrenal gland, testis, thymus and macrophages (32); relatively low expression levels were observed in the liver or small intestine. *In situ* hybridization experiments in mice (13) have confirmed the high ACAT-1 mRNA levels in the preputial gland, sebaceous glands of the skin, adrenal cortex, and aortic atherosclerotic lesions, and a relative lack of signal in the liver and small intestine. An ACAT-1-specific antiserum generated against the amino-terminus of the mouse protein detected a ~46-kDa protein in mouse adrenal, testis, kidney, brain, and peritoneal macrophages (13); only a small amount of ACAT-1 was detected in mouse liver, and none was detected in the small intestine. In rabbit liver, nonparenchymal cells contained 30-fold more ACAT-1 mRNA than parenchymal cells (34). Taken together, these studies suggest that

ACAT-1 may have a prominent role in sebaceous glandular tissues, steroidogenic tissues, and in macrophages, and a less prominent role in the liver and small intestine.

ACAT activity increases markedly in cells cultured in the presence of added cholesterol or oxysterols (41, 42). Previous studies have demonstrated that ACAT activity is regulated chiefly in a posttranslational manner, largely through substrate availability or by substrate-induced modulation of enzyme activity (for a review, see ref. 6). In agreement with these earlier findings, the activity of human ACAT-1 expressed in insect cells, which have low endogenous levels of membrane cholesterol, can be markedly activated by the presence of added cholesterol or 25-hydroxycholesterol in cell-free assays (31). The kinetic data from these experiments suggest that cholesterol, in addition to being a substrate, may activate the enzyme through an allosteric mechanism (6). Changes in mRNA expression also may play some role in the regulation of ACAT-1 activity as ACAT-1 mRNA levels increased ~twofold in livers of mice (32) or livers and aortas of rabbits (34) fed diets rich in fat and cholesterol. However, ACAT-1 protein levels were unchanged in human fibroblasts, HepG2 cells, and Chinese hamster ovary cells cultured in the presence or absence of sterols (36).

MUTATIONS IN ACAT-1

Although humans with a deficiency in ACAT-1 have not been identified, targeted and naturally occurring ACAT-1 mutations in mice have provided an opportunity to explore physiological roles for this enzyme. The mouse ACAT-1 gene (*Acact*) was disrupted in the 5' coding region through homologous recombination in mouse embryonic stem cells (14). Mice homozygous for the gene disruption (*Acact⁻ᐟ⁻* mice) are viable and healthy. Plasma total and HDL cholesterol levels were increased in *Acact⁻ᐟ⁻* mice of a mixed C57BL/6 and 129/Sv genetic background as compared with wild-type control mice (14). However, when the gene disruption was studied in a pure 129/Sv background, no effect on plasma cholesterol levels was observed (E. Sande and R.V. Farese, Jr., unpublished observations). It is likely that the original observations in the mixed background were due to linkage of a 129/Sv gene allele influencing plasma HDL levels (such as *Apoa2*, which is located close to *Acact*) to the *Acact⁻* allele (for a discussion of this phenomenon, see ref. 43).

The phenotype of *Acact⁻ᐟ⁻* mice is that of tissue-specific reductions in ACAT activity and tissue cholesterol ester content (14). In *Acact⁻ᐟ⁻* adrenal glands, cholesterol esters and ACAT activity are markedly reduced (Figure 2). Remarkably, despite the near-total adrenocortical cholesterol ester depletion, the ability to produce corticosterone in response to adrenocorticotropic hormone stimulation is normal in *Acact⁻ᐟ⁻* mice (14). Presumably, a redundancy in pathways providing cholesterol for steroidogenesis accounts for the lack of functional impairment. Peritoneal macrophages from *Acact⁻ᐟ⁻* mice also exhibited a marked reduction in cholesterol ester accumulation when cultured in the presence of acetylated low density lipoproteins (LDL) (14) (Figures 2B and 2C). Because ACAT-1 deficiency in macrophages can result in alterations in cellular concentrations of free cholesterol

A

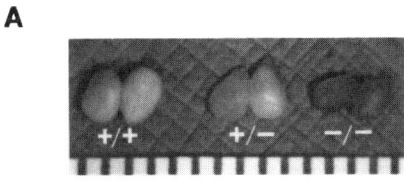

B

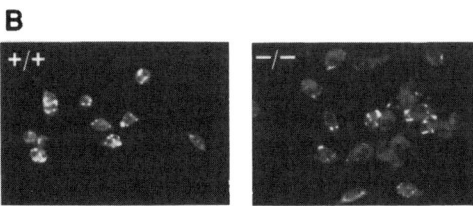

C

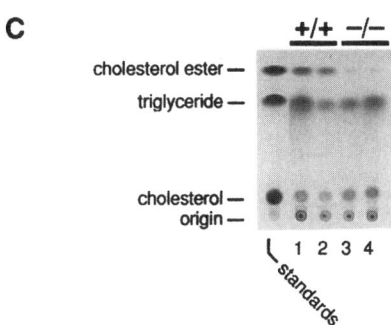

Fig. 2. **Cholesterol ester depletion in adrenal glands and macrophages of ACAT-1-deficient mice.** (A) Gross appearance of adrenal glands from male *Acact$^{+/+}$*, *Acact$^{+/-}$*, and *Acact$^{-/-}$* mice. The adrenals from *Acact$^{-/-}$* mice are darker and more translucent due to the near-total absence of cholesterol esters. (B) Nile Red fluorescence of peritoneal macrophages. Neutral lipid droplets in *Acact$^{-/-}$* mice are diminished in number and intensity, and the droplets have a different pattern of distribution. (C) High-performance thin-layer chromatography of lipid extracts from wild-type or *Acact$^{-/-}$* peritoneal macrophages demonstrating reduced macrophage cholesterol ester accumulation in *Acact$^{-/-}$* mice. Panel B reproduced with permission from *Curr. Opin. Lipidol.* **9**:119–123, 1998; Panels A and C are reproduced with permission from ref. (14).

(29, 30) or oxysterols (S. Cases, manuscript submitted) that may affect macrophage cell biology, *Acact$^{-/-}$* mice could conceivably have alterations in immune function.

Unexpectedly, the livers of *Acact$^{-/-}$* mice contained significant amounts of cholesterol esters (14). Hepatic cholesterol ester levels tended to be lower in *Acact$^{-/-}$* mice than in wild-type mice fed a chow diet, but were similar for mice fed a diet rich in fat and cholesterol. Cholesterol esterification activity was similar for hepatic microsomes from *Acact$^{-/-}$* and wild-type mice fed either diet. Moreover, intestinal cholesterol absorption, a process that is thought to involve cholesterol esterification (8), was not altered in *Acact$^{-/-}$* mice fed either chow (14) or high-fat (E. Sande and R.V. Farese, Jr., unpublished observations) diets. These results indicated that a form of ACAT that was not inactivated by the ACAT-1 gene disruption exists in mice and that this enzyme may be responsible for ACAT act-ivity in tissues such as the liver and small intestine.

A naturally occur-ring mutation in ACAT-1 has been identified in the AKR inbred strain of mice (44). AKR mice carry a recessive allele, *adrenocortical lipid depletion* (*ald*) (45), on chromosome 1 (46), which is associated with a cholesterol ester depletion phenotype in the adrenal cortex (45, 47). The phenotype in AKR mice is conditional: cholesterol esters are normal in the adrenal cortex prior to puberty, but decrease markedly after puberty, especially in males (45). Androgens appear to be a factor in the manifestation of the phenotype. The mapping of *Acact* to a region of chromosome 1 containing *ald* (48) and the adrenocortical cholesterol ester depletion in male and female *Acact*$^{-/-}$ mice (14) implicated *Acact* as a candidate gene for *ald*. Crossing of AKR (*ald/ald*) mice with *Acact*$^{-/-}$ mice did not complement the phenotype of adrenocortical lipid depletion in postpubertal F$_1$ male offspring, indicating that *ald* and *Acact* are allelic (44). The *Acact* cDNA from AKR mice was shown to contain a deletion of the first coding exon and two missense mutations, and the ACAT-1 protein has a reduced molecular mass, consistent with the amino-terminal deletion (44). Despite these differences, ACAT activity in *in vitro* assays of adrenal gland membranes of postpubertal male AKR mice is similar to that in wild-type mice. Also, ACAT activity in ACAT-deficient cells transfected with the *ald* cDNA is similar to those transfected with a wild-type ACAT-1 cDNA (44). These observations indicate that *ald* is indeed a mutant *Acact* allele. However, they suggest that the *ald* allele is necessary, but not sufficient, to cause the adrenocortical lipid depletion phenotype in AKR mice. Another as yet undetermined factor, possibly associated with androgen production at puberty, may contribute to the loss of cholesterol esters in adrenocortical cells.

Because ACAT-1 deficiency causes reduced cholesterol ester accumulation in macrophages but has little effect on processes that influence plasma cholesterol levels (14), the ACAT-1-deficient mice provide an opportunity to selectively study how macrophage ACAT deficiency affects the development of atherosclerosis. In a study performed in rabbits using the ACAT inhibitor CI-976 at doses that did not affect plasma cholesterol levels, atherosclerotic lesion development was inhibited (49). This result and others from animal studies using ACAT inhibitors (24, 50) suggest that macrophage ACAT deficiency may inhibit atherosclerotic lesion development, presumably by inhibiting macrophage foam cell generation. Studies to test this hypothesis by crossing *Acact*$^{-/-}$ mice with mouse models that develop atherosclerosis are in progress.

ACAT-2

A second ACAT gene, ACAT-2, has been identified from the expressed sequence tag (EST) databases, and full-length cDNAs have been cloned from mice (S. Cases et al., manuscript submitted), humans (S. Sturley, personal communication), and primates (L. Rudel, personal communication). The mouse ACAT-2 cDNA encodes a 525–amino acid protein that is 44% identical to mouse ACAT-1 (Figure 3), with the highest regions of homology localized to the carboxyl-termini of the proteins. A serine residue (position 248 of mouse ACAT-2) that is essential for cholesterol esterification activity of the hamster ACAT-1 protein (35) is conserved in ACAT-2. The hydrophobicity plot of ACAT-2 is strikingly similar to that for ACAT-1, and like ACAT-1, the protein may have as many as eight transmembrane domains.

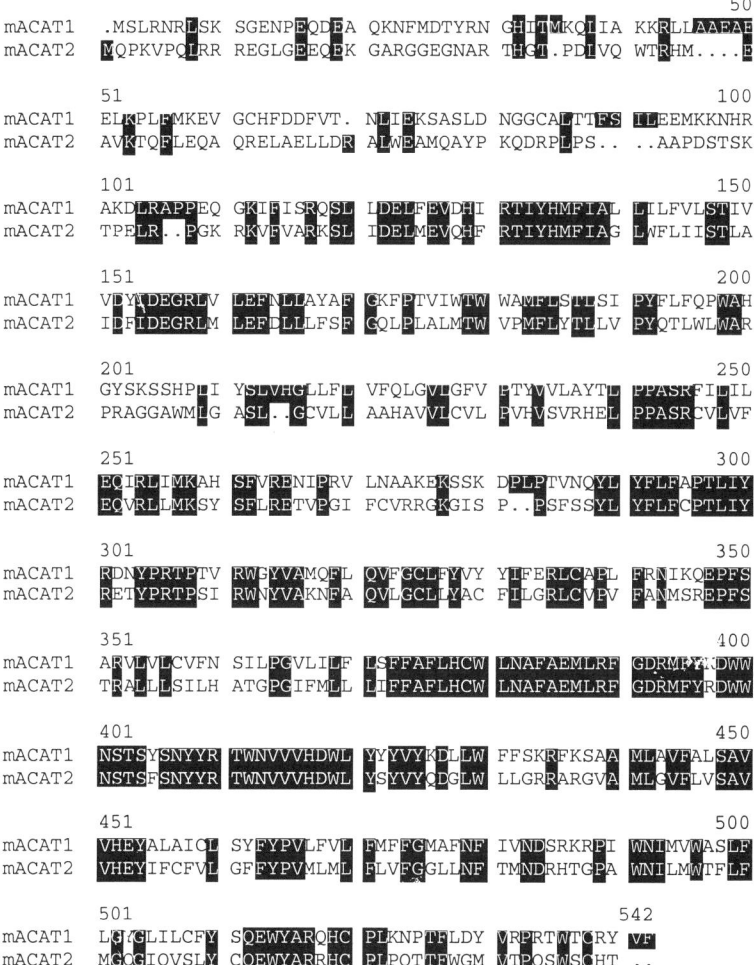

Fig. 3. **Alignment of mouse ACAT-1 (mACAT1) and ACAT-2 (mACAT2) protein sequences**. Identical residues are shaded.

The ACAT-2 protein catalyzes cholesterol esterification at a high specific activity when expressed in insect cell membranes (S. Cases et al., manuscript submitted). Like ACAT-1, ACAT-2 can also use many oxysterols as substrates (S. Cases et al., manuscript submitted). In addition to acting as substrates for the ACAT enzymes, oxysterols can modulate cholesterol esterification activity of ACAT-1 and ACAT-2 using the endogenous cholesterol that is present in insect cell membranes as a substrate. These results suggest that ACAT-2, like ACAT-1 (6), may be allosterically regulated by sterols.

The tissue distribution of ACAT-2 expression has been studied in several species. In mice, ACAT-2 is expressed predominantly in the liver and in the small intestine (S. Cases et al., manuscript submitted). Expression was not detected by northern analysis in brain, heart, lung, spleen, kidney, ovary, testis, or thymus. By using more sensitive techniques, such as RT-PCR, ACAT-2 mRNA can be detected in many mouse tissues at low levels, with relatively higher expression levels in the liver and the small intestine (S. Cases et al., manuscript submitted). In primates, relatively high levels of ACAT-2 expression were observed in the liver and small intestine (L. Rudel, personal communication). In humans, ACAT-2 expression appears to be relatively low in adult tissues (S. Cases and R.V. Farese, Jr., unpublished observations). However, by using RT-PCR techniques, expression can be detected in human liver and small intestine, but not in fibroblasts (S. Cases et al., manuscript submitted).

The role of ACAT-2 in *in vivo* sterol metabolism remains to be determined. The ability of the enzyme to catalyze cholesterol esterification at a high specific activity and its hepatic and intestinal expression pattern make it an attractive candidate for the ACAT enzyme that is important in cholesterol metabolism in these tissues in mice. In support of this possibility, the mouse ACAT-2 gene (*Acact2*) was mapped to a region of chromosome 15 (S. Cases et al., manuscript submitted) that contains a quantitative trait locus (QTL) influencing plasma cholesterol levels in response to a high-fat diet (51). This finding makes *Acact2* a candidate gene for this QTL. The importance of ACAT-2 in mammalian cholesterol metabolism will be best determined by a gene inactivation experiment in mice.

ACAT INHIBITORS

The development of pharmacologic inhibitors of ACAT has received considerable attention (for reviews, see refs. 24, 50). ACAT is a rational target for an anti-atherosclerosis therapy because of the enzyme's role in modulating cholesterol metabolism in the liver and small intestine and because of its role in macrophage foam cell formation in the arterial wall. ACAT inhibitors can be divided into several different classes, such as fatty acyl amides (*e.g.*, SaH 57-118, 58-035, CI-976), urea-based compounds (*e.g.*, CL 277,082, PD 132301-2), and compounds with increased water solubility (*e.g.*, PD 138142-15 or CI-1011).

Many different ACAT inhibitors inhibit intestinal cholesterol absorption and thereby lower plasma cholesterol levels in animals (24). However, despite their success in animal models, several compounds, including CL 277,082 and DuP-128, were not effective in humans (52, 53). The urea inhibitor PD 132301-2 decreased cholesterol absorption in *in vivo* animal models but was found to be extremely adrenotoxic (54, 55), thereby precluding its use in humans.

The inhibition of hepatic ACAT has also been a therapeutic target, based on the rationale that, by limiting cholesterol ester availability, the synthesis and secretion of apo-B-containing lipoproteins can be diminished, which would result in lower plasma cholesterol. Although *in vitro* studies addressing this hypothesis have had mixed results (for a discussion, see ref. 24), the *in vivo* data are promising for some inhibitors. For example, CI-976 decreases cholesterol ester and apo-B secretion in perfused livers from African green monkeys (25), and the intravenous administration of DuP-128 decreased VLDL- and LDL-apo-B production rates in miniature pigs fed a cholesterol-free diet (56). In cholesterol-fed rats, CI-976 administered orally efficiently lowered plasma lipids by inhibiting hepatic ACAT and enhancing biliary bile acid secretion (57). A recently developed water-soluble sulfonamide, CI-1011, is a relatively weak ACAT inhibitor *in vitro*, but causes marked hypocholesterolemia in cholesterol-fed rats (58), possibly through effects on hepatic ACAT.

Macrophage ACAT in atherosclerotic lesions represent another important target for ACAT inhibition. Several inhibitors diminish the cholesterol ester content or extent of atherosclerotic lesions in animals (reviewed in ref. 24). Of note, CI-976 administration to cholesterol-fed rabbits at doses that did not lower plasma cholesterol levels inhibited lesion formation in aortas, suggesting an arterial wall–specific effect (49).

The observation that at least two different ACAT enzymes exist in mammals (S. Cases et al., manuscript submitted) suggests that ACAT inhibition could be directed specifically at either ACAT-1 or ACAT-2. Illustrating this principle, ACAT-1 and ACAT-2 expressed independently in insect cell membranes demonstrated different degrees of inhibition to several of the available inhibitors (S. Cases et al., manuscript submitted). Whereas PD 132301-2 inhibited both ACAT-1 and ACAT-2 to similar degrees, CI-976 and CI-1011 more selectively inhibited ACAT-2 (S. Cases et al., manuscript submitted). However, before embarking on the path of specifically targeting inhibition to ACAT-1 or ACAT-2, more knowledge is needed concerning the *in vivo* function of ACAT-2, particularly in humans.

CONCLUSION

It is now clear that at least two ACAT enzymes exist in mammals. Like the situation for the two sterol esterification enzymes in yeast (11, 12), it is unknown why more than one form of ACAT is necessary for mammalian biology. Possible explanations may relate to: 1) functional redundancy of an important cellular metabolic process, 2) complementary tissue distributions, 3) differences in substrate specificity, or 4) different functions with respect to cell biology. With regard to the latter possibility, it is interesting to note that ACAT-1 deficiency is associated with the loss of cholesterol esters in at least two cell types (adrenocortical cells and macrophages) known to store these molecules in cytosolic droplets. The loss of ACAT-1 had little effect, however, on tissues that export cholesterol esters on lipoproteins. Perhaps different forms of ACAT are involved in cholesterol ester synthesis for storage in the cytosol *vs.* for exportation of cholesterol esters on apo-B-containing lipoproteins. Significant progress in understanding the functions of the different ACAT enzymes should result from better understanding of ACAT-2's role in mammalian biology. Further studies may also provide insights that facilitate the development of more selective, and possibly less toxic, ACAT inhibitors for use in clinical medicine.

40

REFERENCES

1. Bloch K. Cholesterol: Evolution of structure and function. In: *Biochemistry of Lipids, Lipoproteins and Membranes*. DE Vance, J Vance, eds. Amsterdam, The Netherlands: Elsevier, 1991:363–381.
2. Porter JA, Young KE, Beachy PA. Cholesterol modification of hedgehog signaling proteins in animal development. Science 1996;274:255–259.
3. Brown MS, Goldstein JL. The SREBP pathway: Regulation of cholesterol metabolism by proteolysis of a membrane-bound transcription factor. Cell 1997;89:331–340.
4. Chang T-Y, Doolittle GM. Acyl coenzyme A:cholesterol O-acyltransferase. In: *The Enzymes*. PD Boyer, ed. New York: Academic Press, 1983;16:523–539.
5. Suckling KE, Stange EF. Role of acyl-CoA:cholesterol acyltransferase in cellular cholesterol metabolism. J Lipid Res 1985;26:647–671.
6. Chang TY, Chang CCY, Cheng D. Acyl-coenzyme A:cholesterol acyltransferase. Annu Rev Biochem 1997;66:613–638.
7. Billheimer, JT, Gillies, PJ. Intracellular cholesterol esterification. In: *Advances in Cholesterol Research*. M Esfahani, JB Swaney, eds. West Caldwell, NJ: Telford, 1990:7–45.
8. Wilson MD, Rudel LL. Review of cholesterol absorption with emphasis on dietary and biliary cholesterol. J Lipid Res 1994;35:943–955.
9. Dixon JL, Ginsberg HN. Regulation of hepatic secretion of apolipoprotein B-containing lipoproteins: Information obtained from cultured liver cells. J Lipid Res 1993;34:167–179.
10. Chang TY, Chang CCY, Cadigan KM. The structure of acyl coenzyme A–cholesterol acyltransferase and its potential relevance to atherosclerosis. Trends Cardiovasc Med 1994;4:223–230.
11. Yang H, Bard M, Bruner DA, et al. Sterol esterification in yeast: A two-gene process. Science 1996;272:1353–1356.
12. Yu C, Kennedy NJ, Chang CCY, Rothblatt JA. Molecular cloning and characterization of two isoforms of *Saccharomyces cerevisiae* acyl-CoA:sterol acyltransferase. J Biol Chem 1996;271:24157–24163.
13. Meiner V, Tam C, Gunn MD, et al. Tissue expression studies of mouse acyl CoA:cholesterol acyltransferase gene (*Acact*): Findings supporting the existence of multiple cholesterol esterification enzymes in mice. J Lipid Res 1997;38:1928–1933.
14. Meiner VL, Cases S, Myers HM, et al. Disruption of the acyl-CoA:cholesterol acyltransferase gene in mice: Evidence suggesting multiple cholesterol esterification enzymes in mammals. Proc Natl Acad Sci USA 1996;93:14041–14046.
15. Chang CCY, Huh HY, Cadigan KM, Chang TY. Molecular cloning and functional expression of human acyl-coenzyme A:cholesterol acyltransferase cDNA in mutant Chinese hamster ovary cells. J Biol Chem 1993;268:20747–20755.
16. Goodman DS. Cholesterol ester metabolism. Physiol Rev 1965;45:747–839.
17. Hürthle, K. Ueber die Fettsäure-Cholesterin-Ester des Blutserums. Z Physiol Chem 1895; 21:332.
18. Windaus, A. Über den Gehalt normaler und atheromatøoser Aorten an Cholesterin und Cholesterinestern. Z Physiol Chem 1910;67:174.
19. Glomset JA. The plasma lecithin:cholesterol acyltransferase reaction. J Lipid Res 1968;9:155–167.
20. Sperry, WM. Cholesterol esterase in blood. J Biol Chem 1935;111:467–478.
21. Goodman DS, Deykin D, Shiratori T. The formation of cholesterol esters with rat liver enzymes. J Biol Chem 1964;239:1335–1345.
22. Mukherjee, S, Kunitake, G, Alfin-Slater, RB. The esterification of cholesterol with palmitic acid by rat liver homogenates. J Biol Chem 1958;230:91–96.
23. Brown MS, Goldstein JL, Krieger M, Ho YK, Anderson RGW. Reversible accumulation of cholesteryl esters in macrophages incubated with acetylated lipoproteins. J Cell Biol 1979;82:597-613.
24. Krause BR, Bocan TMA. ACAT inhibitors: Physiologic mechanisms for hypolipidemic and anti-atherosclerotic activities in experimental animals. In: *Inflammation. Mediators and Pathways*. RR Ruffolo, Jr., MA Hollinger, eds. Boca Raton, FL: CRC Press, 1995:173–198.
25. Carr TP, Hamilton RL, Jr., Rudel LL. ACAT inhibitors decrease secretion of cholesteryl esters and apolipoprotein B by perfused livers of African green monkeys. J Lipid Res 1995;36:25–36.
26. Yasuhara M, Ohama T, Matsuki N, et al. Deficiency of apolipoprotein B synthesis in Suncus murinus. J Biochem 1991;110:751–755.

27. Nagayoshi A, Matsuki N, Saito H, et al. Deficiency of acyl CoA cholesterol acyl transferase activity in Suncus liver. J Biochem 1994;115:858–861.
28. Brown MS, Goldstein JL. Lipoprotein metabolism in the macrophage: Implications for chole.,terol deposition in atherosclerosis. Annu Rev Biochem 1983;52:223–261.
29. Tabas I, Marathe S, Keesler GA, Beatini N, Shiratori Y. Evidence that the initial up-regulation of phosphatidylcholine biosynthesis in free cholesterol-loaded macrophages is an adaptive response that prevents cholesterol-induced cellular necrosis. Proposed role of an eventual failure of this response in foam cell necrosis in advanced atherosclerosis. J Biol Chem 1996;271:22773–22781.
30. Warner GJ, Stoudt G, Bamberger M, Johnson WJ, Rothblat GH. Cell toxicity induced by inhibition of acyl coenzyme A:cholesterol acyltransferase and accumulation of unesterified cholesterol. J Biol Chem 1995;270:5772–5778.
31. Cheng D, Chang CCY, Qu X-M, Chang T-Y. Activation of acyl-coenzyme A:cholesterol acyltransferase by cholesterol or by oxysterol in a cell-free system. J Biol Chem 1995;270:685–695.
32. Uelmen PJ, Oka K, Sullivan M, Chang CCY, Chang TY, Chan L. Tissue-specific expression and cholesterol regulation of acylcoenzyme A:cholesterol acyltransferase (ACAT) in mice. Molecular cloning of mouse ACAT cDNA, chromosomal localization, and regulation of ACAT *in vivo* and *in vitro*. J Biol Chem 1995;270:26192–26201.
33. Green S, Steinberg D, Quehenberger O. Cloning and expression in Xenopus oocytes of a mouse homologue of the human acylcoenzyme A:cholesterol acyltransferase and its potential role in metabolism of oxidized LDL. Biochem Biophys Res Commun 1996;218:924–929.
34. Pape ME, Schultz PA, Rea TJ, et al. Tissue specific changes in acyl-CoA:cholesterol acyltransferase (ACAT) mRNA levels in rabbits. J Lipid Res 1995;36:823–838.
35. Cao G, Goldstein JL, Brown MS. Complementation of mutation in acyl-CoA:cholesterol acyltransferase (ACAT) fails to restore sterol regulation in ACAT-defective sterol-resistant hamster cells. J Biol Chem 1996;271:14642–14648.
36. Chang CCY, Chen J, Thomas MA, et al. Regulation and immunolocalization of acyl-coenzyme A:cholesterol acyltransferase in mammalian cells as studied with specific antibodies. J Biol Chem 1995;270:29532–29540.
37. Khelef N, Buton X, Beatini N, et al. Immunolocalization of ACAT in macrophages. J Biol Chem 1998; 273:11218–11224.
38. Yang H, Cromley D, Wang H, Billheimer JT, Sturley SL. Functional expression of a cDNA to human acyl-coenzyme A:cholesterol acyltransferase in yeast. Species-dependent sub,trate specificity and inhibitor sensitivity. J Biol Chem 1997;272:3980–3985.
39. Janowski BA, Willy PJ, Devi TR, Falck JR, Mangelsdorf DJ. An oxysterol signalling pathway mediated by the nuclear receptor LXRα. Nature 1996;383:728–731.
40. Cullen P, Fobker M, Tegelkamp K, et al. An improved method for quantification of cholesterol and cholesteryl esters in human monocyte-derived macrophages by high performance liquid chromatography with identification of unassigned cholesteryl ester species by means of secondary ion mass spectrometry. J Lipid Res 1997;38:401–409.
41. Brown MS, Dana SE, Goldstein JL. Cholesterol ester formation in cultured human fibroblasts. Stimulation by oxygenated sterols. J Biol Chem 1975;250:4025–4027.
42. Doolittle GM, Chang T-Y. Acyl-CoA:cholesterol acyltransferase in Chinese hamster ovary cells. Enzyme activity determined after reconstitution in phospholipid/cholesterol liposomes. Biochim Biophys Acta 1982;713:529–537.
43. Smithies O, Maeda N. Gene targeting approaches to complex genetic diseases: Atherosclerosis and essential hypertension. Proc Natl Acad Sci USA 1995;92:5266–5272.
44. Meiner VL, Welch CL, Cases S, et al. Adrenocortical lipid depletion gene (*ald*) in AKR mice is associated with an acyl-CoA:cholesterol acyltransferase (ACAT) mutation. J Biol Chem 1998;273:1064–1069.
45. Arnesen K. Constitutional difference in lipid content of adrenals in two strains of mice and their hybrids. Acta Endocrinol 1955;18:396–401.
46. Taylor BA, Meier H. Mapping the adrenal lipid depletion gene of the AKR/J mouse strain. Genet Res 1976;26:307–312.
47. Arnesen K. The cytology of the adrenal cortex in mice with spontaneous adrenocortical lipid depletion. Acta Pathol Microbiol Scand 1963;58:212–218.
48. Welch CL, Xia Y-R, Shechter I, et al. Genetic regulation of cholesterol homeostasis: Chromosomal organization of candidate genes. J Lipid Res 1996;37:1406–1421.
49. Bocan TMA, Mueller SB, Uhlendorf PD, Newton RS, Krause BR. Comparison of CI-976, an ACAT inhibitor, and selected lipid-lowering agents for antiatherosclerotic activity in iliac–

femoral and thoracic aortic lesions. A biochemical, morphological, and morphometric evaluation. Arterioscler Thromb 1991;11:1830–1843.

50. Sliskovic DR, White AD. Therapeutic potential of ACAT inhibitors as lipid lowering and anti-atherosclerotic agents. Trends Pharmacol Sci 1991;12:194–199.

51. Purcell-Huynh DA, Weinreb A, Castellani LW, Mehrabian M, Doolittle MH, Lusis AJ. Genetic factors in lipoprotein metabolism. Analysis of a genetic cross between inbred mouse strains NZB/BINJ and SM/J using a complete linkage map approach. J Clin Invest 1995;96:1845–1858.

52. Hainer JW, Terry JG, Connell JM, et al. Effect of the acyl-CoA:cholesterol acyltransferase inhibitor DuP 128 on cholesterol absorption and serum cholesterol in humans. Clin Pharmacol Ther 1994;56:65–74.

53. Harris WS, Dujovne CA, von Bergmann K, et al. Effects of the ACAT inhibitor CL 277,082 on cholesterol metabolism in humans. Clin Pharmacol Ther 1990;48:189–194.

54. Dominick MA, Bobrowski WA, MacDonald JR, Gough AW. Morphogenesis of a zone-specific adrenocortical cytotoxicity in guinea pigs administered PD 132301-2, an inhibitor of acyl-CoA:cholesterol acyltransferase. Toxicol Pathol 1993;21:54–62.

55. Dominick MA, McGuire EJ, Reindel JF, Bobrowski WF, Bocan TMA, Gough AW. Subacute toxicity of a novel inhibitor of acyl-CoA:cholesterol acyltransferase in beagle dogs. Fundam Appl Toxicol 1993;20:217–224.

56. Huff MW, Telford DE, Barrett PHR, Billheimer JT, Gillies PJ. Inhibition of hepatic ACAT decreases apoB secretion in miniature pigs fed a cholesterol-free diet. Arterioscler Thromb 1994;14:1498–1508.

57. Krause BR, Anderson M, Bisgaier CL, et al. In vivo evidence that the lipid-regulating activity of the ACAT inhibitor CI-976 in rats is due to inhibition of both intestinal and liver ACAT. J Lipid Res 1993;34:279–294.

58. Lee HT, Sliskovic DR, Picard JA, et al. Inhibitors of acyl-CoA:cholesterol O-acyl transferase (ACAT) as hypocholesterolemic agents. CI-1011: An acyl sulfamate with unique cholesterol-lowering activity in animals fed noncholesterol-supplemented diets. J Med Chem 1996;39:5031–5034.

STEROL ESTERIFICATION AND HOMEOSTASIS IN A MODEL EUKARYOTE.

Peter Oelkers[1], Stephen L. Sturley[1,2]* and Arthur Tinkelenberg[1] [a]

[1]Institute of Human Nutrition, [2]Departments of Pediatrics; Physiology and Molecular Biophysics, Columbia University College of Physicians and Surgeons, 630 W168th St. New York, NY 10032
e-mail: sls37@columbia.edu

KEY WORDS: yeast, acyl-CoA cholesterol acyl transferase (ACAT), Niemann Pick type C (NPC), sterol homeostasis, sterol transport.

ABSTRACT

The intracellular esterification of sterols by all eukaryotes represents a critical homeostatic response to excess sterol. This process is mediated by a multigene family in several organisms, including the unicellular eukaryote, *Saccharomyces cerevisiae* (yeast). We describe here our recent efforts to identify molecular components of sterol esterification and sterol transport in this model system. This includes homologs to the acyl-CoA cholesterol acyl transferase (ACAT) gene, a putative homolog to the Niemann Pick C (NPC) gene, and novel yeast genes with human counterparts.

CONTENTS.

INTRODUCTION.

BUDDING YEAST, A MODEL SYSTEM FOR STEROL HOMEOSTASIS.

A FAMILY OF STEROL ESTERIFICATION GENES IN YEAST AND OTHER EUKARYOTES.

----

[a] All authors contributed equally to this paper.
* To whom correspondence should be addressed.

GENETIC SCREENS FOR STEROL TRANSPORTERS.

CANDIDATE GENES FOR STEROL TRANSPORTERS.

CONCLUSIONS.

INTRODUCTION

Free sterols are important components of all eukaryotic membranes. The esterification of sterols with fatty acids is a critical homeostatic response to excess intracellular free sterol and the end-point of a poorly understood transport phenomenon. In mammals this reaction is mediated by acyl-CoA cholesterol acyl transferase (ACAT) where the acyl donor is acyl CoA. This distinguishes it from extracellular reactions (e.g. by lecithin cholesterol acyl transferase) which use phosphatidylcholine as the acyl donor. Moreover, the ACAT reaction has been characterized as exclusively intracellular, predominantly occurring in the endoplasmic reticulum. This cellular domain is relatively sterol-poor, implying insulated transport pathways for delivery of sterols specifically to ACAT. The sterol esterification reaction is ubiquitous in eukaryotes, from yeast to humans, and is frequently mediated by multigene families. This laboratory has defined the molecular components of an ACAT-like reaction in the model eukaryote, yeast (*Saccharomyces cerevisiae*). We subsequently identified two novel members of the human ACAT gene family. We also extended this genetic analysis to investigate components of intracellular sterol transport, one of which may be directly homologous to the human Niemann Pick C gene.

BUDDING YEAST, A MODEL SYSTEM FOR STEROL HOMEOSTASIS.

Sterols (primarily cholesterol in mammals and ergosterol in yeast) are essential components of all eukaryotic cellular membranes (1) (2-4) and obligate precursors for steroidogenesis (5). They maintain membrane fluidity (6) and modulate the activity of a number of membrane proteins (7-11). The homeostatic control of this metabolite has been strongly conserved between yeast and mammalian cells. For example, sterol esterification, a critical step in the storage of excess cholesterol, is mediated by a two gene family in yeast. Both genes, *ARE1* and *ARE2* (ACAT-Related Enzymes 1 and 2) are structural and functional homologs of human ACAT1 (12-14). Similarly, a gene family mediates esterification reactions in human, mouse and simian cells as demonstrated by this laboratory and others (Oelkers et al. in press, (15), Farese R. and Rudel L. personal communications). Other components of mammalian sterol homeostasis are also conserved in yeast [for review see (16)]. For example, the biosynthesis of cholesterol or ergosterol is mediated by a pathway of enzymes with high homology between these organisms [e.g. squalene synthase (17), mevalonate kinase (18) and others reviewed in (19)]. HMG-CoA reductase, the rate limiting enzyme of sterol biosynthesis, is represented by two isoforms in yeast

(Hmg1 and Hmg2) with 65% and 93% identity to the COOH-terminal 400 amino acids of the human enzyme (20). Moreover, the human HMG-CoA reductase cDNA functionally complements *hmg1 hmg2* deletion mutants for viability as effectively as yeast *HMG1* itself (21). In an almost identical fashion to the situation in mammalian cells, the regulation of at least one of these gene products, Hmg2, is achieved at the level of proteasomal hydrolysis via feedback control from the mevalonate pathway (22, 23). Yeast also regulates sterol biosynthesis at the transcriptional level, possibly through Sterol Regulatory Element (SRE)-related enhancers (24, 25).

Subcellular sterol distribution is remarkably similar in yeast and mammalian cells. Sterols are generally maintained at a high concentration in the plasma membrane (PM) relative to other subcellular membranes, including the ER, which is thought to be a major site of synthesis (26-29). Sterol biosynthesis and sterol storage reside at distinct subcellular locations necessitating specific trafficking systems to maintain cellular sterol homeostasis (30-33). A number of mechanisms have been proposed for sterol transport, including aqueous diffusion, vesicle-mediated pathways, and soluble carrier molecules (34), however, a comprehensive paradigm to describe sterol trafficking and distribution remains elusive.

A FAMILY OF STEROL ESTERIFICATION GENES IN YEAST AND OTHER EUKARYOTES.

The utility of yeast as a model eukaryote to study basic cellular processes is powerfully exemplified by the plethora of information arising from completion of the yeast genome project (16). By screening the yeast genome for sequences similar to the human ACAT1 enzyme (35), we successfully identified the yeast equivalents, *ARE1* and *ARE2* (12, 14). Strains deleted for both *ARE* genes are viable, contain no steryl ester, and have reduced numbers of neutral lipid droplets (12). Further, *are1 are2* null strains down-regulate sterol biosynthesis, indicating the central role played by these ACAT homologs in cellular sterol homeostasis (12, 36).

In mammals, several studies of ACAT inhibitors have shown that microsomal preparations from various tissues have different sensitivities (37-39). This suggested that either more than one enzyme catalyzes ACAT activity or that different membrane milieus in different cell types effect sensitivity to the various inhibitors. More recently, an induced mutant mouse for ACAT1 was created in which cholesterol esters were barely detectable in macrophages, fibroblasts, and cortical adrenal cells (40, 41). However, normal esterification was observed in ACAT1 mutant hepatocytes and intestinal absorption of cholesterol was indistinguishable from control litter-mates. This strongly suggests that multiple enzymes mediate ACAT activity in mice. By extension from the situation we found in yeast, we hypothesized that an ACAT gene family would be a feature in humans and that these additional ACAT enzyme(s) would share sequence similarity to ACAT1 and the ACAT-related genes from yeast.

Using a region conserved between yeast Are1 and Are2, and human ACAT1 to screen the database of expressed sequence tags, two independent human genes which encode for proteins with sequence similarity to ACAT1 were identified (Fig. 1). These two genes were entitled ACAT Related Gene Product (ARGP) 1 and 2. ARPG1 is a 488 amino acid, transmembrane protein with a predicted molecular mass of 55.2 kDa. Over the entire molecule it is 22% identical to ACAT1; when the comparison is restricted to the COOH-terminal half of the proteins they are 28% identical. Among the conserved motifs is the tetrapeptide D-W-W-N and a tyrosine phosphorylation site, found in all cloned ACATs (Fig. 1). The limited sequence identity with ACAT1 makes it unclear whether ARGP1 shares evolutionary lineage with ACAT1 or whether the similarity results from convergent evolution.

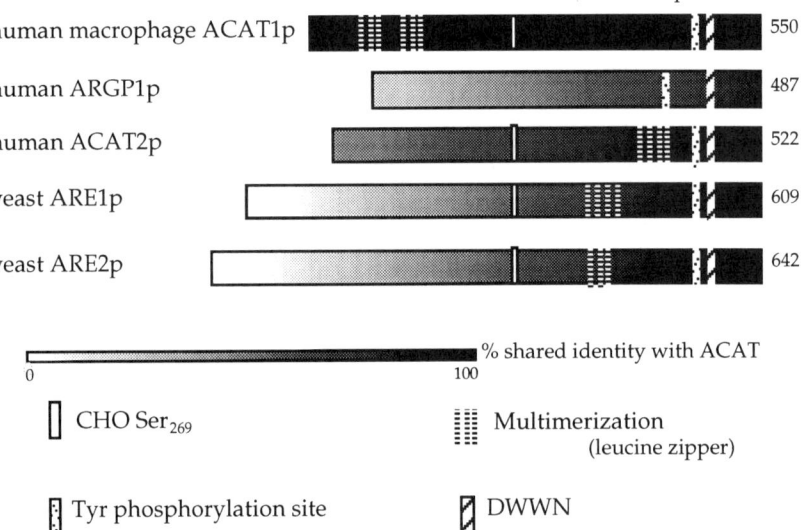

Figure 1: Protein sequence conservation within the eukaryotic ACAT gene family. Conserved sequences may represent functionally critical regions of the molecule, particularly across organisms. Shown here are multimerization motifs (leucine zippers), tyrosine phosphorylation consensus sites, and two regions shown in CHO cells or yeast to be required for enzymatic activity, i.e. Serine corresponding to residue 269 of human ACAT and a DWWN tetrapeptide motif common to all ACATs identified.

When overexpressed in an ACAT deficient yeast strain and assayed for ACAT activity, ARGP1 showed no activity. ARGP1 transcripts were found in most human tissues as shown by northern blotting but qualitatively high amounts were detected in adrenal cortex, adrenal medulla, small intestine, and testes. This is consistent with ARGP1 not acting as an ACAT since the ACAT1 induced mutant mice had negligible ACAT activity in the adrenal cortex, a tissue which in humans has abundant amounts of ARGP1 message. Other potentially ARGP1 mediated esterification reactions that use fatty-acyl CoA include methyl ester formation, triterpene esterification, monoacylglycerol transferase, and diacylglycerol transferase.

ARPG2 is a 522 amino acid protein with a predicted molecular mass of 59.9 kDa and multiple predicted transmembrane domains. Over the entire molecule it is 48% identical to ACAT1 and when the comparison is restricted to the COOH-terminal half of the proteins, they are 60% identical. We have demonstrated that this gene is expressed in a more limited array of tissues and cell types than ARGP1. When overexpressed in an ACAT deficient yeast strain, ARGP2 conferred a significant amount of cholesterol esterification. Thus ARGP2 has been renamed ACAT2 and is a likely candidate for the liver and intestinal forms of this enzyme.

GENETIC SCREENS FOR STEROL TRANSPORTERS.

Our approach to studying the conserved mechanisms underlying eukaryotic sterol trafficking and homeostasis in *Saccharomyces cerevisiae* has been a comprehensive application of molecular and classical genetics. At present, a major limitation in the study of sterol trafficking has been the lack of a genetic paradigm; for example, such as that developed by Novick and Schekman for eukaryotic protein secretion (using the yeast system). Our aim is to develop such a paradigm for sterol trafficking using genetic screens to identify conserved proteins regulating sterol trafficking and distribution in eukaryotic cells. One of the strengths of the yeast system is that mutations disrupting sterol homeostasis can be easily combined with mutations affecting other aspects of membrane physiology (e.g., protein secretion, lysosomal function, sphingolipid and phospholipid metabolism) in order to test functional relationships. We have also utilized genetic approaches which assume that components of this process will physically interact with the ACAT protein. These yeast two-hybrid approaches are described elsewhere (16, 42).

The basis of our current screen relies on the role played by ACAT enzymes (in yeast and higher eukaryotes) in maintaining cellular sterol levels and mitigating toxicity of high sterol levels. We reasoned that mutations rendering sterol trafficking less efficient might lead to toxic accumulations of free sterol in the absence of sterol esterification. Mutations in at least four genes have been recovered that render an *are1 are2* strain (deleted for the yeast ACAT homologs) inviable unless *ARE2* is expressed; these mutations are termed *'arv'* for *'ARE2* Required for Viability'. The gene corresponding to the *arv1* mutation, the most common class of mutations in the screen so far, was isolated. Its amino acid sequence predicts a novel transmembrane protein with similarity to uncharacterized proteins reported in the database of human expressed sequence tags and the nematode *C. elegans*.

CANDIDATE GENES FOR STEROL TRANSPORTERS.

Niemann-Pick Type C (NPC) disease, an autosomal recessive neurodegenerative disorder, is currently understood as a secondary sphingomyelin storage disease. The cellular defect in NPC fibroblasts is a block in lysosomal efflux of LDL-derived cholesterol leading to accumulation of free cholesterol in that organelle (43, 44). The gene defective in NPC was recently cloned and found to be homologous to an uncharacterized open reading frame in yeast as well as the morphogen receptor and tumor suppressor, Patched (45, 46). The NPC disease gene (NPC1) and its yeast homologue, NCR1 (Niemann Pick C Related gene, 42% amino acid identity with NPC1), share a short region of homology (termed the NPC domain) that is lacking in Patched, even though the full length similarity between all three is quite high.

The finding that yeast has a gene homologous to both NPC1 and Patched provides an opportunity to use molecular genetics to determine the conserved functions (if any) this class of genes share. If the yeast *NCR1* gene is functionally related to the NPC1 gene, then yeast genetics can be exploited to identify the pathways and other proteins that NPC1 may interact with in human cells. The NPC1/NCR1/Patched story is intriguing and presents a dilemma that the yeast system may be ideal to address: i.e. what is the conserved cellular function of these genes? Yeast has neither lipoprotein uptake nor multicellular development; yet, it has a gene with striking similarity to genes from humans involved in both processes. At the very least, these observations suggest some primordial function for the Ncr1 protein that is shared by NPC1 and Patched, perhaps in addition to other roles taken on during the evolution of multicellular organisms.

CONCLUSIONS.

We have taken a genetic approach to understanding the molecular nature of sterol esterification and transport in the model eukaryote, yeast. We have studied the orthologous genes in this organism to the human ACAT and Niemann Pick type C genes. Furthermore, using our yeast system as a paradigm, we established that the ACAT reaction is mediated by a multigene family in human cells. Subsequently, we have established genetic assays of ACAT interactions resulting in the isolation of novel components of sterol homeostasis in yeast and their human counterparts.

ACKNOWLEDGEMENTS.

The work was supported in part by a Grant-in-Aid/Investigatorship from the American Heart Association, (NYC Affiliate) and by the Ara Parseghian Medical Research Foundation to SLS. SLS is an established investigator of the National American Heart Association. PMO was supported by an NHLBI training grant fellowship #HL07343 in Arteriosclerosis. AT was supported by an NHLBI training grant fellowship #DK07715 in Nutrition.

REFERENCES.

1. Bloch KE. Sterol structure and membrane function. CRC Crit. Rev. Biochem. 1983;14(1):47-92.
2. Parks LW, Casey WM. Physiological implications of sterol biosynthesis in yeast. Annu Rev Microbiol 1995;49:95-116.
3. Rothblat GH, Mahlberg FH, Johnson WJ, Phillips MC. Apolipoproteins, membrane cholesterol domains, and the regulation of cholesterol efflux. J. Lipid Res. 1992;33(8):1091-1097.

4.   Waldeck AR, Nouri-Sorkhabi MH, Sullivan DR, Kuchel PW. Effects of cholesterol on transmembrane water diffusion in human erythrocytes measured using pulsed field gradient NMR. Biophys. Chem. 1995;55(3):197-208.

5.   Shepherd J. Lipoprotein metabolism. An overview. Drugs, Suppl. 2 1994;47:1-10.

6.   Jackson RL, Gotto AMJ. Hypothesis concerning membrane structure, cholesterol, and atherosclerosis. Atherosclerosis Reviews 1976;1:1-21.

7.   Lijnen P, Petrov V. Cholesterol modulation of transmembrane cation transport systems in human erythrocytes. Biochem. and Mol. Med. 1995;56(1):52-62.

8.   Yeagle PL. Lipid regulation of cell membrane structure and function. FASEB J. 1989;3:1833-1842.

9.   Szolderits G, Hermetter A, Paltauf F, Daum G. Membrane properties modulate the activity of a phosphatidylinositol transfer protein from the yeast, *Saccharomyces cerevisiae*. Biochim. Biophys. Acta 1989;986:301-309.

10. Wang X, Sato R, Brown MS, Hua X, Goldstein JL. SREBP-1, a Membrane-Bound Transcription Factor released by Sterol-Regulated Proteolysis. Cell 1994;77:53-62.

11. Brasitus TA, Dahiya R, Dudeja PK, Bissonnette BM. Cholesterol modulates alkaline phosphatase activity of rat intestinal microvillus membranes. J. Biol. Chem. 1988;263:8592-8597.

12. Yang H, Bard M, Bruner DA, et al. Sterol Esterification in Yeast: A two gene process. Science 1996;272:1353-1356.

13. Yu C, Kennedy NJ, Chang CCY, Rothblatt JA. Molecular Cloning and Characterization of Two Isoforms of Saccharomyces cerevisiae Acyl-CoA:Sterol Acyltransferase. J. Biol. Chem. 1996;271:24157-24163.

14. Sturley SL. Molecular Aspects of Intracellular Sterol Esterification: The Acyl Coenzyme A:Cholesterol Acyltransferase (ACAT) reaction. Curr. Opin. Lipidol. 1997;8:167-173.

15. Oelkers P, Behari A, Sturley SL. Isolation And Characterization of Two Human Genes Encoding Enzymes Related to Acyl Coenzyme A-Cholesterol Acyltransferase. Circulation 1997;96:I-411.

16. Sturley SL. A molecular approach to understanding human sterol metabolism using yeast genetics. Curr. Opin. Lipidol. 1998;9:85-91.

17. Robinson GW, Tsay YH, Kienzle BK, Smith-Monroy CA, Bishop RW. Conservation between human and fungal squalene synthetases: similarities in structure, function, and regulation. Mol Cell Biol 1993;13(5):2706-17.

18. Tsay YH, Robinson GW. Cloning and characterization of ERG8, an essential gene of Saccharomyces cerevisiae that encodes phosphomevalonate kinase. Mol Cell Biol 1991;11(2):620-31.

19. Lees ND, Skaggs B, Kirsch DR, Bard M. Cloning of the late genes in the ergosterol biosynthetic pathway of Saccharomyces cerevisiae--a review. Lipids 1995;30(3):221-6.

20. Basson ME, Thorsness M, Rine J. Saccharomyces cerevisiae contains two functional genes encoding 3-hydroxy-3-methylglutaryl-coenzyme A reductase.

Proceedings of the National Academy of Sciences of the United States of America 1986;83(15):5563-7.

21. Basson ME, Thorsness M, Finer-Moore J, Stroud RM, Rine J. Structural and functional conservation between yeast and human 3-hydroxy-3-methylglutaryl coenzyme A reductases, the rate-limiting enzyme of sterol biosynthesis. Mol. Cell. Biol. 1988;8:3797-3808.

22. Hampton RY, Gardner RG, Rine J. Role of 26S proteasome and HRD genes in the degradation of 3-hydroxy-3-methylglutaryl-CoA reductase, an integral endoplasmic reticulum membrane protein. Mol Biol Cell 1996;7(12):2029-44.

23. Hampton R, Dimster-Denk D, Rine J. The biology of HMG-CoA reductase: the pros of contra-regulation. Trends Biochem Sci 1996;21(4):140-5.

24. Dimster-Denk D, Rine J. Transcriptional regulation of a sterol-biosynthetic enzyme by sterol levels in *Saccharomyces cerevisiae*. Mol. Cell. Biol. 1996;16(8):3981-3989.

25. Smith SJ, Crowley JH, Parks LW. Transcriptional regulation by ergosterol in the yeast Saccharomyces cerevisiae. Mol Cell Biol 1996;16(10):5427-32.

26. Lange Y. Disposition of intracellular cholesterol in human fibroblasts. J. Lipid Res. 1991;32:329-339.

27. Zinser E, Sperka-Gottlieb CDM, Fasch E-V, Kohlwein SD, Paltauf F, Daum G. Phospholipid synthesis and lipid composition of subcellular membranes in the unicellular eukaryote *Saccharomyces cerevisiae*. J. Bacteriol. 1991;173:2026-2034.

28. Reinhart MP, Billheimer JT, Faust JR, Gaylor JL. Subcellular Localization of the enzymes involved in cholesterol biosynthesis and metabolism in rat liver. J. Biol. Chem. 1987;262:9649-9655.

29. Paltauf F, Kohlwein SD, Henry SA. Regulation and Compartmentalization of Lipid Synthesis in Yeast. In: Strathern JN, Jones EW, Broach JR, eds. The Molecular and Cellular Biology of the Yeast *Saccharomyces*: Metabolism and Gene Expression. New York: Cold Spring Harbor Laboratory, 1992:415-500.

30. Zinser E, Paltauf F, Daum G. Sterol composition of yeast organelle membranes and subcellular distribution of enzymes involved in sterol metabolism. J. Bacteriol. 1993;175:2853-2858.

31. Miller WL. Mitochondrial specificity of the early steps in steroidogenesis. J. Steroid Biochem. & Mol. Biol. 1995;55:607-616.

32. Liscum L, Underwood KW. Intracellular cholesterol transport and compartmentation. J. Biol. Chem. 1995;270(26):15443-15446.

33. Frolov A, Woodford JK, Murphy EJ, Billheimer JT, Schroeder F. Spontaneous and protein-mediated sterol transfer between intracellular membranes. J. Biol. Chem. 1996;271:16075-16083.

34. Reinhart MP. Intracellular sterol trafficking. Experientia 1990;46:599-611.

35. Chang CCY, Huh HY, Cadigan KM, Chang TY. Molecular Cloning and Functional Expression of Human Acyl Coenzyme A:Cholesterol Acyltransferase cDNA in Mutant Chinese Hamster Ovary Cells. J. Biol. Chem. 1993;268(28):20747-20755.

36. Arthington-Skaggs BA, Crowell DN, Yang H, Sturley SL, Bard M. Positive and negative regulation of a sterol biosynthetic gene (ERG3) in the post-squalene portion of the yeast ergosterol pathway. FEBS Letts. 1996;392:161-165.

37. Maduskie T, Billheimer J, Germain S, et al. Design, synthesis and structure-activity relationship studies for a new Imidazole series of J774 macrophage specific Acyl-CoA: cholesterol acyltransferase (ACAT) inhibitors. J. Med. Chem. 1995;38:1067-1083.

38. Kinnunen PM, Spilburg CA, Lange LG. Chemical modification of acyl-CoA:cholesterol O-acyltransferase. 2. Identification of a coenzyme A regulatory site by p-mercuribenzoate modification. Biochemistry 1988;27(19):7351-6.

39. Kinnunen PM, DeMichele A, Lange LG. Chemical modification of acyl-CoA:cholesterol O-acyltransferase. 1. Identification of acyl-CoA:cholesterol O-acyltransferase subtypes by differential diethyl pyrocarbonate sensitivity. Biochemistry 1988;27(19):7344-50.

40. Meiner VM, Cases S, Myers H, et al. Disruption of the acyl-CoA:cholesterol acyltransferase gene in mice: Evidence suggesting multiple cholesterol esterification enzymes in mammals.. Proc. Natl. Acad. Sci. USA 1996;93:14041-14046.

41. Farese RV, Jr. Acyl CoA:cholesterol acyltransferase (ACAT) genes and knockout mice. Curr. Opin. in Lipidol. 1998;9(2):in press.

42. Guo Z, Yang H, Sturley SL. Multimerization of the human ACAT enzyme and mutagenesis of the genes required for sterol esterification in yeast. Circulation 1996;94:I-35.

43. Pentchev PG, Brady RO, Blanchette-Mackie EJ, et al. The Niemann-Pick C lesion and its relationship to the intracellular distribution and utilization of LDL cholesterol. Biochim. Biophys. Acta 1994;1225:235-243.

44. Liscum L, Klansek JJ. Niemann-Pick disease type C. Current Opinion in Lipidology 1998;9:131-135.

45. Carstea ED, Morris JA, Coleman KG, et al. Niemann-Pick C1 disease gene: homology to mediators of cholesterol homeostasis. Science 1997;277(5323):228-31.

46. Loftus SK, Morris JA, Carstea ED, et al. Murine model of Niemann-Pick C disease: mutation in a cholesterol homeostasis gene [see comments]. Science 1997;277(5323):232-5.

CHOLESTEROL DISTRIBUTION IN GOLGI, LYSOSOMES AND ENDOPLASMIC RETICULUM

E. Joan Blanchette-Mackie* and Peter G. Pentchev[+]

National Institutes of Diabetes, Digestive and Kidney Diseases* and Neurological Disorders and Stroke[+], National Institutes of Health, Bethesda, MD 20892. e-mail: Joanbm@bdg8.niddk.nih.gov

KEY WORDS: Golgi, *trans* Golgi cisternae, lysosomes, endoplasmic reticulum, intracellular cholesterol

ABSTRACT

Insight into the effect of exogeneously derived lipoprotein cholesterol on distribution of intracellular membrane cholesterol has been gained from structural studies on normal and Niemann Pick Type C human fibroblasts. Endocytic uptake of LDL enriches Golgi cholesterol in both normal and NPC cells. However, the NPC mutation and treatment of normal cells with progesterone during LDL uptake produces abnormal accumulation of cholesterol in lysosomes and *trans* Golgi cisternae. This lysosomal/Golgi block in cholesterol trafficking results in the inability of endocytosed cholesterol to induce cellular homestatic responses. In addition to lysosomes and Golgi the endoplasmic reticulum can also be a site along the intracellular cholesterol transport pathway that becomes a temporary depot for cholesterol. Specific inhibition of acyl CoA:cholesterol acyltransferase with S-58035 during endocytic uptake results in a reversable accumulation of cholesterol in membranes of ER. Thus the ER, normally low in intracellular cholesterol, has the capacity to act as a sink for endocytosed cholesterol when esterification is blocked. In contrast to the lysosomal/Golgi cholesterol sequestration, ER accumulation of cholesterol does not compromise but appears to enhance the induction of cellular homeostatic responses.

CONTENTS

GOLGI AND LYSOSOMAL CHOLESTEROL DISTRIBUTION AFTER LDL
UPTAKE IN NIEMANN PICK TYPE C FIBROBLASTS AND NORMAL FIBRO-
BLASTS TREATED WITH PROGESTERONE

ENDOPLASMIC RETICULUM CHOLESTEROL DISTRIBUTION IN NORMAL
FIBROBLASTS AFTER LDL UPTAKE DURING ACAT INHIBITION

CONCLUSIONS

INTRODUCTION

The heterogeneous distribution of cholesterol in cellular membranes indicates that
cholesterol transport in cells is a regulated process. The highest concentration of
cholesterol is found in plasma membrane and the remainder is distributed in
decreasing amounts between lysosomal, Golgi and endoplasmic reticulum
membranes (1). Exogeneous cholesterol provided to cells through the endocytic
uptake of cholesteryl ester rich lipoproteins such as LDL is first transfered to
lysosomes where hydrolysis of the cholesteryl esters releases free cholesterol for
subsequent distribution to other intracellular sites (2,3). Normally the transfer of
cholesterol from lysosomes is an efficient process since studies have shown that very
little endocytosed cholesterol accumulates in these organelles during a period when
cellular cholesterol can be enriched four fold (4). Documentation of defective
lysosomal cholesterol egress in fibroblasts derived from patients with the genetic
disorder, Niemann-Pick C (NP-C) disease (5,6), strongly suggests that this
pathway of cholesterol translocation depends on specific protein factors. Important
insight into the post-lysosomal trafficking of cholesterol in normal cells was
obtained from the initial structural studies comparing cholesterol distribution in
normal cells with that in fibroblasts with the NP-C lesion (7). Fluorescence
visualization of unesterified cholesterol in Golgi of both normal and mutant human
fibroblasts after LDL uptake suggested that the Golgi complex plays a critical role
in the relocation of endocytosed cholesterol to other cellular membranes (7). We
have subsequently shown that enrichment of the cholesterol content of cells with
LDL modifies the fluorescent properties of the specific Golgi marker, C_6-NBD
ceramide (8). Changes in the fluorescent behavior of this Golgi reporter molecule
(9) during incubation of cells with LDL indicate that cholesterol in the Golgi can be
derived from the endocytosed lipoprotein. The concept that the Golgi plays an
active role in directing the relocation of endocytosed lysosomal cholesterol to the
plasma membrane is supported by the biochemical studies of Neufeld et al., using
cyclodextrin to probe cholesterol transport (10). Our present studies show that
endocytic uptake of LDL in the presence of progesterone produces in normal cells a
lysosomal/Golgi cholesterol lipidosis similar to that induced by LDL uptake in NPC
cells. Additionally, we have gained information on the ability of the endoplasmic
reticulum to accumulate cholesterol from studies on normal fibroblasts incubated
with LDL during drug induced inhibition of acyl CoA: cholesterol acyltransferase.

GOLGI CHOLESTEROL DISTRIBUTION AFTER LDL UPTAKE IN NORMAL FIBROBLASTS

Hydrolysis of endocytosed LDL cholesteryl esters releases free cholesterol for subsequent distribution to other intracellular sites (2,3). Cholesterol accumulation in Golgi of normal fibroblasts incubated with LDL suggested that endocytosed lipoprotein cholesterol may be transported from lysosomes to other cellular destinations through the Golgi (7). The initial demonstration of Golgi involvement in intracellular cholesterol transport was documented with fluorescence microscopy using filipin as a specific cytochemical probe for unesterified cholesterol. After uptake of LDL the Golgi in normal human fibroblasts was illuminated with filipin fluorescence when compared to that in normal cells maintained in lipid depleted serum (7 and Fig.1).

Fig.1. Normal human fibroblast incubated with LDL for 24 hours. Fluorescence staining with filipin shows unesterified cholesterol in Golgi membranes (G). X 500

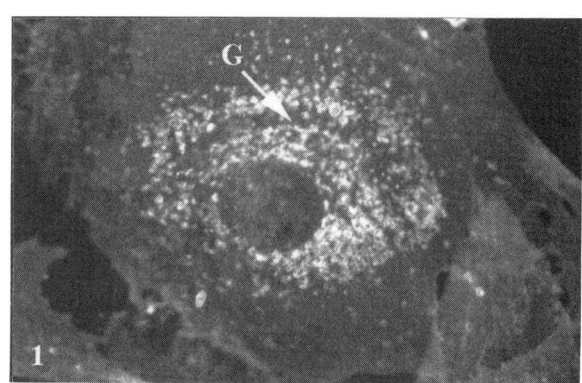

The Golgi is a complex polarized organelle consisting of a collection (stacks) of flattened membrane bounded cisternae. The functional compartmentalization of the Golgi proceeds from the *cis* (entry) cisternae through the medial cisternae to the *trans* (exit) cisternae with the cis and trans side having different capacities for directing cellular components (11). In fibroblasts, there may be several Golgi stacks near the cell nucleus, each consisting of three to five cisternae, small vesicles (20-100 nm) associated with the cisternae and vacuoles (100-350 nm) located on the *trans* side of the cisternal stack. The lumen of the *trans* cisternae is continuous with the *trans* Golgi network composed of vesicles and vacuoles in transit to the plasma membrane. In fibroblasts, the vacuoles have been characterized as exocytic *trans* Golgi vacuoles by wheat germ agglutinin staining and immunostaining for fibronectin (12).

The distribution of cholesterol in Golgi compartments of normal fibroblasts and the enrichment of cholesterol in the various compartments after uptake of LDL was determined with freeze-fracture electron microscopy and filipin staining. Filipin complexes with unesterified cholesterol in membranes forming distinct structures, termed filipin cholesterol deformations (FCD) that are visualized as pits or protruberances in the hydrophobic planes of Golgi membranes (Fig. 2).

56

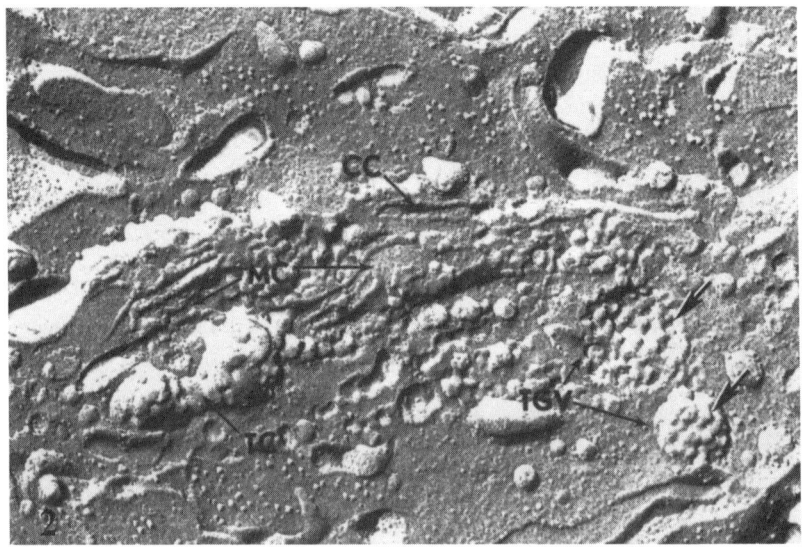

Fig.2 Electron micrograph of replica of freeze-fractured, filipin treated Golgi in normal fibroblasts incubated with LDL for 24 hours. *Trans* Golgi vacuoles (TGV) of the Golgi are "saturated" with filipin-cholesterol deformations (large arrows). *Trans* cisterna (TC) has a few filipin-cholesterol deformations. *Cis* cisternae, CC; *medial* cisternae, MC. X 60,000. Reproduced with permission J. Lipid Res. (13).

EFFECT OF LDL UPTAKE ON DISTRIBUTION AND
DENSITY OF FILIPIN-CHOLESTEROL DEFORMATIONS
IN GOLGI MEMBRANES OF NORMAL FIBROBLASTS

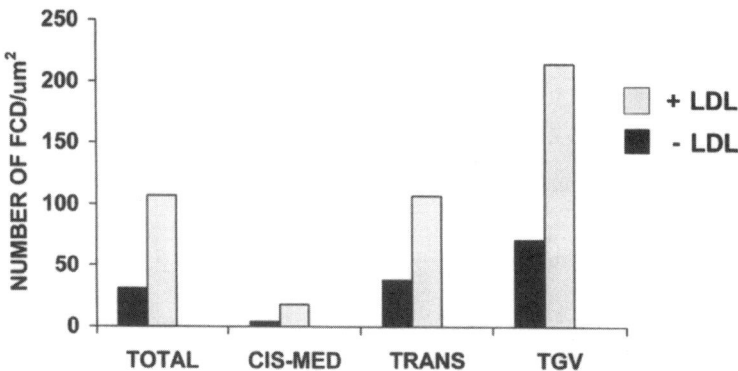

Fig. 3. Normal fibroblasts were grown on lipid depleted serum without (-LDL) or with (+LDL 50μg/ml) for 24 hours (13). Total = Total Golgi Membranes (*cis,medial trans* cisternae and *trans* Golgi vacuoles).

Evaluation of the effect of LDL uptake on Golgi cholesterol distribution in normal fibroblasts was accomplished by quantitation of the number $FCD/\mu m^2$ in Golgi membranes of *cis, medial* and *trans* cisterane and trans Golgi vacuoles (Fig.3).

Methods for selection of samples, preparing and visualization of replicas, and recording and analysis of quantitative data are reported in Coxey et al. (13).

Uptake of LDL for 24 hours in normal fibroblasts increased the cholesterol concentration in Golgi (Total Golgi, Fig. 3). The concentration of cholesterol increases in a cis-trans direction across the Golgi cisternae. A similar gradient was reported for cholesterol concentration across the Golgi in pancreatic acinar cells using freeze fracture analysis and filipin cytochemistry (14). The cis-trans gradient was maintained irrespective of this LDL-induced increase in Golgi cholesterol levels. The membrane cholesterol content increased in both cis/medial and trans Golgi vacuoles with the greater cholesterol increase in trans Golgi vacuoles (statistical data in ref. 13).

GOLGI AND LYSOSOMAL CHOLESTEROL DISTRIBUTION AFTER LDL UPTAKE

Niemann Pick Type-C Fibroblasts

The close link between lysosomal egress of cholesterol and its appearance in the Golgi complex has come forth from the characterization of the sites of intracellular cholesterol accumulation in mutant NP-C fibroblasts. Abnormal accumulation of cholesterol during the endocytic uptake of LDL by mutant cells was not only associated with cholesterol sequestration in lysosomes but also with a premature buildup of cholesterol in the Golgi complex (7, & Fig. 4).

Fig. 4. Nieman Pick Type C fibroblast incubated with LDL for 24 hours. Fluorescence staining with filipin shows accumulation of unesterified cholesterol in Golgi (G) and Lysosomes (L) X 450. Reproduced with permission, Science (15)

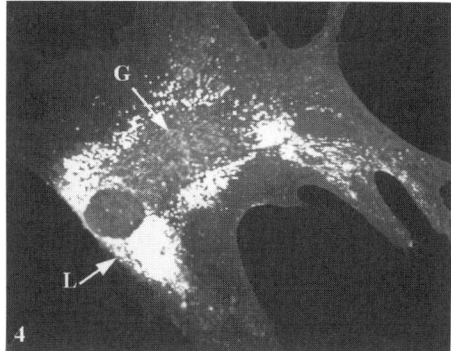

Golgi

Freeze-fracture electron microscopic analyses showed that uptake of LDL for 24 hours in mutant fibroblasts induced an abnormal accumulation of cholesterol in Golgi. (13, Figs. 5 & 6) in addition to its buildup in lysosomes (7, Fig. 7).

58

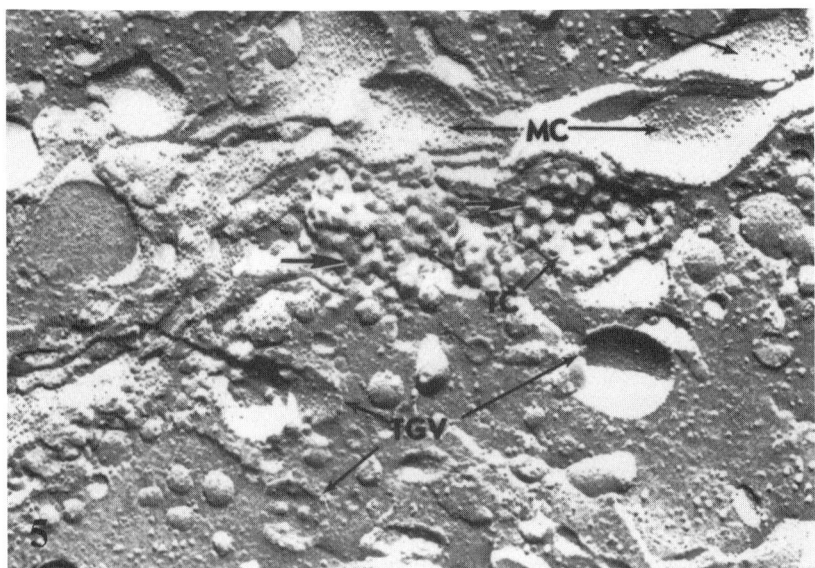

Fig.5. Electron micrograph of replica of freeze-fractured Golgi in Niemann Pick C fibroblast incubated with LDL for 24 hours. *Trans* Golgi cisternae (TC) are "saturated" with filipin-cholesterol deformations (large arrows). *Trans* Golgi vacuoles (TGV) have few filipin cholesterol deformations. *cis* cisternae, CC; *medial* cisternae, MC. X 60,000. Reproduced with permission J. Lipid Res. (13)

**EFFECT OF LDL UPTAKE ON DISTRIBUTION AND
DENSITY OF FILIPIN-CHOLESTEROL DEFORMATIONS
IN GOLGI MEMBRANES OF NP-C FIBROBLASTS**

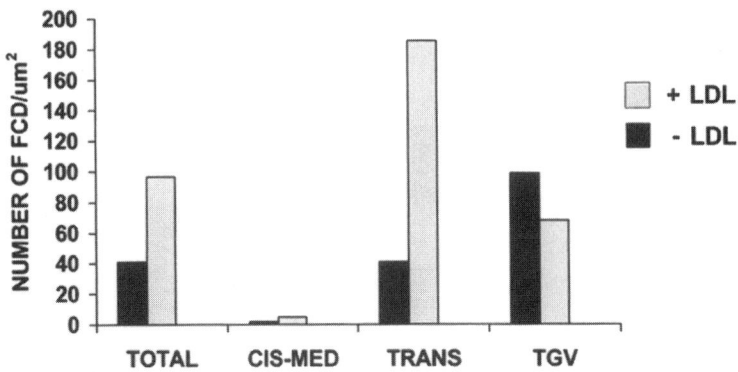

Fig. 6. NPC fibroblasts were grown on lipid depleted serum without (-LDL) or with (+LDL 50µg/ml) for 24 hours (13). Total = Total Golgi Membranes (*cis, medial trans* cisternae and *trans* Golgi vacuoles).

Evaluation of the effect of LDL uptake on Golgi cholesterol distribution in Niemann Pick Type C fibroblasts was accomplished by quantitation of the number of FCD/μm^2 in Golgi membranes of *cis, medial* and *trans* cisternae and *trans* Golgi vacuoles (Fig. 6).

Uptake of LDL for 24 hours by NPC fibroblasts increased the concentration of cholesterol in the Golgi similar to the increase in normal cells (Fig. 3). However the NPC mutation causes an abnormal accumulation of cholesterol in *trans* Golgi cisternae.

Lysosomes

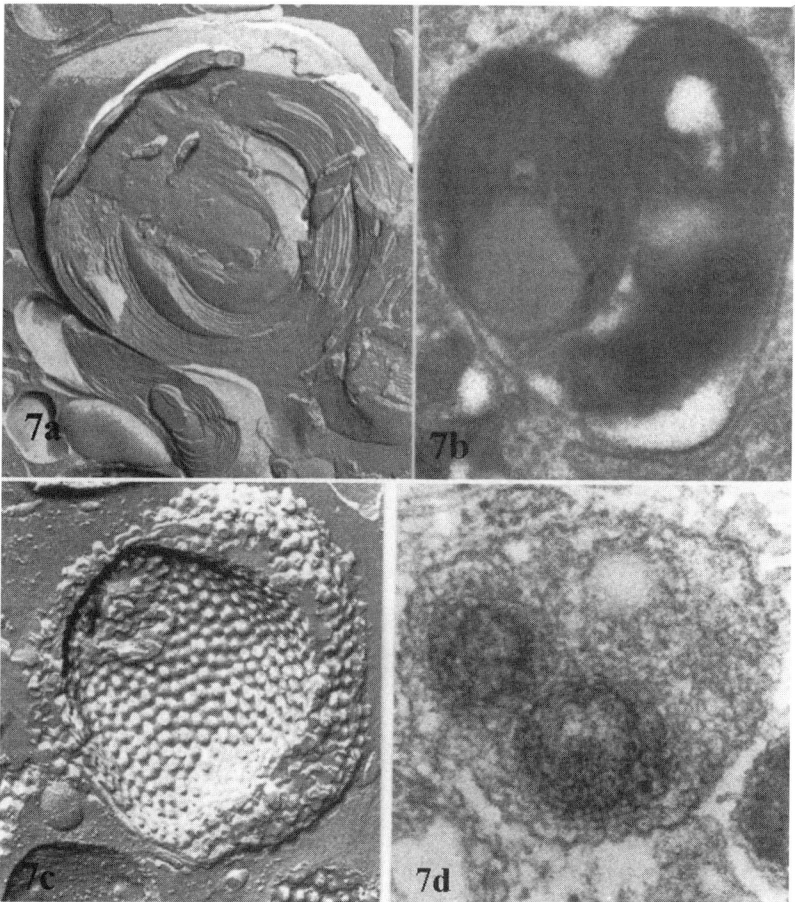

Fig. 7. Lysosomes contain multilamellar cores (a, freeze fracture) and (b, section). Multilamellae are cholesterol rich as shown by filipin treatment. Filipin cholesterol deformations are seen as protuberances in the core lamellae in freeze fracture replicas (c) and as scalloped deformations of core lamellae and outer membrane of the lysosome in sections (d). X 33,000

60

Cholesterol accumulation in lysosomes resulting from uptake of LDL for 24 hours by Niemann Pick C fibroblasts was shown previously with immunostaining (7 and Fig. 4). Cholesterol loaded lysosomes are enlarged (mean of 2μ) as compared to lysosomes before loading (mean of 1μ). The cores of lysosomes are packed with cholesterol enriched multilamellae (Fig. 7).

Since lysosomes are the doorway through which LDL-derived cholesterol must pass prior to its transfer to other intracellular sites, accumulation of cholesterol in the *trans* Golgi cisternae suggested that the lysosomes of these mutant cells are not inherently limited in their ability to release a portion of their cholesterol to this distal organelle. However, transfer of endocytosed lysosomally derived cholesterol to other organelles such as the ER (4) and plasma membrane (6, 10) has been shown to be blocked in these mutant cells. These data collectively suggest that the Golgi complex may serve as an intermediate in the intracellular distribution of cholesterol and that the NP-C block could limit both movement of cholesterol through the Golgi complex as well as egress of cholesterol from lysosomes.

Normal Fibroblasts Treated with Progesterone

The normal intracellular distribution of endocytosed cholesterol in cultured fibroblasts can be disrupted with progesterone which induces an intracellular accumulation of cholesterol (16).

Cytochemical Characterization of the Progesterone Induced Golgi and Lysosomal Cholesterol Lipidosis.

Previous fluorescence studies showed cholesterol accumulates in lysosomes in normal human fibroblasts treated with LDL and progesterone for 24 hours (16) and (Fig. 8a). Subsequent washout of progesterone allows egress of cholesterol from lysosomes resulting in a burst of cholesterol esterification (16). Clearance of cholesterol from lysosomes during progesterone washout is retarded at lower temperatures (18-20°C) strongly suggesting post lysosomal egress of cholesterol is mediated by energy dependent vesicular transport (Figs. 8b & 8c).

Fig. 8. Fluorescence cytochemistry of cultured fibroblasts incubated with LDL and progesterone for 24 hours.
(a) Cells were cultured in the presence of LDL (50µg/ml) and progesterone (10µg/ml) for 24 hours. The intense intracellular fluorescence visible in the perinuclear region of the cells marks cholesterol loaded lysosomes. X 190.

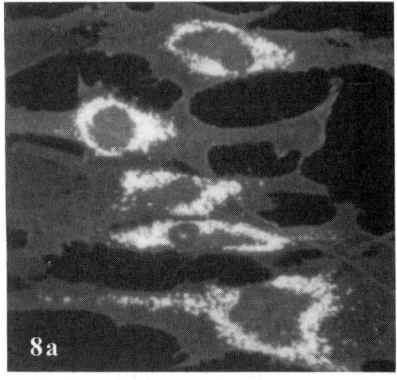

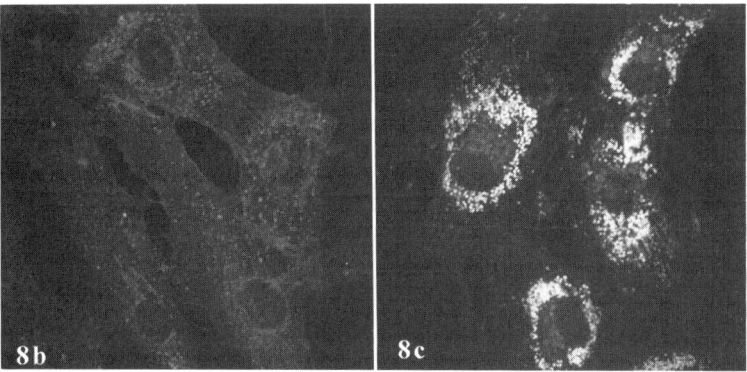

Fig. 8 cont. (b and c) Cells were incubated for an additional 24 hours in the absence of both progesterone and LDL at 37°C (b) or 20°C (c), followed by filipin staining as above. After this washout period only the cells incubated at 37°C (b) and not those incubated at 20°C (c), contain less intracellular filipin-cholesterol fluorescence when compared with cells in (a). X 285.

Electron Microscopic Freeze-Fracture Analysis of Golgi Cholesterol Distribution.

The ultrastructural distribution of cholesterol in the Golgi apparatus was analyzed by quantitative freeze-fracture and filipin cytochemistry.

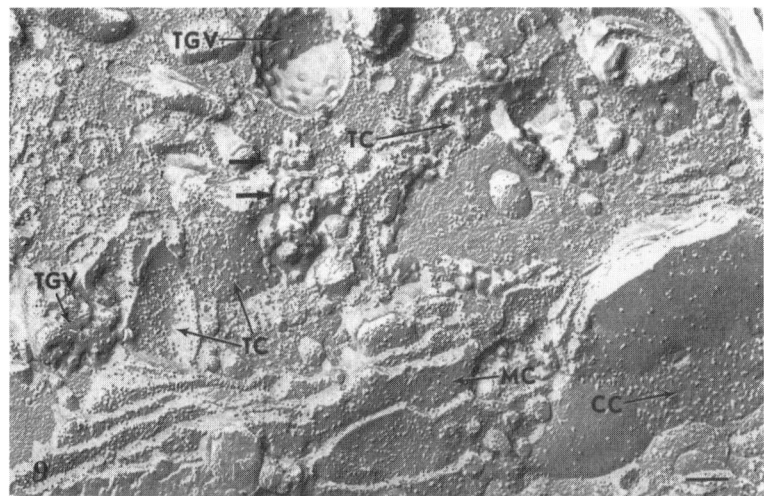

Fig. 9. Electron micrograph of replica of freeze-fractured Golgi in Normal fibroblast incubated with LDL and progesterone for 24 hours. *Trans* Golgi cisternae (TC) are "saturated" with filipin-cholesterol deformations (large arrows). *Trans* Golgi vacuoles (TGV) have few FCD's. *cis* cisternae, CC; *medial* cisternae, MC. X58.000.

Quantitative analyses of the filipin-cholesteol deformations shows that uptake of LDL, by itself, increases the density of filipin-cholesterol deformations in the total

Golgi membrane population (combined values obtained for each Golgi compartment) from 20 ± 7 FCD per μm2 to 86 ± 17 FCD per μm2 (Table 1). Cholesterol in the Golgi cisternal compartments increases in a cis to trans direction as previously described (13). Analyses of cells cultured in the presence of both LDL and progesterone also reveal an increase in total Golgi FCD from 13 ± 5 FCD/μm2 to 80 ± 19 FCD/μm2. However in contrast to cells treated with only LDL, cells exposed to both LDL and progesterone contain a significant increased accumulation of FCD in *trans* Golgi cisternae. The average density of FCD in the *trans* Golgi cisternae of these latter cells increased from the level of 19 ± 10 FCD/μm2 without LDL to 186 ± 30 FCD/μm2. Furthermore, in these cells the fraction of *trans* Golgi cisternal membrane area saturated with FCD (>500FCD/μm2) increased from 0% to 33%. No significant increase in the cholesterol content of *trans* Golgi cisternal membranes was detectable in cells after incubation with either progesterone or LDL alone (Table 1).

Table 1. Effect of progesterone on the enrichment of the Golgi with cholesterol during LDL uptake in cultured fibroblasts

Cholesterol depleted human fibroblasts were seeded in 150 mm glass culture dishes as described. Cells were cultured with or without LDL (50μg/ml), in the presence or absence of progesterone (10μg/ml) for 24 hours and then fixed in the presence of filipin and processed for freeze-fracture analysis.

Golgi Compartments	Density of filipin-cholesterol deformations (number FCD/μm2 ±SEM)			
	(-) Progesterone		(+)Progesterone	
	(-) LDL	(+) LDL	(-) LDL	(+)LDL
Cis/Medial Cisternae	3 ±2 (0%)[1]	18 ±12 (0%)	0 (0%)	5 ±3[d] (0%)
Trans Cisternae	16 ±6 (0%)	66 ±24 (0%)	19 ±10 (0%)	186 ±50[e] (33%)[f,g]
Trans Golgi Vacuoles	76 ±37 (0%)	194 ±43[c] (18%)	14 ±7 (0%)	68 ±18 (0%)
Total Golgi	20 ±7 ----	86 ±17[a] ----	13 ±5 ----	80 ±19[b] ----

(1) percent of membranes saturated (>500 FCD/μm2)
(a) different from Golgi of cells incubated without LDL, P=<0.01
(b) different from Golgi of cells incubated with progesterone only, P=<0.01
(c) different from cis/medial cisternae of cells incubated with LDL only, P=<0.01
(d) different from trans cisternae (P=<0.01) and trans Golgi vacuoles (P=<0.05) of cells incubated with LDL and progesterone
(e) different from trans cisternae of cells incubated with progesterone alone, P=<0.01
(f) different from cis/medial cisternae (P=<0.01) and trans Golgi vacuoles (P=<0.05) of cells incubated with LDL and progesterone
(g) different from trans cisternae of cells incubated with progesterone only (P=<0.01), without LDL (P=<0.01) and with LDL only (P=<0.05)

Evaluation of the effect of progesterone on Golgi cholesterol distribution in normal human fibroblasts incubated with LDL shows progesterone does not effect the amount of cholesterol enrichment of the Golgi due to LDL uptake but does cause an abnormal accumulation of cholesterol in *trans* Golgi cisternae (Fig.10).

**EFFECT OF PROGESTERONE ON THE
ENRICHMENT OF THE GOLGI WITH CHOLESTEROL
DURING LDL UPTAKE IN NORMAL FIBROBLASTS**

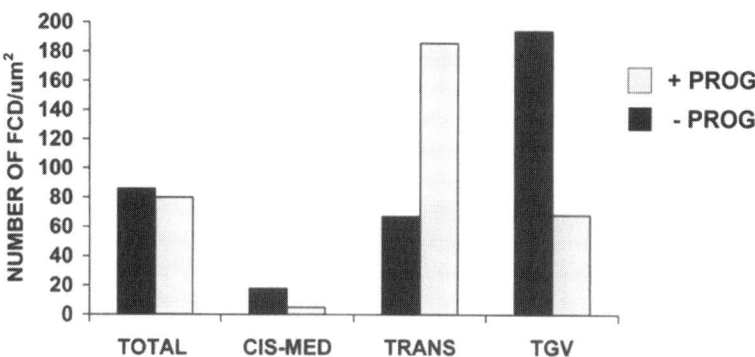

Fig. 10. Normal fibroblasts were grown with LDL for 24 hours with and without progesterone (10μg/ml). Total = Total Golgi Membranes (*cis,medial trans* cisternae and *trans* Golgi vacuoles.

Uptake of LDL for 24 hours in the presence of progesterone did not alter the increase in Golgi cholesterol induced by LDL uptake alone, previously shown for normal cells and NPC cells (13). However progesterone did cause an abnormal accumulation of cholesterol in *trans* Golgi cisternae similar to that caused by the NPC mutation.

Effect of Secondary Drug treatments On Lysosomal Cholesterol Transport During Progesterone Washout.

The ability of cells to reestablish cholesterol egress from lysosomes during progesterone removal was monitored in the presence of agents such as NH_4CL, monensin, (17), C_6-ceramide (18), brefeldin-A (19) which perturb various functions of the cellular vacuolar system. A qualitative evaluation of the effect of these drugs on depletion of cholesterol from lysosomes and Golgi of cultured fibroblasts was carried out by fluorescence cytochemistry. Filipin-cholesterol fluorescence in the aggregated perinuclear lysosomes, (shown in Fig. 8a) is diminished after a 3 hr period of progesterone washout. However, the loss of perinuclear fluorescence during progesterone-washout (Fig. 11a) is retarded by NH_4Cl (Fig. 11b), C_6-ceramide (Fig. 11c) or monensin (Fig. 11d) with filipin-cholesterol fluorescence remaining in structures consistent with the Golgi complex and lysosomes.

64

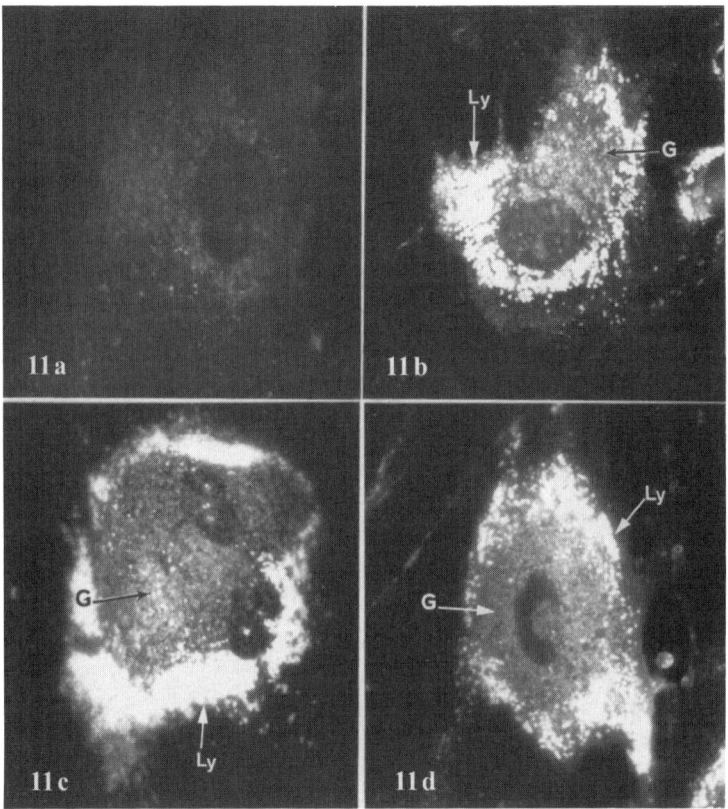

Fig. 11. Fibroblasts were cultured in the presence or absence of LDL (50µg/ml) and progesterone (10µg/ml) for 24 hours. Cells were subsequently washed free of the lipoprotein and steroid and incubated an additional 3 hours in (a) culture medium without additives or with (b) 15mM NH$_4$Cl, (c) 10 µM C$_6$-ceramide or (d) 10µM monensin. Cells were fixed and treated with filipin. Filipin-cholesterol fluorescence is lost from cells if the progesterone washout is carried out in medium alone (a) but is retained if either NH$_4$Cl (b), C$_6$-ceramide (c) or monensin (d) are added to the medium. X 470. The fluorescence is retained in perinuclear organelles that were identified cytochemically as lysosomes (Ly) and Golgi (G) (data not shown). X 470

The reduced mobilization of cholesterol from lysosomes and Golgi of cells treated with NH$_4$Cl, monensin, or C$_6$-ceramide, shown above, does not reflect a drug-induced delay in the removal of progesterone from the cultures. Following 24 hours of incubation with [^{3}H]-progesterone (10µg/ml) and LDL, clearance of the labeled steroid from cell cultures by washing occurred with a half life of 30 to 40 min. This rate of progesterone removal was not significantly affected when the drugs indicated above were present during the washout procedure. Movement of cholesterol to endoplasmic reticulum from lysosomes as monitored by a burst in cholesterol esterification (16) is also retarded by the presence of NH$_4$CL, monensin and C$_6$-ceramide during progesterone removal (Table 2). Although NH$_4$CL and monensin can more directly affect ACAT activity somewhat as detected by their ability to partially block cellular activation of ACAT with 25-hydroxysterol (Table 2), this mode of action is less marked than the block in cholesterol esterification induced by these same drugs during progesterone washout (Table 2). This suggests that these drugs primarily affect cholesterol esterification during washout by delaying the reversal of the sterol sequestration and supports the cytochemical evidence that they retard mobilization of lysosomal cholesterol. C_6-ceramide strongly affected esterification associated with both oxysterol treatment as well as progesterone washout. These drugs do not appear to effect the endoplasmic reticulum in a non-specific fashion since triglyceride synthesis is not disturbed (Table 2). Brefeldin A does not appear to substantially affect movement of lysosomal cholesterol to the ER as determined by its minimal suppression of cholesterol esterification. This may be due to the BFA induced redistribution of Golgi elements to the ER (19).

Table 2: Cellular cholesterol esterification associated with progesterone washout and 25-hydroxycholesterol treatment

Cholesterol depleted human fibroblasts were seeded in standard 6 well plates and cultured with 10% LPDS-EMEM media for two days followed by an additional two days with 1% bovine serum albumin in EMEM. Cells were subsequently cultured with LDL (50µg/ml) in the presence or absence of progesterone (10µg/ml) for 24 hours. All cultures were quickly washed three times with 10% LPDS containing 100µM (^{3}H)-oleate (200dpm/nmol). Cells not treated with progesterone during the LDL-uptake phase were exposed to 25-hydroxycholesterol (5µg/ml) during the addition of (^{3}H)-oleate. Cultures were concurrently exposed to secondary drugs as indicated for 6 hours at 37° C. Cellular synthesis of cholesteryl-(^{3}H)oleate and (^{3}H)-triglyceride was measured as described in methods. The data represent the average value ±S.D. determined for six separate culture wells.

Secondary drug exposures	Cholesteryl ester synthesis		Triglyceride synthesis	
	Steroid washout	25-OH Chol addition	Steroid washout	25-OH Chol addition
	(nmoles/mg protein/6 hours)			
None	4.2 ±0.6	6.2 ±0.7	3 ±1	10 ±2
15 mM NH$_4$CL	0.4 ±0.1	4.0 ±0.3	7 ±2	11 ±1
10 µM Monensin	0.7 ±0.1	3.7 ±0.4	5 ±1	7 ±1
10µM C$_6$-Ceramide	0.2 ±0.0	0.5 ±0.0	4 ±0	2 ±0
10 µM Brefeldin-A	2.0 ±0.1	4.2 ±0.3	8 ±1	3 ±0

The mode by which progesterone blocks intracellular trafficking of lysosomal cholesterol is not known. It has been reported that inhibition of vacuolar H+-ATPase blocks cholesterol egress from lysosomes (20). The anomalous accumulation of cholesterol in lysosomes and the *trans* cisternae of the Golgi complex induced by progesterone, however, does not appear to be a result of pH-mediated perturbations of the vacuolar transport pathway since the receptor-mediated endocytosis and subsequent degradation of ^{125}I-LDL are unaffected in progesterone treated cells. Similarly the ability C$_6$-ceramide to block cholesterol mobilization from lysosomes and Golgi also does not appear to involve pH perturbations (unpublished data). Based on the abnormal accumulation of cholesterol in lysosomes and the Golgi complex it is tempting to speculate that the site of progesterone interdiction lies within these organelles. pH perturbations can not distinguish between these sites since both have acidic compartments that can be simultaneously neutralized by drugs (17). The observed proximity of lysosomes and Golgi during LDL uptake (Figure 4) may provide for efficient transfer of cholesterol between these organelles. and progesterone may block efficient transfer of endocytosed cholesterol from both lysosomes and Golgi. It is also possible that progesterone blocks transfer of LDL-derived cholesterol through the Golgi rather than directly affecting the initial egress from lysosomes. C_6-ceramide, which has been shown to disrupt the Golgi apparatus and Golgi-mediated transport processes (18), may likewise block cholesterol transport because of Golgi perturbations. The Golgi complex may have limited capacity to accommodate excess cholesterol in its membranes as a consequence of its relatively limited membrane surface area (21) and low capacity for phospholipid biosynthesis (22). This saturation of Golgi membrane with cholesterol could block further cholesterol accumulation in the Golgi resulting in accumulation of LDL-derived cholesterol in lysosomes.

ENDOPLASMIC RETICULUM CHOLESTEROL DISTRIBUTION IN NORMAL CELLS AFTER LDL UPTAKE DURING ACAT INHIBITON

Our knowledge of the ability of endoplasmic reticulum membranes to accommodate unesterified cholesterol is derived from studies on normal fibroblasts incubated with a specific inhibitor of acyl-CoA:cholesterol acyltransferase (ACAT), S58035, during 24 hours of LDL uptake. This treatment induces intracellular accumulation of unesterified cholesterol (23). Filipin cytochemical evaluation of cellular sites of cholesterol accumulation with fluorescence microscopy showed cholesterol in discrete cytoplasmic domains (Fig. 12a). Subsequent removal of S58035 from the incubation medium releases ACAT inhibiton resulting in a burst in cholesterol esterification (16) which is accompanied by disappearance of filipin positive domains and appearance of neutral lipid (cholesteryl ester) droplets in the cells. Electron microscopic evaluation of cultured fibroblasts showed that the endoplasmic reticulum was the site of cholesterol accumulation induced by LDL uptake during ACAT inhibition. Treated cells accumulated aggregates of smooth-membrane bounded tubules and lamellae continuous with endoplasmic reticulum membranes and enriched in cholesterol as determined with filipin (Fig. 13). Continuity between the smooth cholesterol rich membranes and ER was also discerned by freeze-fracture analysis. In untreated cells the hydrophobic surfaces of membrane leaflets of endoplasmic reticulum contain intramembranous particles (IMP's), (13,24). However, in fibroblasts treated with both S58035 and LDL there were large expanses of ER membrane leaflets devoid of IMP's (Fig. 14a) and in cross-fractures these accumulations of particle free ER membrane leaflets appeared lamellar in structure (Fig. 14a, arrowhead). After filipin treatment, regions of ER membrane leaflets without IMP's became filled with filipin cholesterol deformations visualized in freeze fracture replicas as pits and protuberances on the hydrophobic face of the membrane leaflet (Fig. 14b). Thus the tubular and lamellar membrane extensions of the RER are cholesterol-enriched domains of the ER that are present only in cells treated with both S 58035 and LDL and they correspond to the filipin fluorescent structures visualized by light microscopy (Fig. 12a).

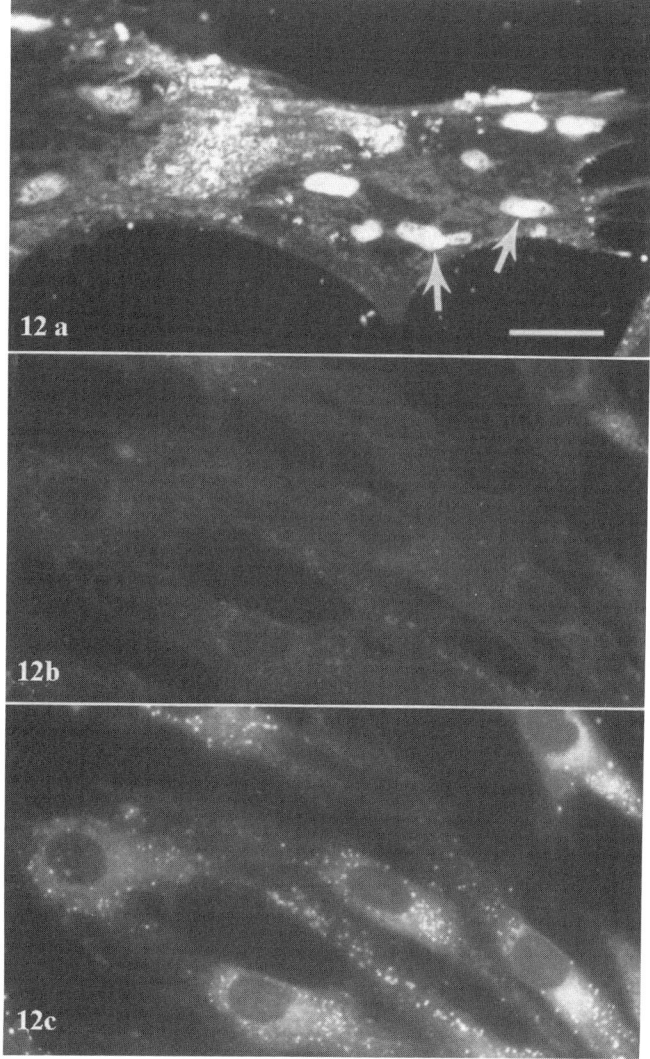

Figure 12. Fluorescence cytochemical documentation of intracellular cholesterol accumulation in fibroblasts incubated with LDL and the acyl-CoA:cholesterol acyltransferase inhibitor, S 58035. Cells were incubated either with (a) both LDL (50mg/ml) and S 58035 (2mg/ml) for 24 hours or with (b & c) both LDL and S 58035 followed by a 24 hr incubation without LDL and S 58035 and subsequently fixed and stained with filipin. Cells incubated with both LDL and S 58035 (a) display extensive intracellular filipin-positive areas (arrows) in the cytoplasm. Fibroblasts incubated for an additional 24 hr in LPDS after removal of LDL and S 58035 (b) do not contain filipin-positive areas but (c) do contain numerous Nile Red positive lipid droplets (b and c show the same cells). Bar=50μ

68

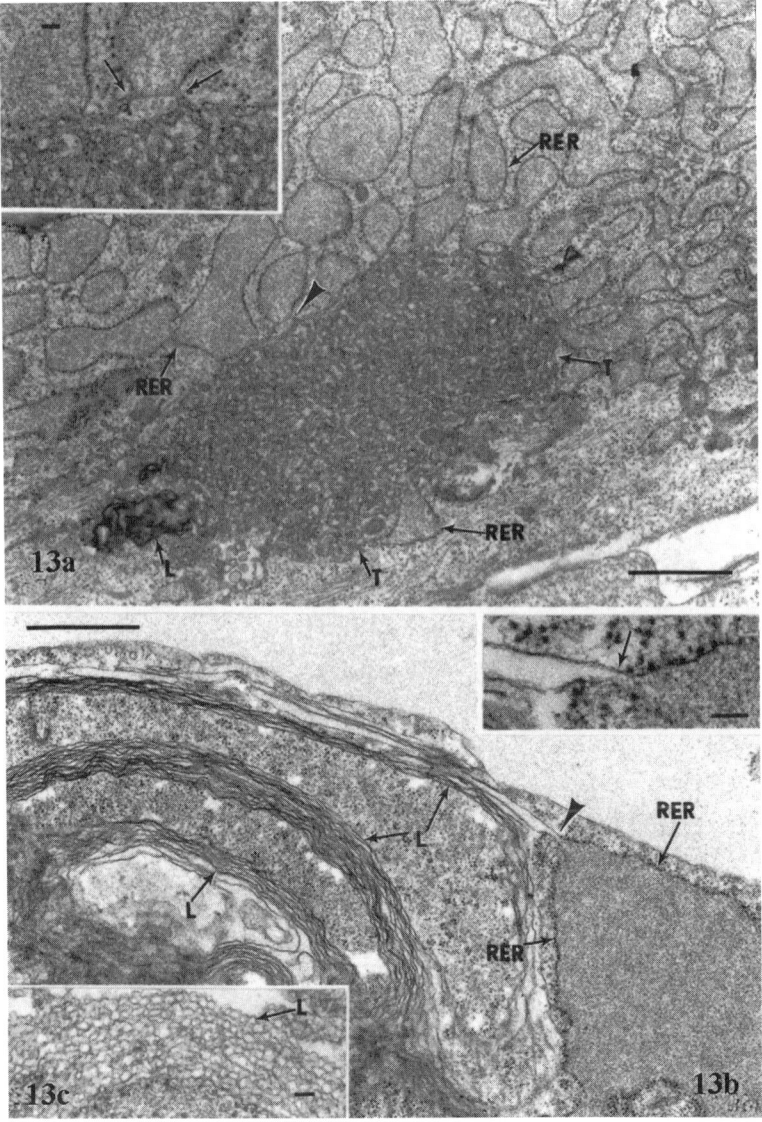

Figure 13. Structure of endoplasmic reticulum in fibroblasts incubated with LDL and 58035. Cells were incubated with LDL (50mg/ml) and S 58035 (2mg/ml) for 24 hours followed by fixation (a,b) and exposure to filipin ©. (a) Large aggregates of membranous tubules (T) are interspersed among profiles of rough endoplasmic reticulum (RER). A smaller lamellar structure (L) can also be seen in the area of the tubular tangle. Inset: Enlarged area of (a) at arrowhead showing continuity (arrows) between smooth membranous tubules and rough endoplasmic reticulum. (b) Larger aggregates of lamellae (L) are associated with membranes

69

of rough endoplasmic reticulum (RER). Inset: Enlarged area of (b) at arrowhead showing continuity (arrow) between membranes of smooth lamellae and rough endoplasmic reticulum. (c) lamellae in cells treated with filipin show scalloped appearance indicative of filipin-cholesterol deformations and cholesterol enrichment. Bars in (a) and (b) = 1.0μ; Bars in insets = 0.1μ.

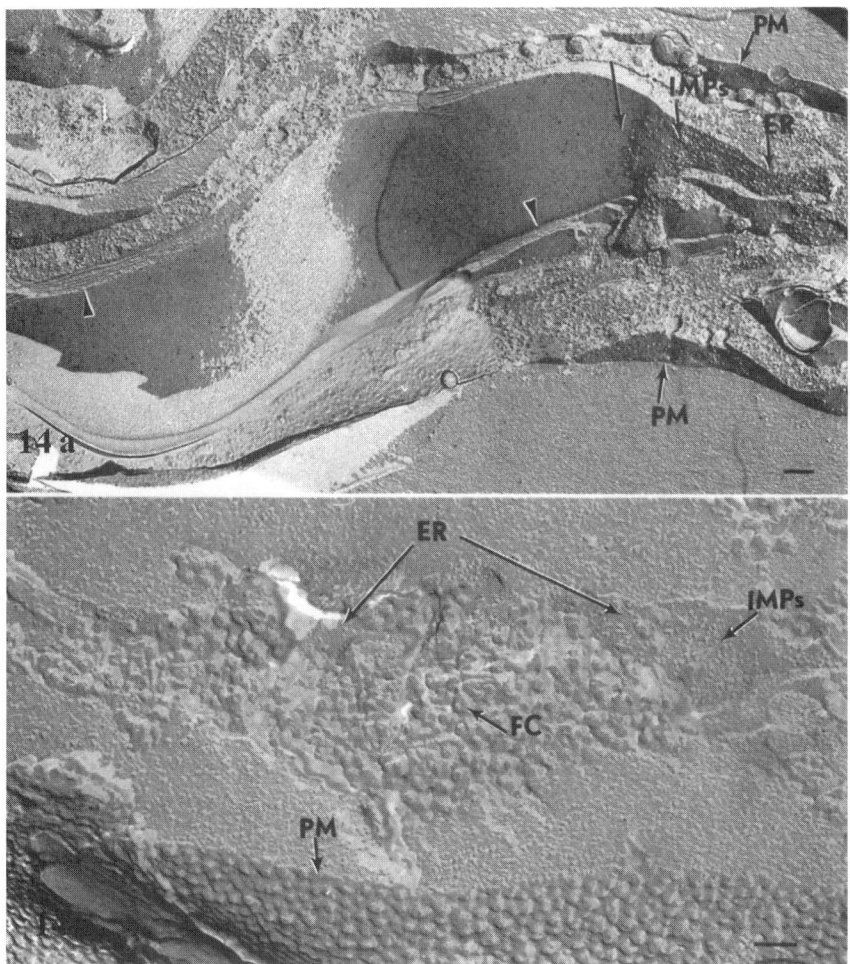

Figure 14.   Freeze-fracture documentation of cholesterol accumulation in endoplasmic reticulum. Cells were incubated with LDL (50mg/ml) and S 58035 (2mg/ml) for 24 hours followed by fixation (a) and exposure to filipin (b). (a) The cytoplasmic leaflet of endoplasmic reticulum (ER) studded with intramembranous particles (IMP's) is continuous with the particle-free leaflet (at arrow) and is associated with lamellae (arrowheads). Plasma membrane, PM. (b). Cells exposed to filipin show deformations characteristic of filipin-cholesterol complexes (FC) in membranes of endoplasmic reticulum (ER) that in adjacent areas contain IMP's.   Filipin-cholesterol deformations are also present in the plasma membrane (PM). Bar =0.1μ

Although the cholesterol content of the endoplasmic reticulum is normally low (23, 25), direct inhibition of cholesterol esterification by blocking ACAT activity with S58035 (26) during the cellular processing of LDL induces extensive accumulation of unesterified cholesterol at the RER. In such cells, even though the cellular content of unesterified cholesterol can reach levels seen in both NP-C and progesterone treated cells (16) there is no discernable sequestration of cholesterol in lysosomes (16). We now find that direct inhibition of cholesterol esterification can disrupt the transport pathway of LDL cholesterol at yet another cellular location, the endoplasmic reticulum itself. Our results show that cells can accumulate cholesterol-enriched domains within membrane leaflets of the endoplasmic reticulum. The RER appears to possess the capacity to sustain incorporation of cholesterol from other cellular membranes. The expansive membrane area (21) and high capacity for phospholipid synthesis (22) of the RER may facilitate its ability to accumulate high levels of membrane cholesterol in expanding domains visualized as the membrane whorls lacking intramembranous particles (IMP's) in freeze fracture replicas (Fig. 3). These IMP's, a unique feature of the hydrophobic faces of leaflets of cellular membranes representing the presence of integral membrane proteins (24) are also cleared from other lipid domains, such as those formed by accumulations of fatty acids in cellular membranes (27). The notion that the endoplasmic reticulum can act as a cholesterol sink is supported by a recent study showing stimulation of phospholipid biosynthesis when ACAT is inhibited in murine macrophages processing exogeneous cholesterol (28). The presence of multilamellar whorls in these macrophages may correspond to the multilamellar cholesterol enriched extensions of the endoplasmic reticulum identified in the current study after incubation of fibroblasts with LDL and 58035. In our studies, both the phospholipid and cholesterol content of the fibroblast lamellar whorls and tubules have been verified. The accumulation of phospholipid in these membranes was confirmed by the fact that the lamellar and tubular structures were partially preserved by permanganate fixation (results not shown) specific for phospholipid (29). The cholesterol content of the structures was in turn confirmed by their reaction with the specific cholesterol marker, filipin (7,13).

CONCLUSIONS

The Golgi complex of normal and NPC cells is enriched with cholesterol by endocytic uptake of LDL. Cholesterol accumulation in the Golgi complex of cells with the NP-C lesion (7,13) and the progesterone-induced lipidosis (16) identify that this organelle is a pivitol component of post-lysosomal cholesterol transport to the plasma membrane. To date, accumulation of cholesterol in the *trans* Golgi cisternae of cells occurs in tandem with lysosomal accumulation. Block in lysosomal/Golgi transport of LDL derived cholesterol results in lack of cellular homeostatic responses. Inhibition of the cells ability to esterify cholesterol with S 58035 does not appear to disrupt those segments of the intracellular cholesterol transport pathway that lie proximal to the RER and does not compromise the induction of homeostatic responses. Release of this esterification block upon removal of the drug indicates that ACAT can access the extensive accumulations of endoplasmic reticulum membrane cholesterol. It has been shown that specific inhibition of ACAT during LDL uptake can enhance down-regulation of both the LDL-receptor and 3-hydroxy-3-methylglutaryl-CoA-reductase in murine macrophages (30). The sterol composition of membranes can play a critical role in regulation of gene expression (31). A transcription factor that regulates expression of the genes for the LDL receptor and 3-hydroxy-3-methylglutaryl CoA synthetase translocates to the nucleus following proteolytic release from a larger inactive precursor protein located in the nuclear envelope and endoplasmic reticulum. Sterol inhibits proteolytic release of active transcription factor thereby blocking expression of the genes (31). These findings and our demonstration of cholesterol accumulation in the RER during LDL uptake may explain the enhanced down regulation of the LDL receptor and *de novo* cholesterol synthesis that occurs in ACAT inhibited cells.

REFERENCES

1. Wattenberg BW, Silbert DF. Sterol partitioning among intracellular membranes. J Biol. Chem 1983; 258; 2284-89
2. Brown MS, Goldstein JL. A receptor-mediated pathway for cholesterol homeostasis. Science 1986; 232: 34-47.
3. Liscum L, Dahl NK. Intracellular cholesterol transport J. Lipid Res. 1992; 33: 1239-54
4. Sokol J, Blanchette-Mackie EJ, Kruth HS, Dwyer NK, Amende LM, Butler JD, Robinson E, Patel S, Brady RO, Comly ME, Vanier MT, Pentchev PG: Type-C Niemann-Pick Disease: Lysosomal accumulation and defective intracellular mobilization of LDL-cholesterol. J. Biol. Chem. 1988 263: 3411-17
5. Pentchev PG, Comly ME, Kruth HS, Tokoro T, Butler JD, Sokol J, Filling-Katz M, Quirk, J.M, Marshall DC, Patel S, Vanier MT, Brady RO. Group C Niemann-Pick disease: Faulty regulation of low-density lipoprotein uptake and cholesterol storage in cultured fibroblasts. FASEB J. 1987; 1: 40-45
6. Liscum L, Ruggiero RM, Faust JR. The intracellular transport of low density lipoprotein-derived cholesterol is defective in Niemann-Pick type C fibroblasts. J. Cell Biol. 1989; 108:1625-16367.
7. Blanchette-Mackie EJ, Dwyer NK, Amende LM, Kruth HS, Butler JD, Sokol J, Comly ME, Vanier MT, August JT, Brady RO, Pentchev PG. Type C Niemann-Pick Disease: low density lipoprotein uptake is associated with premature cholesterol accumulation in the Golgi complex and excessive cholesterol storage in lysosomes. Proc Natl Acad. Sci. USA 1988; 85: 8022-26
8. Martin OC, Comly ME, Blanchette-Mackie EJ, Pentchev PG, Pagano RE. Cholesterol deprivation affects the fluorescence properties of a ceramide analog at the Golgi apparatus of living cells. Proc. Natl. Acad. Sci. 1993; 90: 2661-65.
9. Pagano RE, Sepanski MA, Martin OC. Molecular trapping of a fluorescent ceramide analogue at the Golgi apparatus of fixed cells: interaction with endogeneous lipids provides a trans-Golgi marker for both light and electron microscopy. J. Cell Biol. 1989; 109: 2067-79

10. Neufeld EB, Cooney AM, Pitha J, Dawidowitcz EA, Dwyer NK, Pentchev PG, Blanchette-Mackie EJ. Intracellular trafficking of cholesterol monitored with a cyclodextrin, J Biol Chem 1996; 271: 21604-13.

11. Alcade J, Bonay P, Roa A, Vilaro S, Sandoval V. Assembly and disassembly of the Golgi complex: two processes arranged in a cis-trans direction. J Cell Biol. 1992; 6: 69-83

12. Anderson RGW, Pathak RK: Vesicles and cisternae in the trans Golgi apparatus of human fibroblasts are acidic compartments. Cell 1985; 40: 635-43

13. Coxey RA, Pentchev PG, Campbell G, Blanchette-Mackie EJ. Differential accumulation of cholesterol in Golgi compartments of normal and Niemann-Pick type C fibroblasts incubated with LDL: a cytochemical freeze-fracture study. J. Lipid Res. 1993; 34: 1165-1176

14. Orci L, Montesano R, Meda P, Malaisse-Lagae F, Brown A, Perrelet A, Vassalli P. Heterogeneous distribution of filipin-cholesterol complexes across cisternae of the Golgi apparatus. Proc. Natl. Acad. Sci USA. 1981; 78: 293-97.

15. Carstea E.D., Morris JA, Coleman KG, Loftus SK, Zhang D, Cummings C, Gu J., Rosenfeld   MA, Pavan WJ, Krizman DB, Nagle J, Polymeropoulos MH, Sturley SL, Ioannou YA,   Higgens ME, Comly M, Cooney A, Brown A, Kaneski CR, Blanchette-Mackie EJ, Dwyer   NK, Neufeld EB, Chang T-Y, Liscum L, Strauss JF, Ohno K, Zeigler M, Garmi R, Sokol J,   Markie   D, O'Neill RR, van Diggelen OP, Elleder M, Patterson MC, Brady, Vanier MT, Pentchev PG and DA Tagle. Niemann-Pick C1 disease gene:homology to mediators of cholesterol homeostasis. Science 1997 Jul 11; 277(5323):228-31

16. Butler J, Blanchette-Mackie EJ, Goldin E, O'Neill RR, Carstea GD, Roff CF, Patterson   MC, Patel S, Comly ME, Cooney AM, Vanier MT, Brady RO, Pentchev PG: Progesterone blocks cholesterol translocation from lysosomes. J. Biol. Chem. 1992; 267: 23797-805

17. Mellman I, Fuchs R, Helenius A. Acidification of the endocytic and exocytic pathways. Ann. Rev. Biochem. 1986; 55: 663-700

18. Rosenwald AG, Pagano RE: Inhibition of glycoprotein traffic through the secretory pathway by ceraqmide. J. Biol. Chem. 1993; 268: 4577-79

19. Klausner RD, Donaldson JG, Lippincott-Schwartz J: Brefeldin A: Insights into the control of membrane traffic and organelle structure. J. Cell Biol. 1992; 116: 1071-80

20. Furuchi T, Aikawa K, Arai H, Inoue K. Bafilomycin A1, a specific inhibitor of vacuolar H+-ATPase, blocks lysosomal cholesterol trafficking in macrophages. J. Biol. Chem 1993; 268: 27345-48

21. Bolender RP. Correlation of morphometry and stereology with biochemical analysis of cell fractions. Int. Rev. Cytology 1978; 55: 247-89

22. Voelker DR.   Organelle biogenesis and intracellular lipid transport in eukaryotes. Microbiol. Rev. 1991; 55: 543-60

23. Lange Y. Disposition of intracellular cholesterol in human fibroblasts. J. Lipid Res. 1991; 32: 329-39

24. Rash JE, Johnson TA, Dinchuk JE, Duch DS, Levinson SR (in *Freeze-Fracture Studies of Membranes.*) 1989; ed. S.W. Hui. CRC Press, Boca Raton, Florida. 41-59.

25. Urbani L, Simoni RD. Cholesterol and vesicular stomatitis virus G protein take separate routes from the endoplasmic reticulum to the plasma membrane. J. Biol. Chem. 1990; 265: 1919-23,

26. Jamal Z, Siffolk A, Boyd GS, Suckling KE. Metabolism of cholesterol ester in monolayers of bovine adrenal corticla cells. Effect of an inhibitor of acyl-CoA:cholesterol acyltransferase. Biochim. Biophys. Acta 1985; 834:230-37

27. Amende LM, Blanchette-Mackie, E.J, Scow RO. Demonstration of fatty acid domains in membranes produced by lipolysis in mouse adipose tissue: A freeze-fracture study. *Cell Tissue Res.* 1986; 246:495-508

28. Shiratori Y, Okwu AK, Tabas I. Free cholesterol loading of macrophages stimulates phosphatidylcholine biosynthesis and up-regulation of CTP: phosphocholine cytidyltransferase. *J. Biol. Chem.* 1994; 269:11337-48

29. Hayat M.A. (in *Fixation for Electron Microscopy*) 1981; ed Hayat, Academic Press, New York, N.Y. 1-50

30. Tabas I, Rosoff WJ, Boykow GC. Acyl-coenzymeA: cholesterol acyltransferase in macrophages utilizes a cellular pool of oxidase accessable cholesterol as substrate. J. Biol. Chem. 1988; 263: 1266-72

31. Wang X, Sato R, Brown MS, Hua X, Goldstein JL, SREBP-1, a membrane-bound transcription factor released by sterol regulated proteolysis. Cell 1994; 77: 53-62

Analysis of Somatic Cell Mutants That Express Defective Intracellular Cholesterol Transport

Laura Liscum

Department of Physiology, Tufts University School of Medicine
Boston, MA 02111; e-mail: lliscum@opal.tufts.edu

KEY WORDS: cholesterol, intracellular transport, somatic cell mutant, Niemann-Pick type C

ABSTRACT

The distribution of cholesterol in mammalian cells is tightly controlled. However, the factors involved in shuttling cholesterol among cellular membranes have not been well defined. Our experimental approach employs somatic cell mutants and pharmacological agents to investigate the routes and mechanisms of intracellular cholesterol transport. Two complementation groups of Chinese hamster ovary cell mutants have been isolated, with gene defects that impair separate cholesterol transport pathways. Identification of genes that correct the cholesterol transport defects is in progress. Pharmacological inhibitor studies have revealed a pathway that lipoprotein-derived cholesterol takes to reach the endoplasmic reticulum. The mechanism of that transport pathway is under investigation. In coming years, our studies should turn from biochemical description of transport pathways to molecular analysis of candidate transport proteins.

CONTENTS

INTRODUCTION

In mammalian cells, most cellular cholesterol is localized in the plasma membrane (1). There, cholesterol's modulation of membrane fluidity is likely to have effects on membrane receptor, channel and transport functions. However, cholesterol is not static in the plasma membrane. Instead, it is dynamic, continuously cycling between the cell interior and cell surface (2).

Cultured mammalian cells acquire cholesterol by two coordinately regulated processes, *de novo* synthesis in the endoplasmic reticulum (ER) and internalization of cholesterol rich plasma lipoproteins, such as low density lipoprotein (LDL) (3). Cholesterol obtained by either mechanism becomes available for incorporation into membranes, efflux to extracellular acceptors, and metabolism to cholesteryl esters, bile acids, and steroid hormones.

While much exciting progress has been made in defining the mechanisms by which cholesterol homeostasis is regulated (4), our knowledge of the mechanisms and pathways of cholesterol movement to regulatory sites has lagged. One traditional approach to identification of lipid transport factors has been to isolate soluble proteins that facilitate lipid movement in *in vitro* assays. However, subsequent demonstration of the physiological significance of such proteins in intact cells has, in many cases, been problematic.

One clue that the delivery of LDL-cholesterol to regulatory sites must be protein-mediated is the existence of the human genetic disease, Niemann-Pick type C (NPC). Cultured skin fibroblasts from individuals with NPC exhibit defective movement of LDL-cholesterol due to a single gene mutation (5, 6). The recent cloning of the NPC1 gene (7) provides the first window on a *bona fide* cholesterol transport molecule (E. Neufeld and P. Pentchev, this volume).

Our approach has centered on isolating and characterizing a family of somatic cell mutants with single gene defects that impair intracellular cholesterol transport. Examination of these mutants is helping us to evaluate cholesterol transport pathways in normal cells.

ISOLATION OF CHOLESTEROL TRANSPORT MUTANTS

Two complementation groups of Chinese hamster ovary (CHO) cell mutants were isolated using amphotericin B as the selection agent. Amphotericin B is a polyene antibiotic that complexes with cholesterol and forms pores, lysing cells that have a threshold level of plasma membrane cholesterol (8, 9). Our selection protocol was designed to isolate cells with low plasma membrane cholesterol due to defects in LDL-cholesterol transport to the cell surface (10, 11). Each mutant class that we obtained exhibits profoundly defective LDL-mediated stimulation of cholesterol esterification. Thus, the mutant classes are termed *c*holesterol *e*sterification *d*efective, ced-1 and ced-2.

Ced-1 mutants have an NPC phenotype

The ced-1 mutant class is characterized by expression of the classical NPC biochemical phenotype (10). Ced-1 mutants show lysosomal storage of LDL-cholesterol due to impaired transport of cholesterol from lysosomes to other cell membranes (Figure 1). The transport defect is accompanied by abnormal LDL-mediated regulation of cellular cholesterol homeostasis. Our finding that homeostatic responses are elicited normally by oxysterols, such as 25-hydroxycholesterol, indicates that the gene mutation causes a sterol trafficking defect and not a defect in signaling.

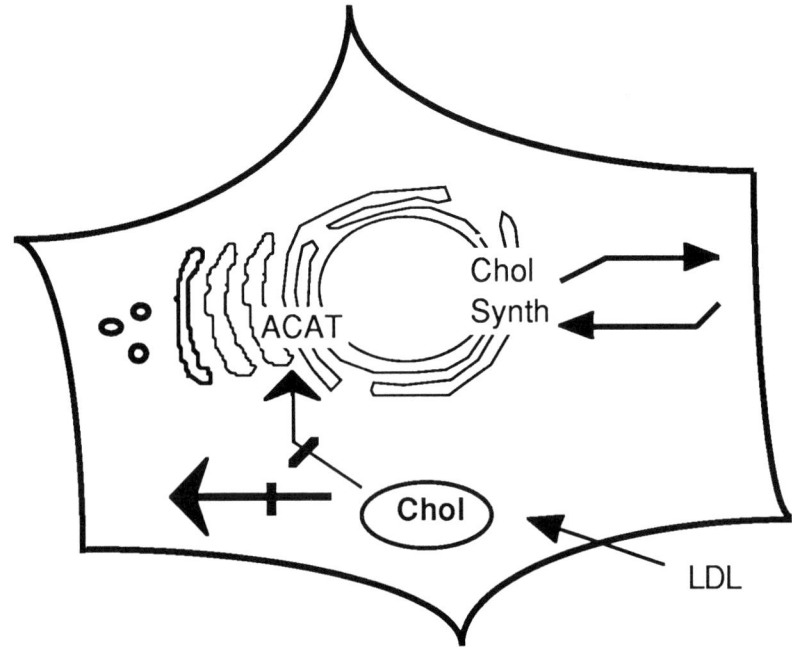

Figure 1. Intracellular cholesterol transport and regulation in ced-1 mutants. The internalization and lysosomal hydrolysis of LDL is normal. However, the movement of LDL-cholesterol to the plasma membrane and regulatory sites, such as ACAT in the ER, is greatly delayed. Cholesterol cycling from the ER to the plasma membrane and back appears normal.

Like mutant CT-60 isolated by Cadigan et al. (12), ced-1 mutants represent a somatic cell model for NPC. The recent utilization of CT-60 in the positional cloning of NPC1 illustrates the power of such model systems (13).

Our hypothesis is that ced-1 cells harbor a mutation in the NPC1 gene, although the genetic evidence in support of this is still indirect. Complementation analysis showed that ced-1 mutants are in the same class as mutant CT-60 (11), while CT-60 appears to be in the same class as NPC1 (13). Examination of the NPC1 locus in ced-1 cells is in progress.

Ced-2 mutation affects mobilization of plasma membrane cholesterol to the ER

Among our collection of cholesterol transport mutants, one cell line, mutant 3-6, belongs to a separate complementation class (11). Early analysis of this mutant was quite puzzling because it showed that the movement of LDL-derived as well as endogenously synthesized cholesterol to the plasma membrane is normal (11, 14). Despite the movement of cholesterol to the plasma membrane, mutant 3-6 is resistant to amphotericin B-mediated cell killing. The plasma membrane cholesterol content was probed by cholesterol oxidase treatment and Semliki Forest virus infection (14), as well as plasma membrane isolation (J. Reodica, N. Jacobs, and L. Liscum, unpublished observations). All methods showed the cholesterol content to be the same in parental CHO and mutant 3-6 cells. Nevertheless, the mutant cells show strong resistance to amphotericin B, whether the source of cholesterol is LDL, endogenously-synthesized cholesterol, or cholesterol sulfate added to the medium. We conclude that the amphotericin B resistance is the result of an altered plasma membrane cholesterol organization, rather than cholesterol content.

The defective plasma membrane cholesterol organization is linked to one defective cholesterol transport pathway, the movement of plasma membrane cholesterol to the cell interior (Figure 2). Arrival of cholesterol in the cell interior is measured by its ACAT-catalyzed esterification to cholesteryl esters. Both the basal movement of plasma membrane cholesterol to ACAT (L. Liscum, unpublished observation) and the movement stimulated by sphingomyelinase treatment of cells (14) is profoundly defective.

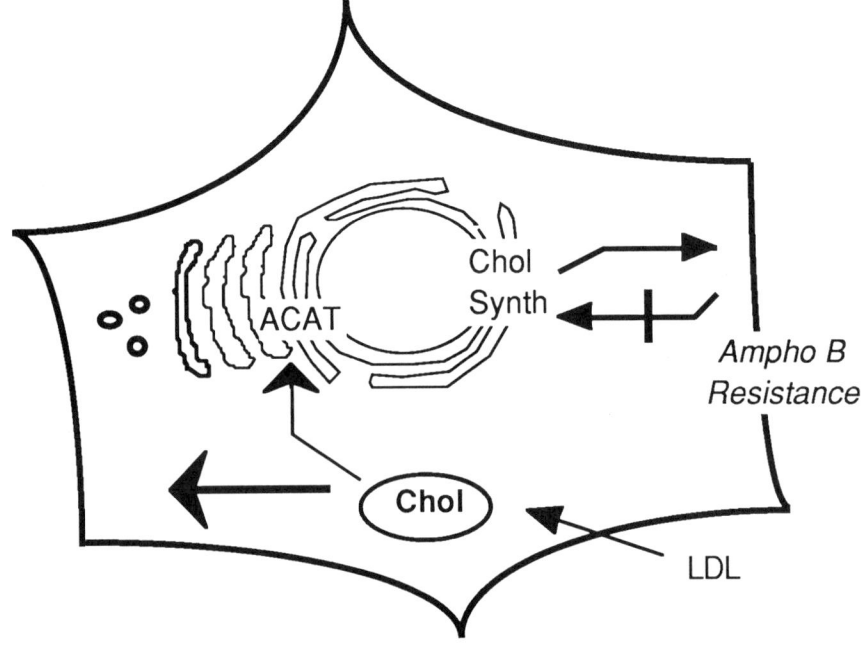

Figure 2. Intracellular cholesterol transport and regulation in ced-2 mutant 3-6. The internalization and lysosomal hydrolysis of LDL is normal. The movement of endogenously synthesized and LDL-cholesterol to the plasma membrane is normal. However, cholesterol cycling from the plasma membrane to the ER is impaired. Despite the normal plasma membrane cholesterol content, the cells are amphotericin B resistant.

Plasma membrane lipid domains

Defective influx of cholesterol could be due to mutation of a soluble cholesterol carrier, but it is hard to envision how a defective cytoplasmic transport factor could cause amphotericin B resistance as well. Our hypothesis is that the 3-6 mutation alters the domain structure of the plasma membrane, which prevents cholesterol from becoming accessible to the transport mechanism.

Lateral domains of phospholipids have been well-documented in biological membranes (15). Evidence is accruing for heterogeneity in the cholesterol distribution as well (reviewed in 15-18).

One possible plasma membrane domain that might be required for amphotericin B sensitivity and cholesterol influx is the caveolae. Caveolae are flask-shaped microdomains of the plasma membrane that are rich in cholesterol, sphingolipids and caveolin protein (19, 20). Cholesterol is important for maintaining caveolar shape and function (21). While only a fraction of plasma membrane cholesterol is localized to caveolae, links to cholesterol transport have been established (C. Fielding and P. Fielding, this volume; E. Smart, this volume). Both LDL-derived and nascent cholesterol arriving at the plasma membrane are first incorporated into the caveolae domain (22, 23).

Sphingolipid-cholesterol microdomains also organize into "rafts" on the exoplasmic leaflet of cellular membranes in the absence of caveolin (24). Rafts are proposed to play a role in trafficking of glycophosphatidylinositol-anchored proteins and specific membrane proteins to the apical surface of polarized cells and other cell types that lack caveolae. However, in many cell types, the distinct roles of rafts and caveolae appear somewhat blurred.

How might lateral lipid domains be linked to impaired cholesterol influx? Cholesterol movement to the cell interior may require its initial localization in a specific plasma membrane lateral domain, such as caveolae or rafts. However, we find normal levels of caveolin-1 and caveolar cholesterol in 3-6 cells (14). Furthermore, purified plasma membranes from parental and 3-6 cells show identical protein and phospholipid contents, by two-dimensional gel electrophoresis and two-dimensional thin layer chromatography, respectively (J. Reodica, N. Jacobs, and L. Liscum, unpublished observations). Nevertheless, probing of plasma membrane lipid domains in mutant 3-6 remains an important area for future work.

Cell growth

Analysis of ced-2 mutant 3-6 has given us one new insight into cell growth. Mammalian cell growth requires a source of cholesterol, presumably for membrane biogenesis. Where is the sensor that tells a cell that it has enough cholesterol to proceed through cell division? Our work with mutant 3-6 suggests that the sensor must be in the ER rather than the plasma membrane (Figure 3).

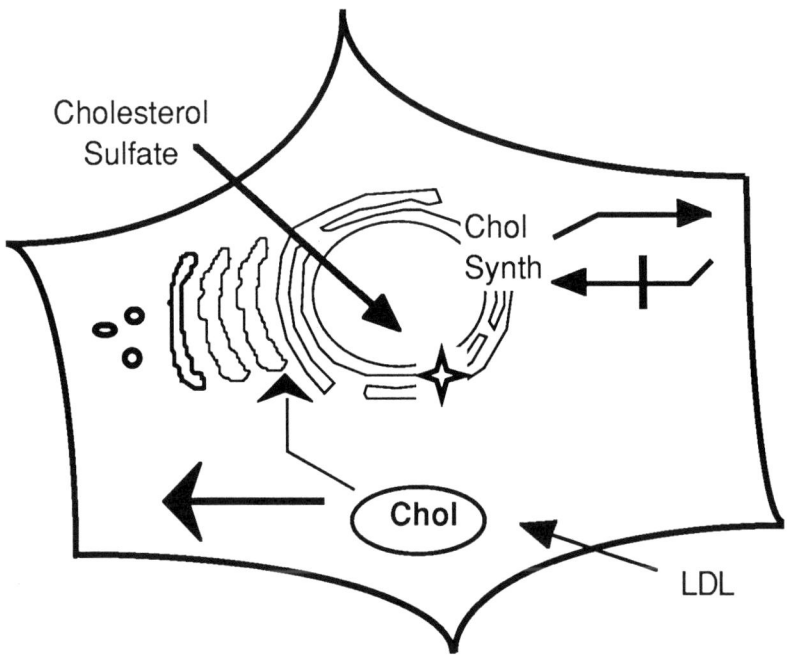

Figure 3. Control of cell growth in mutant 3-6. Endogenously-synthesized and LDL-cholesterol are transported to the plasma membrane, filling that cholesterol pool. However, 3-6 cells do not grow as readily as parental cells using those sources of cholesterol. Mutant 3-6 cells do grow when supplied with cholesterol sulfate, which is hydrolyzed in the ER. Thus the sensor that detects cellular cholesterol and allows cells to proceed through cell division is postulated to be in the ER.

Parental CHO cells grow when supplied with either endogenously synthesized cholesterol or LDL-derived cholesterol. Mutant 3-6 exhibits lagged growth with either source of cholesterol, despite normal movement of the cholesterol to the plasma membrane (14). This suggests that merely filling the plasma membrane cholesterol pool is not sufficient to sustain cell growth. Instead, cholesterol movement or signaling to the cell interior must be required. In support of this model, mutant 3-6 growth is supported by the addition of cholesterol sulfate to the medium (14). Cholesterol sulfate is hydrolyzed by an ER sulfatase. Our hypothesis is that cholesterol derived in the ER from cholesterol sulfate triggers an ER sensor that is not reached by nascent cholesterol or low concentrations of LDL-cholesterol.

Higher concentrations of LDL do promote the growth of 3-6 cells to normal levels. This is consistent with our finding that a portion of LDL cholesterol is transported to the ER by a plasma membrane independent pathway, and that this lysosome to ER pathway is intact in mutant 3-6 (25). When higher concentrations of LDL are internalized, a threshold amount of LDL-cholesterol must be transported along this minor pathway, filling the ER pool.

PATHWAYS OF LDL-CHOLESTEROL MOVEMENT

The routes of cholesterol transport is an area of intense investigation. Several studies have provided evidence that LDL-cholesterol is rapidly transported from lysosomes to the plasma membrane. Brasaemle and Attie measured the kinetics of LDL-cholesterol transport to the plasma membrane, assessing arrival at the plasma membrane with a brief cholesterol oxidase treatment of intact fibroblasts (26). They found that LDL-cholesterol arrived at the plasma membrane minutes after hydrolysis, with a $T_{1/2}$ of 42 min. We now know that cholesterol oxidase treatment of intact cells preferentially oxidizes caveolar cholesterol (27). Therefore, Brasaemle and Attie were likely measuring the arrival of LDL-cholesterol at the caveolar domain.

Johnson et al. (28) measured a similar T1/2 for LDL-cholesterol movement to the plasma membrane (40-50 min). They provided additional information that the kinetics are unaffected by ACAT activity, suggesting that LDL-cholesterol does not pass through the ER on its route to the plasma membrane. Liscum et al. (29) also presented evidence that LDL-cholesterol is transported to the plasma membrane, while Sokol et al. (30) showed that LDL incubation raises the cholesterol content of the plasma membrane.

While these studies provided evidence for a lysosome-to-plasma membrane pathway, they did not eliminate an alternate lysosome-to-ER route. Indeed, Spillane et al. (31) presented results consistent with such a route. They subjected mutant CT-60 to DNA transfection and isolated a cell line that exhibited greatly increased LDL-cholesterol transport to ACAT. This increased pathway was not accompanied by increased plasma membrane to ER cholesterol movement, suggesting that the lysosome-to-ER route does not go through the plasma membrane.

Hydrophobic amine inhibition of cholesterol transport

Analysis of pharmacological inhibition of cholesterol transport has complemented the study of mutant cells and allowed us (25, 32) and Neufeld et al. (33) to define further the pathways that LDL-cholesterol takes after its egress from lysosomes. When LDL-cholesterol exits the lysosomes, the cellular cholesterol pool expands and ACAT is activated in the ER. How is ACAT signaled that excess cellular cholesterol is available for esterification? LDL-cholesterol may be transported to the plasma membrane, where the rising cholesterol content increases the amount of cholesterol cycling into the cell interior. Alternatively, LDL-cholesterol may be transported along multiple pathways.

Our studies have utilized the hydrophobic amines U18666A and imipramine, drugs that inhibit specific cholesterol transport pathways (Figure 4).

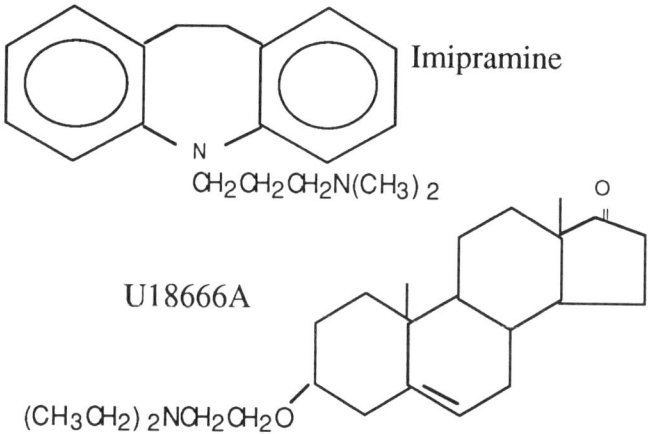

Figure 4. Structures of the hydrophobic amines U18666A and imipramine.

U18666A came to our attention from work of Panini et al. (34), who showed that it blocks the LDL-mediated suppression of cholesterol synthesis. We found that this effect is likely due to U18666A's inhibition of LDL-cholesterol transport (35). U18666A-treated CHO cells have the biochemical phenotype of NPC fibroblasts. LDL-cholesterol accumulates in lysosomes of U18666A-treated cells, causing delays in transport to other cell membranes and regulatory sites. Rodriguez-Lafrasse et al. (36) and Roff et al. (37) showed that imipramine and other related hydrophobic amines have a similar effect. Finally, Harmala et al. (38) found that the plasma membrane to ER pathway is also inhibited by U18666A. One explanation for these findings is that the drugs inhibit the action of a protein or lipid that facilitates cholesterol movement.

Route of LDL-cholesterol transport to ACAT

We utilized hydrophobic amines to test the hypothesis that while the bulk of LDL-cholesterol is delivered to the plasma membrane, a portion is transported to ACAT in the ER by a route that is independent of the plasma membrane. Three lines of evidence support the existence of such a pathway (25).

First, LDL-cholesterol is transported from lysosomes to ACAT via a pathway that is exquisitely sensitive to hydrophobic amine inhibition (32) (Figure 5). U18666A and imipramine inhibit LDL-cholesterol movement to ACAT, as measured by re-esterification, at concentrations that have little or no effect on bulk cholesterol pathways through the plasma membrane. This suggests that a separate pathway exists, distinguished by its sensitivity to hydrophobic amines.

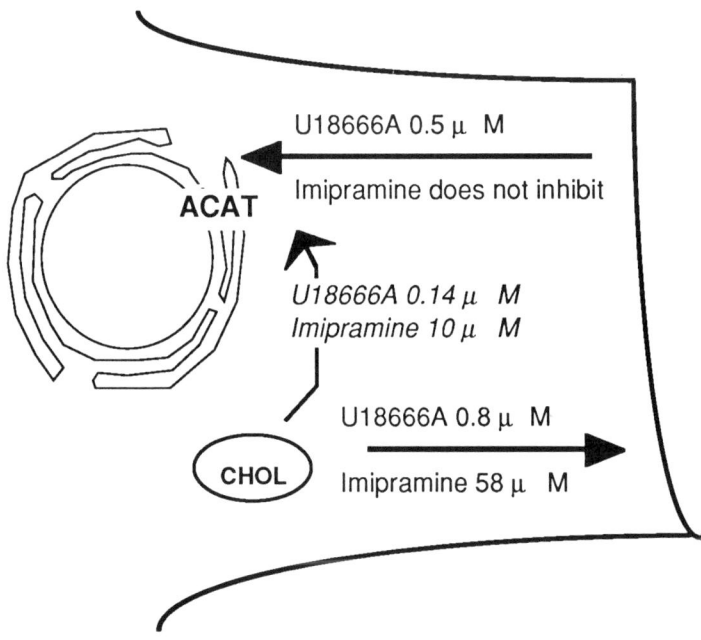

Figure 5. Hydrophobic amine inhibition of cholesterol transport pathways. The cholesterol transport pathways assessed and the IC_{50}s of U18666A and imipramine action are indicated.

Second, LDL-cholesterol transport from lysosomes-to-ACAT is normal in mutant 3-6. This indicates that the lysosome-to-ACAT pathway does not involve movement of cholesterol from the plasma membrane.

Third, the kinetics of LDL-cholesterol versus cellular cholesterol incorporation into cholesteryl esters indicates that it is LDL-cholesterol that activates ACAT. Upon ACAT activation, the first cholesteryl esters are synthesized predominantly from LDL-cholesterol. Once ACAT is activated, cellular cholesterol is incorporated into cholesteryl esters.

These findings are consistent with the study of Neufeld et al. (33), which used a different approach to arrive at the same model (E. Neufeld and P. Pentchev, this volume).

Our study did not resolve whether the Golgi apparatus is involved in LDL-cholesterol movement, although we did show that an intact Golgi complex is not necessary (25). These experiments utilized brefeldin A, which causes the cis and medial Golgi to fuse with the ER and also disrupts the delivery of internalized ligands to lysosomes. Brefeldin A reduced LDL internalization and delivery to lysosomes; however, the residual LDL-cholesterol released was efficiently transported to ACAT and reesterified. This does not eliminate the Golgi as a player since any transport "vehicle" that normally docks at the Golgi is likely capable of finding its docking site in the merged Golgi/ER.

LDL-cholesterol transport from lysosomes to ER is likely to be a vesicle-mediated event. Transport was inhibited by ionophores, such as monensin and nigericin, that affect vesicle sorting through the Golgi complex and endosomes. Transport was also disrupted by cytochalasin D, suggesting that actin filaments are required, while colchicine had no effect, eliminating a role for microtubules.

FUTURE DIRECTIONS

Our emphasis will be on the following questions.

What are the ced-1 and ced-2 genes? Molecular analysis will soon reveal if ced-1 is the NPC1 gene. If so, the ced-1 mutants can serve as recipient cell lines for NPC1 expression studies and domain mapping. Their importance is heightened by the inability of our laboratory to develop stable immortalized NPC fibroblasts. If ced-1 is not the NPC1 gene, the possibility exists that ced-1 is a genetic model for NPC2 instead. NPC2 is clinically and biochemically indistinguishable from NPC1; however, the NPC2 gene does not map to the region of chromosome 18 containing NPC1 (39). Molecular strategies to identify NPC2 are lacking and a somatic cell model would play a valuable role in that effort.

Expression cloning strategies are being implemented by our laboratory to isolate genes whose expression corrects the defective phenotype in ced-1 and ced-2 cells. This effort should identify the ced-1 and ced-2 genes, in addition to genes encoding modulators of cholesterol transport or distribution in the cell.

What is the biological function of NPC1? The intracellular location of NPC1 and any homeostatic regulation will provide important clues to its function. Does NPC1 participate in a vesicular transport pathway? If so, is the transport pathway specific for cholesterol? How do mutations in NPC1 lead to lysosomal accumulation of cholesterol in peripheral tissues, and gangliosides in the central nervous system? Domain analysis of NPC1 revealed intriguing links, not only to cholesterol homeostasis, but also to a signal transduction pathway important for development. Future work is likely to search for an analogous NPC1 pathway.

Finally, what is the mechanism of hydrophobic amine inhibition of cholesterol transport? Hydrophobic amines inhibit specific cholesterol transport pathways, causing an NPC phenotype in cultured cells. To date, our work has eliminated several possible mechanisms of action but has not provided positive evidence in favor of any mechanism. Is

NPC1 the target of hydrophobic amine action? Or, are there additional cholesterol transport proteins that can be identified by analysis of drug-binding sites? Identification of the binding site and mechanism of hydrophobic amine action is integral to our understanding of NPC1 and intracellular cholesterol transport.

References

1. Warnock DE, Roberts C, Lutz MS, Blackburn WA, Young WW, Baenziger JU. Determination of plasma membrane lipid mass and composition in cultured Chinese hamster ovary cells using high gradient magnetic affinity chromatography. J. Biol. Chem. 1993; 268:10145-10153.

2. Lange Y, Strebel F, Steck TL. Role of the plasma membrane in cholesterol esterification in rat hepatoma cells. J. Biol. Chem. 1993; 268:13838-13843.

3. Brown MS, Goldstein JL. A receptor-mediated pathway for cholesterol homeostasis. Science 1986; 232:34-47.

4. Brown MS, Goldstein JL. The SREBP pathway: regulation of cholesterol metabolism by proteolysis of a membrane-bound transcription factor. Cell 1997; 89:331-340.

5. Pentchev PG, Blanchette-Mackie EJ, Liscum L. Biological implications of the Niemann-Pick C mutation. Sub-Cellular Biochem. 1997; 28:437-451.

6. Pentchev PG, Vanier MT, Suzuki K, Patterson MC. Niemann-Pick Disease type C: Cellular cholesterol lipidosis. In: Scriver CR, Beaudet AL, Sly WS, Valle D, ed. The Metabolic and Molecular Bases of Inherited Disease. 7th ed. New York: McGraw-Hill, 1995: 2625-2639. Vol II.

7. Carstea ED, Morris JA, Coleman KG, et al. Niemann-Pick C1 disease gene: Homology to mediators of cholesterol homeostasis. Science 1997; 277:228-231.

8. Saito Y, Chou SM, Silbert DF. Animal cell mutants defective in sterol metabolism: A specific selection procedure and partial characterization of defects. Proc. Natl. Acad. Sci. U.S.A. 1977; 74:3730-3734.

9. Hidaka K, Endo H, Akiyama S, Kuwano M. Isolation and characterization of amphotericin B-resistant cell lines in Chinese hamster cells. Cell 1978; 14:415-421.

10. Dahl NK, Reed KL, Daunais MA, Faust JR, Liscum L. Isolation and characterization of Chinese hamster ovary cells defective in the intracellular metabolism of LDL-derived cholesterol. J. Biol. Chem. 1992; 267:4889-4896.

11. Dahl NK, Daunais MA, Liscum L. A second complementation class of cholesterol transport mutants with a variant Niemann-Pick type C phenotype. J. Lipid Res. 1994; 35:1839-1849.

12. Cadigan KM, Spillane DM, Chang T-Y. Isolation and characterization of Chinese hamster ovary cell mutants defective in intracellular low density lipoprotein-cholesterol trafficking. J. Cell Biol. 1990; 110:295-308.

13. Gu JZ, Carstea ED, Cummings C, et al. Substantial narrowing of the Niemann-Pick C candidate interval by yeast artificial chromosome complementation. Proc. Natl. Acad. Sci. U.S.A. 1997; 94:7378-7383.

14. Jacobs NL, Andemariam B, Underwood KW, et al. Analysis of a Chinese hamster ovary cell mutant with defective mobilization of cholesterol from the plasma membrane to the endoplasmic reticulum. J. Lipid Res. 1997; 38:1973-1987.

15. Glaser M. Lipid domains in biological membranes. Curr. Op. Struc. Biol. 1993; 3:475-481.

16. Rothblat GH, Mahlberg FH, Johnson WJ, Phillips MC. Apolipoproteins, membrane cholesterol domains, and the regulation of cholesterol efflux. J. Lipid Res. 1992; 33:1091-1097.

17. Liscum L, Faust JR. Compartmentation of cholesterol within the cell. Curr. Op. Lipid. 1994; 5:221-226.

18. Slotte JP. Cholesterol-sphingomyelin interactions in cells - effects on lipid metabolism. Sub-Cellular Biochemistry 1997; 28:277-293.

19. Parton RG. Caveolae and caveolins. Curr. Op. Cell Biol. 1996; 8:542-548.

20. Harder T, Simons K. Caveolae, DIGS, and the dynamics of sphingolipid-cholesterol microdomains. Curr. Op. Cell Biol. 1997; 9:534-542.

21. Rothberg KG, Ying Y-S, Kamen BA, Anderson RGW. Cholesterol controls the clustering of the glycophospholipid-anchored membrane receptor 5-methyltetra-hydrofolate. J. Cell Biol. 1990; 111:2931-2938.

22. Fielding PE, Fielding CJ. Plasma membrane caveolae mediate the efflux of cellular free cholesterol. Biochem. 1995; 34:14288-14292.

23. Smart EJ, Ying Y-S, Donzell WC, Anderson RGW. A role for caveolin in transport of cholesterol from endoplasmic reticulum to plasma membrane. J. Biol. Chem. 1996; 271:29427-29435.

24. Simons K, Ikonen E. Functional rafts in cell membranes. Nature 1997; 387:569-572.

25. Underwood KW, Jacobs NL, Howley A, Liscum L. Evidence for a cholesterol transport pathway from lysosomes to endoplasmic reticulum that is independent of the plasma membrane. J. Biol. Chem. 1998; 273:4266-4274.

26. Brasaemle DL, Attie AD. Rapid intracellular transport of LDL-derived cholesterol to the plasma membrane in cultured fibroblasts. J. Lipid Res. 1990; 31:103-112.

27. Smart EJ, Ying Y-S, Conrad PA, Anderson RGW. Caveolin moves from caveolae to the Golgi apparatus in response to cholesterol oxidation. J. Cell Biol. 1994; 127:1185-1197.

28. Johnson WJ, Chacko GK, Philips MC, Rothblat GH. The efflux of lysosomal cholesterol from cells. J. Biol. Chem. 1990; 265:5546-5553.

29. Liscum L, Ruggiero RM, Faust JR. The intracellular transport of low density lipoprotein-derived cholesterol is defective in Niemann-Pick Type C fibroblasts. J. Cell Biol. 1989; 108:1625-1636.

30. Sokol J, Blanchette-Mackie J, Kruth HS, et al. Type C Niemann-Pick disease. Lysosomal accumulation and defective intracellular mobilization of low density lipoprotein cholesterol. J. Biol. Chem. 1988; 263:3411-3417.

31. Spillane DM, Reagan JW, Kennedy NJ, Schneider DL, Chang T-Y. Translocation of both lysosomal LDL-derived cholesterol and plasma membrane cholesterol to the endoplasmic reticulum for esterification may require common cellular factors involved in cholesterol egress from the acidic compartments (lysosomes/endosomes). Biochim. Biophys. Acta 1995; 1254:283-294.

32. Underwood KW, Andemariam B, McWilliams GL, Liscum L. Quantitative analysis of hydrophobic amine inhibition of intracellular cholesterol transport. J. Lipid Res. 1996; 37:1556-1568.

33. Neufeld EB, Cooney AM, Pitha J, et al. Intracellular trafficking of cholesterol monitored with a cyclodextrin. J. Biol. Chem. 1996; 271:21604-21613.

34. Panini SR, Sexton RC, Rudney H. Regulation of 3-hydroxy-3-methylglutaryl coenzyme A reductase by oxysterol by-products of cholesterol biosynthesis. Possible mediators of low density lipoprotein action. J. Biol. Chem. 1984; 259:7767-7771.

35. Liscum L, Faust JR. The intracellular transport of low density lipoprotein-derived cholesterol is inhibited in Chinese hamster ovary cells cultured with 3-β-[2-(diethylamino)ethoxy]androst-5-en-17-one. J. Biol. Chem. 1989; 264:11796-11806.

36. Rodriguez-Lafrasse C, Rousson R, Bonnet J, Pentchev PG, Louisot P, Vanier MT. Abnormal cholesterol metabolism in imipramine-treated fibroblast cultures. Similarities with Niemann-Pick type C disease. Biochim. Biophys. Acta 1990; 1043:123-128.

37. Roff CF, Goldin E, Comly ME, et al. Type C Niemann-Pick Disease: Use of hydrophobic amines to study defective cholesterol transport. Dev. Neurosci. 1991; 13:315-320.

38. Harmala A-S, Porn MI, Mattjus P, Slotte JP. Cholesterol transport from plasma membranes to intracellular membranes is inhibited by 3β-[2-(diethylamino)ethoxy]androst-5-en-17-one. Biochim. Biophys. Acta 1994; 1211:317-325.

39. Vanier MT, Duthel S, Rodriguez-Lafrasse C, Pentchev P, Carstea ED. Genetic heterogeneity in Niemann-Pick C disease: A study using somatic cell hybridization and linkage analysis. Am. J. Hum. Genet. 1996; 58:118-125.

WHAT THE NIEMANN-PICK TYPE C GENE HAS TAUGHT US ABOUT CHOLESTEROL TRANSPORT

Edward B. Neufeld

Lipid Cell Biology Section, Laboratory of Cell Biochemistry and Biology, National Institute of Diabetes and Digestive and Kidney Diseases, National Institutes of Health, Bethesda, MD 20892; e-mail: neufeld@codon.nih.gov

KEY WORDS: Niemann-Pick Type C disease, NPC1, npc1, lysosomes, Golgi complex, cyclodextrin, brefeldin A, endocytosed cholesterol

---

ABSTRACT

Niemann-Pick Type C disease (NP-C) is characterized by abnormal accumulation of endocytosed cholesterol in lysosomes and the Golgi complex and by delayed induction of cellular homeostatic responses to cholesterol. Positional and complementation cloning has recently identified both the human *NPC1* and the syntenic mouse *Npc1* genes. Complementation analysis has shown that a second mutant human gene, designated *NPC2*, produces an identical phenotype. The notable cellular defect in NP-C disease involves impaired trafficking of lysosomally-derived cholesterol to the plasma membrane and endoplasmic reticulum. Defective enrichment of regulatory sterol pools in the endoplasmic reticulum delays induction of homeostatic responses. Current evidence suggests that the two NPC proteins may: (i) mediate transport of endocytically-derived cholesterol from lysosomes to the Golgi complex and (ii) be linked to regulation of anterograde and retrograde Golgi cholesterol transport. In addition to the disruptions in cholesterol metabolism, impaired trafficking of endocytosed glycolipids along the NPC protein-mediated pathway may play a role in the neuropathology associated with NP-C disease. Future studies of the *NPC* genes and their products will provide insight into the relationship between cholesterol and glycolipid trafficking.

CONTENTS

INTRODUCTION

CLONING OF THE *NPC*1 GENE

FUNCTIONS OF THE *NPC* GENE PRODUCTS

Transport of Endocytosed Cholesterol to the Plasma Membrane and
Endoplasmic Reticulum

Golgi-Mediated Cholesterol Transport

Endocytosed LDL-Cholesterol

Endocytosed Plasma Membrane Cholesterol

Other Pathways of Cholesterol Transport

INTEGRATED MODEL OF NPC PROTEIN FUNCTION

BEYOND CHOLESTEROL

INTRODUCTION

Inherited metabolic diseases have taught us a great deal about the intracellular
trafficking of LDL-cholesterol. The pioneering studies of Brown and Goldstein
using fibroblasts from patients with familial hypercholesterolemia elucidated the
receptor-mediated pathway for LDL-cholesterol delivery to lysosomes (1). The
study of Wolman's syndrome, a lysosomal cholesterol ester (CE) storage disease,
revealed the role of lysosomes and cholesterol ester hydrolase in the processing
of the CE core of LDL (2). Similarly, Niemann-Pick Type C disease (NP-C), a
neurovisceral cellular lipidosis, has demonstrated that transport of LDL-
cholesterol from lysosomes to regulatory sites of cholesterol homeostasis is
mediated by a specific set of proteins. The recent cloning of the NPC1 gene in
man (3) and mouse (4) has expedited our efforts to unravel the molecular
mechanisms underlying intracellular trafficking of LDL-derived cholesterol.
This chapter will (i) review the contributions NP-C disease has made to our
understanding of intracellular cholesterol trafficking and (ii) provide a model
which integrates NPC protein-mediated cholesterol trafficking into the broader
scheme of intracellular trafficking.

CLONING OF THE *NPC*1 GENE

The *NPC1* gene, linked to the pericentromeric region on chromosome 18q11
(3,5) was recently identified by positional and complementation cloning (5).
The same gene in mouse localized to chromosome 18 in a region syntenic to the
human gene locus, thus confirming the validity of the BALB/c *npc*[nih] and
C57BLKS/J *spm* mouse models (4). Complementation analyses have identified
a second human mutant gene, NPC2, which is not linked to 18q11 (6). NPC2 is
likely to be tightly linked to NPC1 functionally since it produces the NPC1
phenotype when mutant.

The *NPC1* gene encodes a protein predicted to have 1278 amino acids. The NPC1 protein sequence reveals several putative domains, each potentially consistent with a role in intracellular trafficking of endocytosed LDL-cholesterol (3). The NPC1 protein appears to be an integral membrane protein (has a signal peptide sequence and more than a dozen predicted transmembrane domains), and, to localize to lysosomes (has a lysosomal targeting motif and numerous conserved potential glycosylation sites). NPC1 protein has a unique, highly conserved (mouse, yeast, worm), soluble amino terminal domain likely to play a role in NPC1-specific functions (transfer of sterol between membranes). A leucine zipper motif in this region could mediate interactions with other proteins. A putative tyrosine phosphorylation site at a different region of NPC1 could potentially serve a regulatory function. Most significantly, NPC1 protein has regions with sequence similarity to sterol-sensing regions in other mediators of cholesterol homeostasis (sterol responsive element binding protein (SREBP)-cleavage activation protein (SCAP), and 3-hydroxy-3-methyl-glutaryl-coenzyme A (HMG-CoA) reductase).

Studies of SCAP provide important clues to potential NPC1 function. SCAP and SREBP form a complex through interaction of their carboxy-terminal domains (7). Formation of this complex allows a protease to generate soluble transcription factors from membrane-bound SREBPs (7). Sterol interaction with sterol-sensing regions of SCAP prevents formation of the SREBP-SCAP complex (7). In like manner, membrane sterol content may regulate the interaction of NPC1 protein with NPC2, or, other proteins involved in cholesterol transport. It is tempting to speculate that sterol-regulated interactions of NPC1 protein mediate transfer of cholesterol from sterol-rich lysosomal membranes to sterol-poor acceptor membranes through the action of the unique NPC1 domain.

Database sequence analysis of NPC1 revealed that Patched (PTC), a morphogenic protein (8), also has sterol-sensing regions, suggesting a novel role for cholesterol in the function of PTC. Cholesterol regulated SCAP function appears to have a more general role in membrane biogenesis (9). Thus, perhaps NPC1 sterol-sensing function, like that of SCAP and PTC, is linked to cellular regulation beyond that of cholesterol alone.

FUNCTIONS OF THE *NPC* GENE PRODUCTS

Transport of Endocytosed Cholesterol to the Plasma Membrane and Endoplasmic Reticulum

NP-C is characterized by a unique error in cellular trafficking of LDL-derived cholesterol associated with abnormal accumulation of unesterified cholesterol in lysosomes and the Golgi complex. The transport defect in NP-C targets pathways involving movement of endocytosed cholesterol out of lysosomes. Transport of LDL-derived cholesterol from lysosomes to the plasma membrane is retarded in human NP-C cells both *in vitro* (10, 11) and *in vivo* (12) and results in lysosomal sequestration and deficient enrichment of the plasma membrane (13). The rate of transport from lysosomes to the plasma membrane has been shown to be 2 to 4-fold slower in NP-C fibroblasts (10, 11).

In normal cells, endocytosed LDL-cholesterol enriches cellular pools of cholesterol in the plasma membrane and endoplasmic reticulum (ER) (10, 11, 14, 15). Acetyl-coenzyme A:cholesterol acyltransferase (ACAT) in the ER is stimulated after the capacity of cellular membranes to accommodate cholesterol enrichment is exceeded (16,17). Plasma membrane cholesterol serves as the principal source of substrate cholesterol for ACAT in the ER (10, 11, 18, 19, 20). In NP-C cells, lysosomal cholesterol does not appear to reach the ER by a plasma membrane-mediated pathway (11). Retarded transport of cholesterol from lysosomes to the plasma membrane does not allow the plasma membrane to be enriched beyond the threshold required for subsequent transport to the ER. Several independent lines of research suggest that endocytosed LDL-derived lysosomal cholesterol can also reach the ER by a pathway that bypasses the plasma membrane (11, 21, 22, 23). In NP-C cells, endocytosed cholesterol appears to reach the ER solely by the plasma membrane-independent pathway (11). Movement of cholesterol along this pathway is severely inhibited in NP-C cells since esterification in mutant cells is only 5% of normal.

The induction of SREBP-regulated cellular homeostatic regulatory responses and esterification depends upon enrichment of ER membranes by endocytosed LDL-derived (9) or PM-derived cholesterol (24). Mutant NPC1 retards transport of endocytosed cholesterol from lysosomes to the ER. Consequently, induction of homeostatic responses normally associated with cellular cholesterol enrichment are uniformly delayed in NP-C (25, 26).

Golgi-Mediated Cholesterol Transport

Endocytosed LDL-Cholesterol

The *NPC* mutation has also provided compelling evidence that the Golgi complex plays an active role in the transport of lysosomal cholesterol to other cellular membranes (see also Blanchette-Mackie, this volume). Premature Golgi cholesterol cytochemical staining during uptake of LDL in NP-C cells provided initial evidence for abnormal sterol transport in the Golgi complex (27). Subsequently, the distribution of endocytosed LDL-cholesterol in specific Golgi subcompartments of NP-C cells was shown also to be abnormal (28). Endocytosed cholesterol increasingly enriches Golgi subcompartments of normal cells in the *cis*-to-*trans* direction (28). In NP-C cells, however, only the *trans*-cisternae of the Golgi apparatus are enriched (28). These observations led to the speculation that transport of LDL-derived cholesterol from the Golgi complex to both the plasma membrane and ER may be impaired in NP-C cells (28).

Further evidence supporting a role for the Golgi complex in the distribution of lysosomal cholesterol was provided using brefeldin A (BFA) to disrupt the structure and function of this organelle (11). BFA causes the *cis*-, *medial*-, and *trans*- cisternae of the Golgi complex to merge with the ER (29). BFA allows retrograde Golgi transport to the ER to proceed in the absence of anterograde Golgi transport to the plasma membrane (30). BFA also appears to block anterograde, but not retrograde, transport of lysosomally-derived cholesterol through the Golgi complex. As described below, BFA has revealed that mutant NPC1 protein impairs both anterograde (Golgi to plasma membrane) and retrograde (Golgi to ER) transport of endocytosed cholesterol.

Normal and NP-C cells were treated with BFA during release of LDL-derived cholesterol from pre-loaded lysosomes (11). Arrival of this sterol at the ER was monitored by measuring formation of cholesteryl [3H]oleate. Participation of the plasma membrane in the intracellular distribution of the lysosomal cholesterol pool was monitored by depletion of cellular sterol (via the plasma membrane) to cyclodextrin (a sterol chelator) in the medium. Under the conditions used, it takes 24 hrs for sterol to clear from pre-loaded lysosomes of normal cells (31). Measurements of esterification were made up to 6 hrs, during which time a large excess of lysosomal cholesterol remains to drive further esterification. Thus, rates of esterification during the 6 hr time period reflect formation and not turnover of cellular pools of esterified cholesterol.

NORMAL CELLS

INTACT GOLGI   DISRUPTED GOLGI

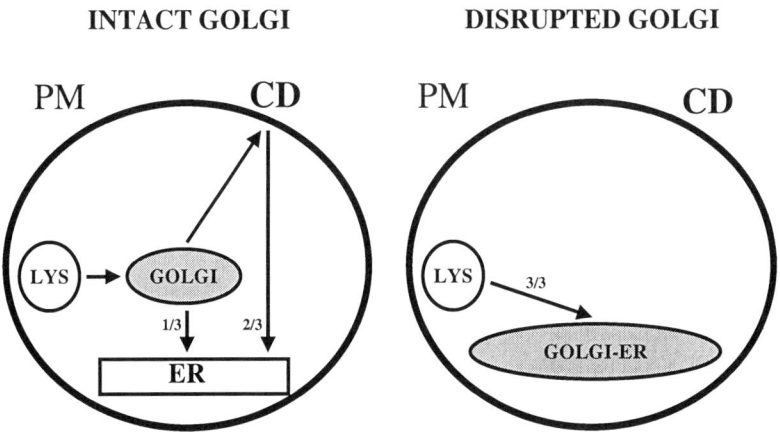

Figure 1. Golgi-mediated pathways of delivery of LDL-cholesterol to the ER in normal fibroblasts. (*Left*) Lysosomal cholesterol is delivered to the ER in part by PM-mediated (2/3) and PM-independent (1/3) pathways. (*Right*) When the Golgi cisternae are absorbed into the ER by BFA, all lysosomal cholesterol is delivered to the ER by the PM-independent pathway. Although the cholesterol content of the Golgi (shaded in grey) is also delivered to the ER by BFA-induced retrograde transport, its contribution to the esterifying pool is too small (relative to the flood of lysosomal cholesterol delivered to the ER) to be detected. PM:plasma membrane; CD:cyclodextrin; LYS:lysosome; Golgi-ER: Golgi cisternae merged with the ER.

In normal cells with intact Golgi, cyclodextrin suppressed esterification by 2/3 (Fig. 1). Thus, it appears that 2/3 of lysosomal cholesterol sent to the ER was first delivered to the plasma membrane, while 1/3 was delivered by a plasma

98

membrane-independent pathway. However, when the Golgi cisternae merged with the ER by the action of BFA, cyclodextrin no longer suppressed esterification, suggesting that all lysosomal cholesterol delivered to the ER now bypasses the plasma membrane. Thus, BFA appears to block anterograde, but not retrograde, Golgi transport of LDL-cholesterol, so that all lysosomal cholesterol is delivered directly to the merged Golgi-ER. The latter finding is readily explained if lysosomal cholesterol is first delivered to a site in the Golgi complex which relocates to the merged Golgi-ER organelle by the action of BFA. Cytochemical studies confirm that BFA redistributes the bulk of lysosomal cholesterol to the merged Golgi-ER (Neufeld et al., in preparation). The contribution of the pre-existing pool of Golgi cholesterol itself to esterification appears to be too small to be discriminated from the incoming flood of lysosomal cholesterol to the ER (since BFA alone does not increase esterification). Inasmuch as BFA appears to block anterograde cholesterol transport in normal cells, this suggests that curtailed transport of lysosomal cholesterol to the plasma membrane in NP-C cells (10, 11) may reflect defective anterograde Golgi transport.

NP-C CELLS

INTACT GOLGI **DISRUPTED GOLGI**

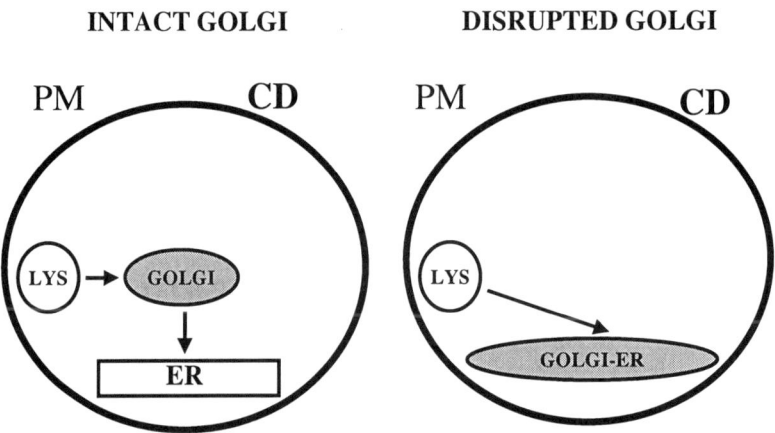

Figure 2. Golgi-mediated pathways of delivery of LDL-cholesterol to the ER in NP-C fibroblasts. (*Left*) Lysosomal cholesterol is delivered to the ER solely by a PM-independent pathway. Esterification by this pathway is 1/5 of normal. Note that when the Golgi complex is intact, cholesterol transported to the PM from the Golgi does not reach the ER. (*Right*) When the Golgi cisternae are absorbed into the ER by BFA, esterification is increased 4X as a result of re-distribution of Golgi cholesterol content to the ER, and, the diversion of cholesterol from the anterograde (Golgi-PM) pathway to the retrograde (Golgi-ER) pathway. PM:plasma membrane; CD:cyclodextrin; LYS:lysosome; Golgi-ER: Golgi cisternae merged with the ER

In NP-C cells with intact Golgi complex, all LDL-derived lysosomal cholesterol is delivered to the ER by a plasma membrane-independent pathway. This pathway appears to be defective in NP-C cells since esterification of lysosomal cholesterol by this pathway is only 1/5 of normal. Studies with BFA suggest that this defective pathway represents retrograde Golgi-ER cholesterol transport (11) (Fig. 2). BFA increases esterification of LDL-derived lysosomal cholesterol in NP-C cells four-fold, consistent with a role for the Golgi in transport of lysosomal cholesterol. This increase most likely represents addition of (i) the pre-existing cholesterol content of the Golgi cisternae, as well as, (ii) lysosomal cholesterol diverted from transport via the anterograde Golgi pathway, to cholesterol transported via the retrograde Golgi pathway to the ER. Thus, when the Golgi is intact in NP-C cells, it appears that only a fraction of the lysosomal cholesterol which reaches the Golgi is transported to the ER by the retrograde pathway. Cytochemical studies in NP-C cells confirm that BFA induces loss of Golgi structure and redistribution of Golgi cholesterol (but little, if any, LDL-derived lysosomal cholesterol) to the merged Golgi-ER organelle (Neufeld et al., in preparation).

Endocytosed Plasma Membrane Cholesterol

Hydrophobic amines, like mutant *NPC,* block intracellular distribution of LDL-derived endocytosed cholesterol resulting in cholesterol accumulation in lysosomes (32, 33). In the absence of lipoprotein uptake, hydrophobic amines have been observed to redistribute cholesterol from the plasma membrane to lysosomes (Pentchev, unpublished results). The latter finding suggests that plasma membrane cholesterol entering the cell via endocytic vesicles traffics, like LDL-cholesterol, to lysosomes. Sphingomyelinase (SMase) digestion of plasma membrane sphingomyelin stimulates plasma membrane endovesiculation (34). Although SMase-induced endovesicles are distinct from receptor-mediated endocytic vesicles, both vesicles appear to deliver endocytic markers to lysosomes by a common pathway (34, 35). SMase-mobilized plasma membrane cholesterol enriches the Golgi complex (11) as well as regulatory pools of cholesterol in the ER (24, 36). Esterification of SMase-mobilized cholesterol in the ER is greatly enhanced in normal cells when Golgi cisternal elements are absorbed into the ER by the action of BFA (11). Taken together, these findings are consistent with delivery of endocytosed plasma membrane cholesterol to the ER by retrograde transport through the Golgi apparatus.

This transport of plasma membrane-derived cholesterol through the Golgi complex appears to involve NPC1 protein (11). Delivery of SMase-mobilized pools of plasma membrane cholesterol to the ER is defective in NP-C cells (11, 37). BFA enhances esterification of SMase-mobilized cholesterol in NP-C cells, consistent with delivery of cholesterol-enriched Golgi cisternae to ACAT in the ER. However, the BFA-induced enhancement of esterification of plasma membrane-derived cholesterol in NP-C cells remains attenuated. NPC-mediated transport of endocytosed plasma membrane cholesterol, like endocytosed LDL-cholesterol, thus seems to be impaired in the BFA-induced ER-Golgi organelle in mutant cells.

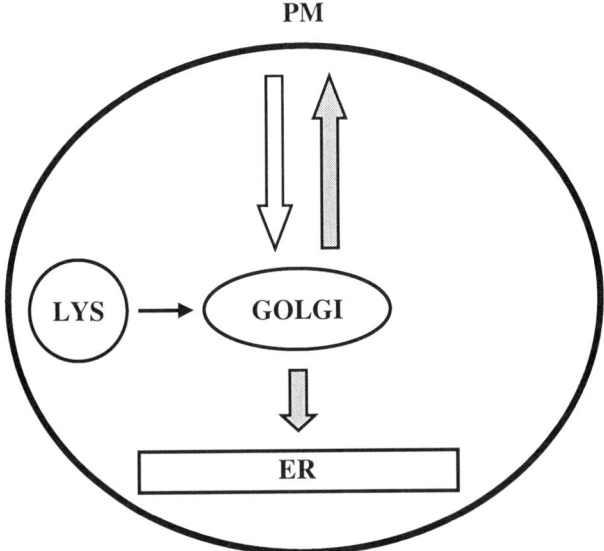

PM

Figure 3. Proposed defective pathways of intracellular cholesterol transport in NP-C fibroblasts. Transport of endocytosed cholesterol out of the Golgi complex is defective (grey arrows). Transport of cholesterol from the plasma membrane to the Golgi complex does not appear to be impaired.

The apparent effect of mutant NPC1 protein on Golgi-mediated distribution of endocytosed cholesterol is shown in Fig. 3. Movement of endocytosed cholesterol from the Golgi complex to both the plasma membrane and ER is defective. This block in Golgi transport results in retention of unesterified cholesterol in *trans*-Golgi cisternae and lysosomes, and, consequent delayed induction of cellular homeostatic responses. The bulk of lysosomal cholesterol delivered to the ER should pass through the Golgi complex twice - once on the way to the plasma membrane, and then a second time, moving from the plasma membrane to the ER through the Golgi. Thus, a Golgi-related defect in NPC protein function would be expected to impair transport of lysosomal cholesterol to the ER more severely than transport of lysosomal cholesterol to the PM, as is observed (10, 11).

Other Cholesterol Transport Pathways

The transport defect in NP-C specifically involves movement of endocytosed LDL-cholesterol. This was confirmed in a recent *in vivo* study of lipoprotein trafficking which showed that intracellular transport of HDL-derived cholesterol is not defective in NP-C patients (12). HDL-derived cholesterol esters were also shown to be hydrolyzed extralysosomally (12). Moreover, the defective transport from lysosomes in NP-C seems to be tightly linked to the Golgi complex, since, to date, pathways known to bypass the Golgi complex are not impaired in NP-C. Transport of newly synthesized cholesterol from the ER to the plasma membrane is not altered by mutant NPC protein (10). Thus, newly synthesized cholesterol and LDL-cholesterol seem to be delivered to the plasma membrane by distinct pathways. Newly synthesized cholesterol was shown to be transported to the plasma membrane by an energy-dependent pathway that bypasses the Golgi (38). More recent evidence suggests that newly synthesized cholesterol is directly transported from the ER to plasma membrane caveolae in a non-vesicular, cytosolic lipoprotein-chaperone complex (39, 40). Recent *in vivo* studies of NP-C further suggest that the intracellular transport of newly synthesized cholesterol also bypasses the lysosome (12). Thus, intracellular transport of newly synthesized cholesterol may function independently of lysosomes, the Golgi complex, and, NPC proteins. To date, little is known about the non-Golgi pathway, which delivers plasma membrane cholesterol to the ER (see Smart, this volume).

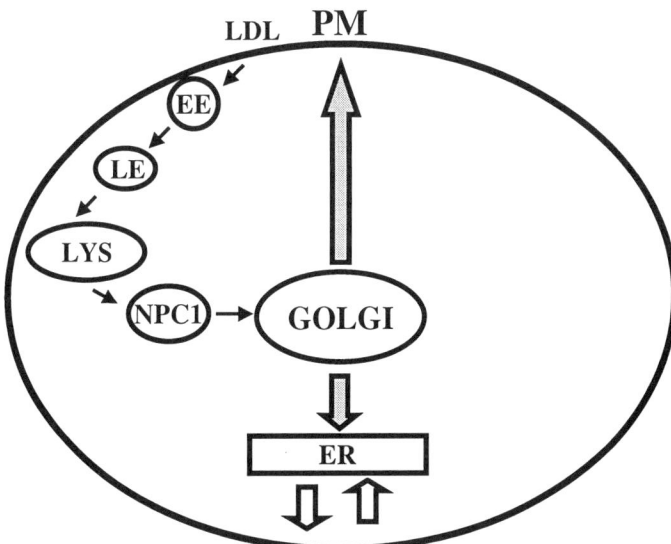

Figure 4. Proposed role of the Golgi complex in NPC-mediated transport of endocytosed cholesterol. Endocytosed LDL or PM-derived cholesterol enters the cell in the lumen, or on the surface, respectively, of early endosomes (EE), via receptor-mediated endocytosis, and is delivered to lysosomes (LYS) via late endosomes (LE). Vesicles containing NPC1 protein mediate transport of cholesterol from lysosomes to the Golgi complex. Intracellular transport pathways that are impaired when NPC1 protein is mutant are shown as grey arrows. Non-Golgi mediated cholesterol transport between the plasma membrane and ER (open arrows) do not appear to be defective when NPC proteins are mutant.

INTEGRATED MODEL OF NPC PROTEIN FUNCTION

The most parsimonious model consistent with cytochemical, biochemical, pharmacological, and subcellular fractionation studies of cholesterol transport in NP-C cells is shown in Fig. 4. Immunocytochemical studies in progress suggest that NPC1 protein resides in a unique vesicular compartment, as shown in Fig. 4 (Neufeld et al., in preparation). The NPC protein-containing compartment is hypothesized to mediate delivery of endocytosed cholesterol (LDL, or plasma membrane-derived) from lysosomes to the Golgi complex. Golgi proteins, which function in NPC protein-mediated transport of cholesterol, remain to be identified. NPC1 protein may be functionally linked to the sorting and transport of endocytosed cholesterol within the Golgi complex itself. Endocytosed cholesterol in the Golgi, like other Golgi proteins (29, 30) and lipids (41), is predicted to be sorted into both the anterograde pathway (delivery to PM via secretory pathway) and the retrograde pathway (delivery to the ER for esterification). The bulk of endocytosed cholesterol delivered to the Golgi complex is delivered to the plasma membrane. Endocytosed cholesterol, together with caveolin protein (delivered from the ER; (42)) and glycosphingolipids (GSLs; synthesized within the Golgi (43)), form membrane microdomains (rafts) which sequester proteins to be delivered in vesicles to the plasma membrane (44, 45). Sorting into the retrograde Golgi-ER pathway regulates the rate of esterification of endocytosed cholesterol.

Together, the ER and Golgi complex generate most cellular membranes. NPC protein-mediated cholesterol transport from lysosomes to the Golgi complex could function to coordinate membrane biogenesis. Enrichment of Golgi membranes with endocytosed cholesterol could serve to adjust the rate of Golgi membrane biogenesis in the anterograde direction. Cholesterol sorted into the retrograde Golgi-ER pathway may represent cholesterol that cannot be accommodated by cellular GSLs in the Golgi complex and plasma membrane. In response to regulatory pools of cholesterol, SREBPs in the ER coordinate the synthesis of two major building blocks of membranes, fatty acids and cholesterol (9). Thus, NPC proteins may also play a role in coordinating cellular membrane biogenesis by modulating Golgi-mediated flow of regulatory cholesterol to the ER.

BEYOND CHOLESTEROL

Clearly, NP-C disease, an autosomal recessive lipid storage disorder, involves defective processing of endocytosed cholesterol. The BALB/c mouse model, originally studied as a neuroviscerally affected strain of unknown etiology, was found to be defective in utilization of LDL cholesterol (46). Subsequently, human NP-C cells were shown to have the same lesion in cholesterol processing (47). Defective utilization of LDL in cultured fibroblasts provides the basis for differential diagnosis of patients presenting with the clinical manifestations of NP-C disease (48). In humans and animal models of NP-C, however, lipid storage in visceral tissues is complex. Unesterified cholesterol, sphingomyelin, lysobisphosphatidic acid, glucosylceramide, and smaller amounts of other phospholipids and glycolipids accumulate in reticuloendothelial tissue (48). Lipid accumulation enlarges the liver and spleen. Other accumulated metabolites in lysosomes, such as endogenous hydrophobic amines (sphingosine (49)) and sphingomyelin may secondarily exacerbate accumulation of cholesterol and other metabolites.

NP-C is a fatal neurodegenerative disease. In the brain, GSLs, not cholesterol, accumulate (48). Glucosylceramide and lactosylceramide levels are elevated manyfold, up to levels seen in Gaucher's disease. Gangliosides GM2 and GM3 are also elevated. The appearance of elevated GM2 in NP-C neurons is linked to ectopic dendritogenesis and correlates with clinical progression of the disease (50, 51). Researchers and clinicians alike have long been puzzled by the lack of an obvious role for cholesterol in the neuropathology of the disorder. Recent findings suggest a plausible explanation for this paradox.

GM2 was shown to accumulate in lysosomes of cultured NP-C fibroblasts, even in the absence of a lysosomal cholesterol load (52). This finding suggests that NPC proteins may play a role in intracellular trafficking of gangliosides as well as cholesterol. Moreover, the release of an endocytosed fluid-phase marker from cholesterol-depleted NP-C fibroblasts is retarded (Neufeld et al., in preparation). This finding suggests that retrograde endocytic transport is defective in NP-C cells. Kinetic modeling of the clearance data suggests that the NPC1-targeted compartment lies between lysosomes and the plasma membrane, consistent with the model in Fig. 4. The NPC protein-mediated vesicular pathway, like all known vesicular pathways, should transport a specific set of lipids and proteins. Thus, perturbations in vesicular trafficking as a result of mutant NPC1 protein function may have consequences well beyond intracellular trafficking of cholesterol alone.

Altered metabolism of cellular GSLs as a result of mutant NPC protein function is expected to be particularly damaging to the brain. Plasma membrane GSLs are known to play critical roles in neuronal development and function. Cholesterol and GSLs are thought to associate to form "rafts" in the trans-Golgi network (TGN) which regulate protein sorting along the apical secretory pathway in both polarized and non-polarized cells (43, 44, 45). Plasma membrane caveolae containing these lipids compartmentalize signal transduction mechanisms, which play critical roles in cellular regulation (53). Cellular sphingomyelin levels modulate the size of regulatory pools of cellular cholesterol (24). The homeostatic mechanisms regulating cellular cholesterol and GSL metabolism may thus be linked so as to maintain proper cellular function. Mutant NPC protein impairs access of endocytosed cholesterol to the sterol-sensing machinery in the ER which regulates cellular sterol homeostasis. Mutant NPC protein might also prevent access of endocytically-processed GSLs to a cellular site involved in regulation of GSL metabolism. Alternatively, cellular cholesterol levels may co-regulate levels of cellular GSLs. Future studies of NP-C disease will clarify the relationship between cholesterol and glycolipid trafficking.

ACKNOWLEDGMENTS

The author wishes to thank Drs. Peter G. Pentchev and Joan Blanchette-Mackie for helpful discussions and critical reading of the manuscript.

REFERENCES

1. Brown ME, Goldstein JL. A receptor-mediated pathway for cholesterol homeostasis. Science 1986;232:34-47

2. Goldstein JL, Dana SE, Faust JR, Beaudet AL, Brown ME. Role of lysosomal acid lipase in the metabolism of plasma low density lipoprotein. Observations in cultured fibroblasts from a patient with cholesteryl ester storage disease. J Biol Chem 1975; 250:8487-95

3. Carstea ED, Morris JA, Coleman KG, Loftus SK, Zhang D, Cummings C, Gu JZ, Rosenfeld MA, Pavan WJ, Krizman DB, Nagle J, Polymeropoulos MH, Sturley SL, Ioannou YA, Higgins, ME, Comly M, Cooney A, Brown A, Kaneski CR, Blanchette-Mackie EJ, Dwyer NK, Neufeld EB, Chang T-Y, Lisum L, Strauss III JF, Ohno K, Zeigler M, Carmi R, Sokol J, Markie D, O'Neill RR, van Diggelen OP, Elleder M, Patterson MC, Brady RO, Vanier MT, Pentchev PG, Tagle DA. Niemann-Pick C1 disease gene: homology to mediators of cholesterol homeostasis. Science 1997;277:228-231

4. Loftus SK, Morris JA, Carstea ED, Cummings C, Brown A, Ellison J, Rosenfeld MA, Tagle DA, Pentchev PG, Pavan WJ. Murine model of Niemann-Pick C disease: mutation in a cholesterol homeostasis gene. Science 1997;277:232-235

5. Carstea ED, Polymeropoulos MH, Parker CC, Detera-Wadleigh SD, O'Neill RR, Patterson MC, Goldin, E, Xiao H, Straub RE, Vanier MT, Brady RO, Pentchev PG. Linkage of Niemann-Pick disease type C to human chromosome 18. Proc Natl Acad Sci 1993;90:2002-2004

6. Vanier MT, Duthel S, Rodriguez-Lafrasse C, Pentchev PG, Carstea ED. Genetic heterogeneity in Niemann-Pick C disease: A study using somatic cell hybridization and linkage analysis. Am J Hum Genet 1996;58:118-125

7. Sakai J, Nohturfft A, Goldstein JL, Brown ME. Cleavage of sterol regulatory element-binding proteins (SREBPs) at site-1 requires interaction with SREBP-cleavage-activating protein. Evidence from in vivo competition studies. J Biol Chem 1998;273:5785-5793

8. Johnson RL, Rothman AL, Xie J, Goodrich LV, Bare JW, Bonifas JM,Quinn AG, Myers RM, Cox DR, Epstein EH Jr, Scott MP. Human homolog of patched, a candidate gene for the basal cell nevus syndrome. Science 1996; 272:1668-16719

9. Brown ME, Goldstein JL. The SREBP pathway: regulation of cholesterol metabolism by proteolysis of a membrane-bound transcription factor. Cell 1997;78:331-340.

10. Liscum L, Ruggiero RM, Faust JR. The intracellular transport of low density lipoprotein-derived cholesterol is defective in Niemann-Pick type C fibroblasts. J Cell Biol 1989;108:1625-1636

11. Neufeld EB, Cooney AM, Pitha J, Dawidowicz EA, Dwyer NK, Pentchev PG, Blanchette-Mackie EJ. Intracellular trafficking of cholesterol monitored with a cyclodextrin. J Biol Chem 1996;271:21604-21613

12. Shamburek RD, Pentchev PG, Zech LA, Blanchette-Mackie EJ, Carstea ED, VandenBroek JM, Cooper PS, Neufeld EB, Phair RD, Brewer Jr HB, Brady RO, Schwartz CC. Intracellular trafficking of the free cholesterol derived from LDL cholesteryl ester is defective in vivo in

Niemann-Pick C disease: insights on normal metabolism of HDL and LDL gained from the NP-C mutation. J Lipid Res 1997;38:2422-2435

13.   Sokol J, Blanchette-Mackie EJ, Kruth HS, Dwyer NK, Amende LM, Butler JD, Robinson E, Patel S, Brady RO, Comley ME, Vanier MT, Pentchev PG. Type C Niemann-Pick disease. Lysosomal accumulation and defective intracellular mobilization of low density lipoprotein cholesterol. J Biol Chem 1988;263:3411-3417

14.   Brasaemle DL, Attie AD. Rapid intracellular transport of LDL-derived cholesterol to the plasma membrane in cultured fibroblasts. J Lipid Res 1990;31:103-112

15.   Johnson WJ, Chacko GK, Phillips MC, Rothblat GH. The efflux of lysosomal cholesterol from cells. J Biol Chem 1990;265:5546-5553

16.   Xu X-X, Tabas I. Lipoproteins activate acyl coenzyme A:cholesteryl acyltransferase in macrophages only after cellular cholesterol pools are expanded to a critical threshold level. J Biol Chem 1991;266:17040-17048

17.   Okwu AK, Xu X-X, Shiratori Y, Tabas I. Regulation of the threshold for lipoprotein-induced acyl-CoA:cholesterol O-acyltransferase stimulation in macrophages by cellular sphingomyelin content. J Lipid Res 1994;35:644-655

18.   Lange Y, Strebel F, Steck TL. Role of the plasma membrane in cholesterol esterification in rat hepatoma cells. J Biol Chem 1993;268:13838-13843

19.   Tabas I, Rosoff WJ, Boykow GC. Acyl coenzyme A:cholesterol acyl transferase in macrophages utilizes a cellular pool of cholesterol oxidase accessible cholesterol as a substrate. J Biol Chem 1988;263:1266-1272

20.   Nagy L, Freeman DA. Cholesterol movement between the plasma membrane and the cholesteryl ester droplets of cultured Leydig tumour cells. Biochem J 1990;271:809-814

21.   Spillane DM, Reagan Jr JW, Kennedy NJ, Schneider DL, Chang T-Y. Translocation of both lysosomal LDL-derived cholesterol and plasma membrane cholesterol to the plasma membrane for esterification may require common cellular factors involved in cholesterol egress from the acidic compartments (lysosomes/endosomes). BBA 1995;1254:283-294

22.   Underwood KW, Andemariam B, McWilliams GL, Liscum, L. Quantitative analysis of hydrophobic amine inhibition of intracellular cholesterol transport. J Lipid Res 1996;37:1556-1568

23.   Underwood KW, Jacobs NL, Howley A, and Liscum L. Evidence for a cholesterol transport pathway from lysosomes to endoplasmic reticulum that is independent of the plasma membrane. J Biol Chem 1998;273:4266-4274

24.   Scheek S, Brown ME, Goldstein, JL. Sphingomyelin depletion in cultured cells blocks proteolysis of sterol regulatory element binding proteins at site 1. Proc Natl Acad Sci 1997;94:11179-11183

25.   Liscum L, Faust JR. Low density lipoprotein (LDL)-mediated suppression of cholesterol synthesis and LDL uptake is defective in Niemann-Pick type C fibroblasts. J Biol Chem 1987; 262:17002-17008

26.   Pentchev PG, Comly ME, Tokoro T, Butler J, Sokol J, Filling-Katz M, Quirk JM, Marshall DC, Patel S, Vanier MT. Group C Niemann-Pick disease: faulty regulation of low-density lipoprotein uptake and cholesterol storage in cultured fibroblasts. FASEB J 1987;1;40-45

27.   Blanchette-Mackie EJ, Dwyer NK, Amende LM, Kruth HS, Butler JD, Sokol J, Comley ME, Vanier MT, August JT, Brady RO, Pentchev PG. Type-C Niemann-Pick disease: low density

lipoprotein uptake is associated with premature cholesterol accumulation in the Golgi complex and excessive cholesterol storage in lysosomes. Proc Natl Acad Sci USA 1988;85:8022-8026

28. Coxey RA, Pentchev PG, Campbell G, Blanchette-Mackie EJ. Differential accumulation of cholesterol in Golgi compartments of normal and Niemann-Pick Type C fibroblasts incubated with LDL: a cytochemical freeze-fracture study. J Lipid Res 1993;34:1165-1176

29. Lippincott-Schwartz J, Yuan LC, Bonifacino JS, Klausner RD. Rapid redistribution of Golgi proteins into the ER in cells treated with brefeldin A: evidence for membrane cycling from Golgi to ER. Cell 1989;56:801-813

30. Lippincott-Schwartz J, Yuan L, Tipper C, Amherdt M, Orci L, Klausner RD. Brefeldin A's effects on endosomes, lysosomes, and the TGN suggest a general mechanism for regulating organelle structure and membrane traffic. Cell 1991;67:601-616

31. Butler JD, Blanchette-Mackie EJ, Goldin E, O'Neill RR, Carstea G, Roff CF, Patterson MC, Patel S, Comley ME, Cooney A, Vanier MT, Brady RO, Pentchev PG. Progesterone blocks cholesterol translocation from lysosomes. J Biol Chem 1992;267:23797-23805

32. Liscum L. Pharmacological inhibition of the intracellular transport of low-density lipoprotein-derived cholesterol in Chinese hamster ovary cells. BBA 1990;1045:40-48

33. Roff CF, Goldin E, Comly ME, Cooney A, Brown A, Vanier MT, Miller, SPF, Brady RO, Pentchev PG. Type C Niemann-Pick disease: use of hydrophobic amines to study defective cholesterol transport. Dev Neurosci 1991;13:315-320

34. Zha X, Pierini LM, Leopold PL, Skiba PJ, Tabas I, Maxfield FR. Sphingomyelinase treatment induces ATP-independent endocytosis. J Cell Biol 1998;140:39-47

35. Skiba PJ, Zha X, Maxfield FR, Schissel SL, Tabas I. The distal pathway of lipoprotein-induced cholesterol esterification, but not sphingomyelinase-induced cholesterol esterification, is energy-dependent. J Biol Chem 1996;271:13392-13400

36. Slotte JP, Bierman EL. Depletion of plasma-membrane sphingomyelin rapidly alters the distribution of cholesterol between plasma membranes and intracellular cholesterol pools in cultured fibroblasts. Biochem J 1988;250:653-658

37. Byers DM, Morgan MW, Cook HW, Palmer FB StC, Spence MW. Niemann-Pick type II fibroblasts exhibit impaired cholesterol esterification in response to sphingomyelin hydrolysis. BBA 1992;1138:20-26

38. Urbani L, Simoni RD. Cholesterol and vesicular stomatitis virus G protein take separate routes from the endoplasmic reticulum to the plasma membrane. J Biol Chem 1990;265:1919-1923

39. Smart EJ, Ying Y-s, Donzell WC, Anderson RGW. A role for caveolin in transport of cholesterol from endoplasmic reticulum to plasma membrane. J Biol Chem 1996;271:29427-29435

40. Uittengogaard A, Ying Y-S, Smart EJ. Characterization of a cytosolic heat-shock protein-caveolin chaperone complex. Involvement in cholesterol trafficking. J Biol Chem 1998;273:6525-6532

41. Hoffmann PM, Pagano RE. Retrograde movement of membrane lipids from the Golgi apparatus to the endoplasmic reticulum of perforated cells: evidence for lipid recycling. Eur J Cell Biol 1993;60:371-375

42. Scheiffele P, Verkade P, Fra AM, Virta H, Simons K, Ikonen E. Caveolin-1 and -2 in the

exocytic pathway of MDCK cells. J Cell Bio 1998;140:795-806

43. Simons K, van Meer, G. Lipid sorting in epithelial cells. Biochemistry 1988;27:6197-6202

44. Keller P, Simons K. Cholesterol is required for surface transport of influenza virus hemagglutinin. J Cell Biol 1998;140:1357-1367

45. Simons K, and Ikonen E. Functional rafts in cell membranes. Nature 1997;387:569-572

46. Pentchev PG, Boothe AD, Kruth HS, Weintroub H, Strivers J, Brady RO. A genetic storage disorder in BALB/c mice with a metabolic block in esterification of exogenous cholesterol. J Biol Chem 1984;259:5784-5791

47. Pentchev PG, Comley ME, Kruth HS, Patel S, Proestel M, Weintroub H. The cholesterol storage disorder of the mutant BALB/c mouse. A primary genetic lesion closely linked to defective esterification of exogenously derived cholesterol and its relationship to human type C Niemann-Pick disease. J Biol Chem 1986;261:2771-2777

48. Pentchev PG, Vanier MT, Suzuki K, Patterson MC. "Niemann-Pick Disease Type C: A Cellular Cholesterol Lipidosis. In *Metabolic and Molecular Bases of Inherited Disease*, Charles R. Sciver, Arthur L. Beaudet, William S. Sly, David Valle, eds. New York, NY: McGraw Hill, 1995.

49. Goldin E, Roff CF, Miller SPF, Rodriguez-Lafrasse C, Vanier MT, Brady RO, Pentchev PG. Type C Niemann-Pick disease: a murine model of the lysosomal cholesterol lipidosis accumulates shingosine and sphinganine in the liver. BBA 1992;1127:303-311

50. Siegel DA, Walkley, SU. Growth of ectopic dendrites on cortical pyramidal neurons in neuronal storage diseases correlates with abnormal accumulation of GM2 ganglioside. J Neurochem 1994;62:1852-1862

51. Goodman LA, Walkley SU. Elevated GM2 ganglioside is associated with dendritic proliferation in normal developing neocortex. Dev Brain Res 1996;93:162-171

52. Yano T, Taniguchi M, Akaboshi S, Vanier MT, Tai T, Sakuraba H, OhnoK. Accumulation of GM2 ganglioside in Niemann-Pick disease type C fibroblasts. Proc Japan Acad 1996;72:214-219

53. Anderson RGW. Caveolae: where incoming and outgoing messengers meet. Proc Natl Acad Sci USA 1993;90:10909-10913

THE MOVEMENT OF PLASMA MEMBRANE CHOLESTEROL THROUGH THE CELL

Yong-Soon Choi and Dale A. Freeman

Department of Medicine, Oklahoma University Health Sciences Center, VA Medical Center, Oklahoma City, Oklahoma 73104; e-mail: dale-freeman @ uokhsc.edu

KEY WORDS: plasma membrane internalization, steroidogenic cholesterol, intracellular pathway

---

ABSTRACT

The cholesterol of the plasma membrane plays a central role as substrate for steroidogenesis in the steroid hormone synthesizing cell. Not only is the cholesterol of this membrane used directly for steroid synthesis but cholesteryl ester cholesterol passes through this membrane prior to utilization as well. Recent work in this laboratory has determined that cAMP-stimulated cells internalize membrane faster than unstimulated cells. Plasma membrane cholesterol enters the cell by means of a vesicular transport pathway with many functional features of the acidic vesicle-lysosome. Anatomically, however, plasma membrane cholesterol is contained in a vesicle population distinct from the acidic vesicle lysosome.

CONTENTS

INTRODUCTION AND PERSPECTIVES

All cells use cholesterol for membrane synthesis. Cholesterol is synthesized or obtained from the outside environment and then inserted into membranes that differ markedly in cholesterol content (1,2). Certain cells such as hepatocytes or cells that produce steroid hormones, have exaggerated needs for cholesterol since these cells synthesize lipoproteins and steroid hormones, respectively. Steroidogenic cells have certain advantages as study cells and have proved very useful for studying lipoprotein metabolism and cholesterol trafficking. Under extreme conditions the small mass of steroidogenic cells can utilize more than one half of the cholesterol turnover in an organism.

Utilization of cellular-free cholesterol is more apparent in Leydig cells (3) or Leydig tumor cells (4). These cells contain much lower cholesteryl ester levels than adrenal cells or corpus luteum cells. The utilization of free cholesterol that is readily apparent in Leydig and Leydig tumor cells can be detected on adrenal tumor cells and

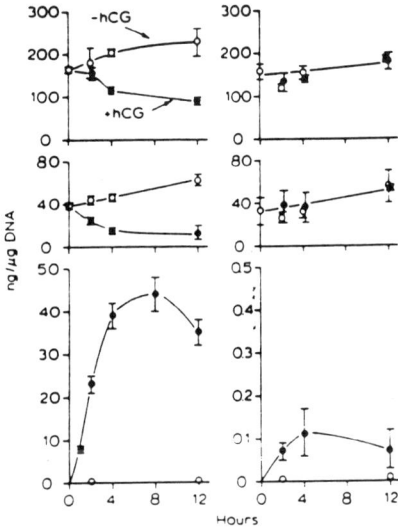

Fig. 1. **Effect of hCG stimulation on the cellular content of free and esterified cholesterol, and progesterone biosynthesis.** Dishes of cells were washed and placed in medium containing 1 mg/ml albumin without (*left*) or with (*right*) aminoglutethimide (50 μM). The dishes then received buffer only (O) or hCG (●) to give a final concentration of 40 ng/ml. The intracellular content of free (*top*) and esterified cholesterol (*middle*) and extracellular progesterone (*lower*) were determined at the times indicated. *Left*, average (± S.E.) of five independent experiments; *right*, average (± range) of two independent experiments. Note the different scales in the ordinates (*lower*).

normal adrenal cells if the cholesteryl ester content of the cells is reduced (5). Figure 1 shows the changes in cholesteryl ester and free cholesterol mass associated with hCG stimulation of MA-10 Leydig tumor cells. Although cholesteryl ester levels decline to a greater extent and somewhat more rapidly than free cholesterol levels, the quantity of free cholesterol utilized by the cell is significantly greater. The changes in cholesterol levels are caused by utilization of cholesterol for steroidogenesis since cholesterol content of stimulated cells treated with aminoglutethimide to block steroidogenesis does not decline (Figure 1 right).

Studies of intact cells treated with cholesterol oxidase indicate that most cellular-free cholesterol is found in the plasma membrane (6). Studies employing completely different techniques reach essentially the same conclusion (7). The location of steroidogenic-free cholesterol is the cholesterol oxidase-sensitive plasma membrane

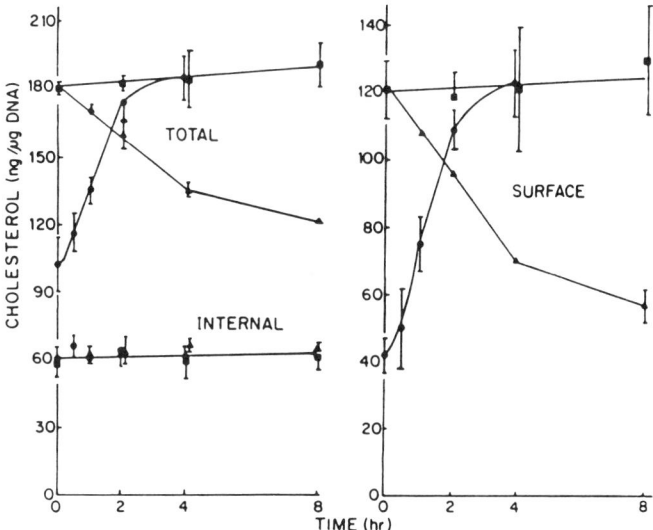

Fig. 2. **Cell surface cholesterol content changes in response to cAMP and cholesterol.** Twenty-four hours prior to the beginning of the experiment ($t = 0$) 100 X 20-mm dishes of cells were placed in 10 ml of medium containing 15% lipoprotein-deficient horse serum. At 14 h prior to $t = 0$, some cells were treated with 1 mM dibutyryl-cAMP (●). At $t = 0$ all dishes were washed twice with 3 ml of fresh assay medium and were placed in either assay medium (■ and ▲) or assay medium containing 50 µM aminoglutethimide and 300 µg/ml LDL cholesterol (●). At the times indicated, dishes were removed and either oxidized or pseudooxidized and the cellular content of cholesterol was determined. Cell surface cholesterol was the difference in total cholesterol and cholesterol post cholesterol oxidase treatment. The values shown are the average ± range of two independent experiments, triplicate points per experiment.

pool (8,9) (Figure 2). This pool makes up about 65% of the total free cholesterol pool of the MA-10 cells, depletes during hCG stimulation and is rapidly repleted by incubating the cells with LDL. Cholesterol depletion occurred both because stimulated cells diverted cholesterol normally incorporated into the membrane into the steroid biosynthetic pathway and because stimulated cells actively catabolize plasma membrane cholesterol. Figure 3 illustrates the effect of stimulation on the arrival of newly synthesized cholesterol at the plasma membrane. In the absence of stimulation most newly synthesized cholesterol is initially cholesterol oxidase resistant, however, with time the label distribution shifts to about the distribution of free cholesterol in the cell. With stimulation less label reaches the plasma membrane pool. Radioactivity retained within the cell is increased compared to that retained intracellularly in unstimulated cells. Studies employing ³H-cholesteryl linoleate labeled LDL showed that stimulated cells exported less LDL cholesterol to the plasma membrane. The data of Figure 4 show that ³H-cholesterol applied to the surface of unstimulated MA-

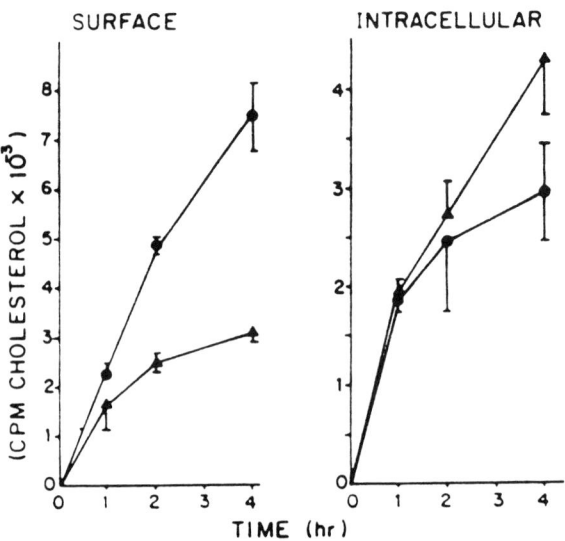

Fig. 3. **cAMP sequesters newly synthesized cholesterol into the cell interior.** Dishes of cells (60 X 15 mm) were incubated for 24 h in 4 ml of medium containing 15% lipoprotein-deficient horse serum. At t = 0, dishes were washed twice with 2 ml of assay medium and then placed in 2 ml of assay medium containing 2.3 μM [1-¹⁴C]sodium acetate (0.10 μCi/nmol) either alone (●) or with 1 mM dibutyryl-cAMP (Radioactivity comigrating with cholestenone (*surface*, ▲) and cholesterol (internal, ●) were quantitated. The results presented are from two experiments, average ± range, triplicate points per experiment.▲). At the times indicated, dishes of cells were removed and treated with cholesterol oxidase. Radioactivity comigrating with cholestenone (*surface*, ▲) and cholesterol (internal, ●) were quantitated. The results presented are from two experiments, average ± range, triplicate points per experiment.

10 cells is only slowly lost from the plasma membrane and that there is little radioactive steroid hormone formed. Stimulated cells, on the other hand, rapidly catabolize cell surface cholesterol and convert it into radioactive steroid hormones. Quantitatively, this is the most important effect of cAMP on plasma membrane cholesterol and can be estimated to account for about 75% of the steroid hormones formed (4).

Studies using inhibitor drugs that are known to inhibit steroidogenesis indicate that cholesterol transport in MA-10 cells can be functionally divided into two categories (10). Transport of newly synthesized or LDL-derived cholesterol to the cell surface is relatively insensitive to microtubule or microfilament inhibitors, proton pump inhibitors, and lysosomotrophic agents. Transport of cell surface cholesterol into the cell, on the other hand, is inhibited by these agents at concentrations active in inhibiting steroidogenesis. A reasonable hypothesis is that transport of plasma membrane cholesterol into the cell occurs by a vesicular transport pathway since vesicular transport would be expected to be sensitive to the inhibitors we tested. Sensitivity to proton pump inhibitors and lysosomotrophic agents might be expected if the acidic vesicle-lysosome pathway is involved in such traffic. The object of the studies reviewed here was to determine if hCG/cAMP stimulated plasma membrane internalization directly and to identify cellular intermediates involved in intracellular transport.

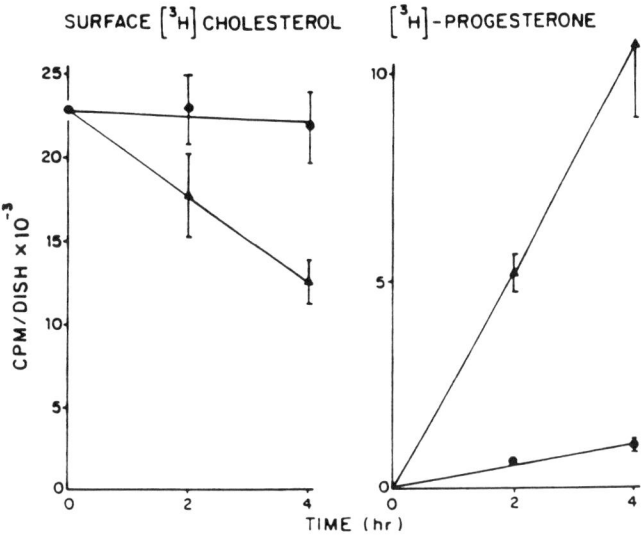

Fig. 4. **cAMP causes cell surface cholesterol to be utilized for steroid hormone synthesis.** At $t = 0$, 60 X 15-mm dishes of cells were washed once with 2 ml of assay medium and placed in 2 ml of assay medium containing 0.5% ethanol and 1 µCi of [³H]cholesterol (1 µCi/nmol). After 2 h at 37°C, dishes were removed, washed four times with warm assay medium, and placed in 4 ml of assay medium alone (●) or in the presence of 1 mM dibutyryl-cAMP (▲). At the times indicated, the medium was collected and saved and the cells were treated with cholesterol oxidase. Cell surface radioactivity (*left panel*) and medium progesterone radioactivity (*right panel*) were quantitated. The data shown are the mean ± S.E. of quadruplicate points from a representative experiment.

THE EFFECT OF cAMP ON INTERNALIZATION OF PLASMA MEMBRANE COMPONENTS

Cyclic-AMP stimulation may facilitate the removal of cholesterol from plasma membrane either by increasing the speed at which plasma membrane is internalized, by increasing the extraction of cholesterol from the membrane or by both mechanisms. The first possibility is measurable since it is possible to measure equilibration half-lives of ligands associated with the cholesterol, protein and phospholipid moieties of the cell.

The simplest measurement to perform is measuring the equilibrium half-life of cholesterol. Addition of ³H-cholesterol to cells at 4°C results in almost all label locating on the plasma membrane (11). With time and warming to 37°, this plasma membrane label enters the cell and becomes esterified. Since all cholesteryl esters are intracellular, measuring radioactivity associated with the esters acts as a measure of

114

internalized plasma membrane cholesterol. Figure 5 shows the time course for accumulation of plasma membrane cholesterol in the cholesterol esters of unstimulated and dibutyryl-cAMP stimulated, aminoglutethimide-treated (to block cholesterol utilization for steroidogenesis) MA-10 cells. It is clear that in each group there is time-dependent accumulation of cholesterol in the ester droplet. Stimulated cells reached equilibrium about 2.5-fold more quickly ($t_{1/2}$ of 84 minutes compared with a $t_{1/2}$ 213 minutes for control cells). Stimulation by cAMP also causes cholesteryl ester hydrolysis to increase. Since increased hydrolysis should facilitate exchange of plasma membrane and ester cholesterol, experiments were next performed with this variable removed.

The data of Figure 6 are from diethylumbelliferyl phosphate-treated MA-10 cells. Diethylumbelliferyl phosphate is an organophosphate compound that inactivates the cholesteryl ester hydrolase enzyme (12) and blocks cAMP-dependent production of the StAR protein (13). This later protein is synthesized in response to cAMP and acts to translocate cholesterol across the mitochondrial membranes. With these proteins inhibited dibutyryl-cAMP-treated cells still internalize plasma membrane cholesterol

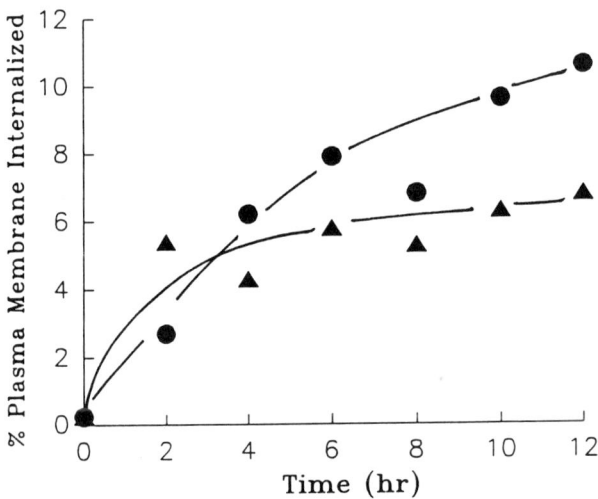

Fig. 5. **Turnover of plasma membrane cholesterol in control and cAMP-stimulated MA-10 cells.** Dishes of MA-10 cells (60 X 15 mm) were labeled with 2 ml of 4°C assay medium containing 1 μCi ^{3}H-cholesterol for 2 h at 4°C. At this time, the dishes were washed twice with 2 ml of 37°C assay medium and placed in assay medium containing 50 μM aminoglutethimide alone (●-●) or with 1 mM dibutyryl-cAMP (▲-▲). At the time indicated, cells were scraped from the dishes, the lipids were extracted and cholesteryl esters separated from cholesterol by thin layer chromatography. Data are expressed as percentage of plasma membrane ^{3}H-cholesteryl internalized and incorporated into cholesteryl ester. The data shown are from triplicate points of a representative experiment.

1.4-fold more rapidly than unstimulated cells ($t_{1/2}$ 132 minutes versus 180 minutes). Stimulated cells also esterified more label than controls. This likely occurs because diethylumbelliferyl phosphate blocks the cAMP-response pathway. A major effect of cAMP is to cause cholesterol movement to the mitochondria. With this effect blocked cholesterol that would have been sequestered near the mitochondria is available for esterification.

To determine whether other plasma membrane components internalized more rapidly in response to stimulation the experiments of Figures 7 and 8 were performed. Cell surface protein was labeled by biotinylation and then recognized by fluorescent strepavidin which was quantitated by flow cytometry (14). As shown on Figure 7, strepavidin-binding sites decrease in both control and in stimulated cells. On average the $t_{1/2}$ for this decrease in control cells was 165 minutes while that of

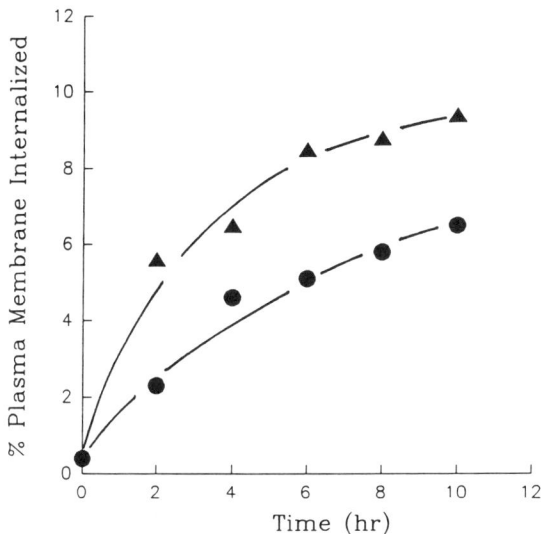

Fig. 6. **Turnover of plasma membrane cholesterol in diethylumbelliferyl phosphate treated control and cAMP-stimulated MA-10 cells.** Dishes of MA-10 cells (60 X 15 mm) were labeled with 2 ml of 4°C assay medium containing 1 μCi ^{3}H-cholesterol for 2 h at 4°C. At this time, the dishes were washed twice with 2 ml of 37°C assay medium containing 10^{-5}M diethylumbelliferyl/phosphate and 50 μM aminoglutethimide alone (●-●) or with 1 mM dibutyryl-cAMP (▲-▲). At the time indicated, cells were harvested and processed exactly as described in the legend of Fig. 5. The data shown are from triplicate points of a representative experiment.

stimulated cells was 90 stimulated minutes, a 1.5-fold decrease with stimulation. The $t_{1/2}$ of protein in either group of cells was similar to the t ½ for cholesterol, the expected finding, if these components were internalizing together.

The data of Figure 8 show the internalization rate of a fluorescent phospholipid analogue; N-[16-{7-Nitrobenz-2-oxa-1,3-diazol-4yl} aminohexanyl] sphingosylphos-phocholine(NBD-sphingomyelin) (15). In both control and stimulated cells, phospholipid analogue enters an intracellular compartment where it is not removed by high protein washes. The $t_{1/2}$ for entry to this compartment is 14.3 minutes for controls and 9.8 minutes for stimulated cells. The stimulated cells internalized label 1.5-fold more rapidly than the control cells. In contrast to the time constants for cholesterol and protein which were measurable in hours, the phospholipid time courses were almost 10-fold faster. The most likely explanation for this finding is that the NBD-sphingomyelin labels a rapid turning over subset of membrane, perhaps a subset like the endosomes which are known to have a very rapid turnover rate (16). The time

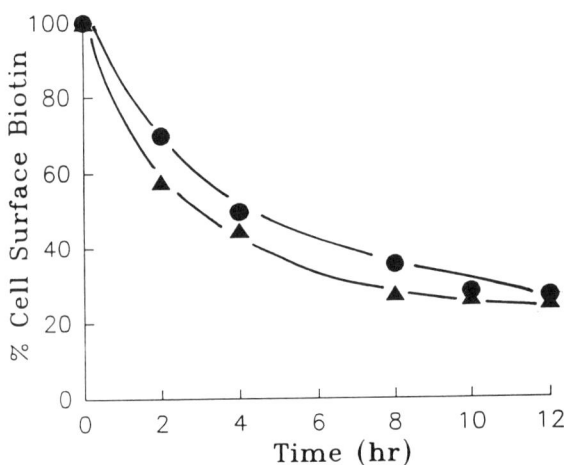

Fig. 7. **Turnover of plasma membrane proteins in control and stimulated MA-10 cells.** Dishes of MA-10 cells (100 X 20 mm) were reacted with NHS-biotin at 4°C. After washing away unbound biotin, cells were either reacted immediately with β-phycoerythrin streptavidin or warmed to 37°C in assay medium alone (●-●) or containing 1 mM dibutyryl-cAMP (▲-▲) for the times indicated and then reacted with streptavidin. After fixation, fluorescence was quantitated by flow cytometry. Data are from triplicate points of a representative experiment.

constants for cholesterol and protein are similar to plasma membrane internalization time constants measured in other cells that are thought to reflect total membrane turnover (17).

MOVEMENT OF PLASMA MEMBRANE CHOLESTEROL THROUGH THE CELL

Nutrients and peptide hormones follow a now well-described pathway from the cell surface to coated vesicles to the lysosome (18). Data employing selective labeling of intralysosomal membrane (19) or plasma membrane (17) indicate that label initially restricted to the lysosome eventually redistributes to the plasma membrane while labeled proteins from the plasma membrane eventually are found on the lysosome. These findings coupled to observations that Niemann-Pick type C cells express a defect in lysosomal release of cholesterol (20) and that cells treated with agents that mimic the Niemann-Pick lesion do not readily esterify sphingomyelinase released plasma membrane cholesterol (21) caused us to systematically explore the

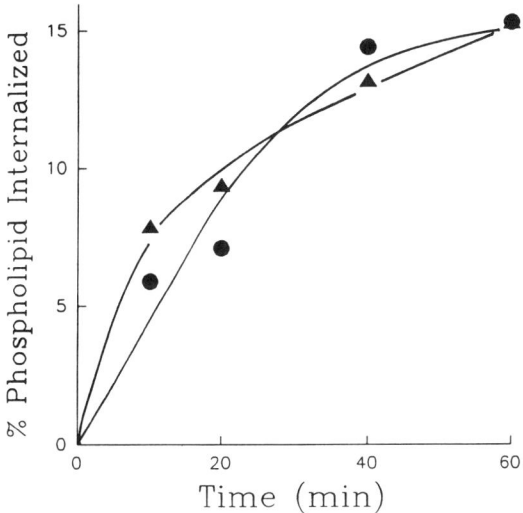

Fig. 8. **Turnover of plasma membrane phospholipid in control and cAMP-stimulated MA-10 cells.** Dishes of MA-10 cells (30 X 15 mm) were incubated in 2 ml of assay medium containing N-[16-(7-nitrobenz-2-oxa-1,3-diazol-4-yl) amino hexanyl] sphingosylphosphocholine (NBD-sphingomyelin) alone (●-●) or containing 1 μM dibutyryl-cAMP (▲-▲). At the times indicated, the dishes were placed at 4° and washed five times with assay medium containing 5% w/v BSA. The cells were then dissolved in 0.01% Triton X-100 and fluorescence quantitated by fluorometer. Data shown are from triplicate points of a representative experiment.

possibility that plasma membrane steroidogenic cholesterol enters through a lysosomal pathway.

To determine whether chloroquine or sphinganine, inhibitors of lysosomal function and lysosomal cholesterol release, respectively, would block lysosomal release of LDL cholesterol, we performed experiments where cells were incubated with LDL and various concentrations of either inhibitor drug (22). Chloroquine maximally increases cell-free cholesterol content 1.5-fold and cholesteryl ester content 1.7-fold when used at 50-100μm. Sphinganine causes greater cholesterol and cholesteryl ester accumulation that was not saturable at 100μm. At this concentration of sphinganine cholesterol increased 2.6-fold while cholesteryl esters increased 7.4-fold. These data indicate that both agents cause intracellular accumulation of cholesterol. Using a sucrose step gradient to isolate lysosomes from the other cholesterol containing membranes of the MA-10 cells, it is possible to assess the effect of each inhibitor on the cholesterol content of lysosomes isolated from control and inhibitor-treated cells. The lysosomal content of cholesterol as well as cholesteryl

118

esters is increased in inhibitor-treated cells. Chloroquine treatment increases lysosomal cholesterol levels 2.4-fold and cholesteryl ester levels 2-fold compared to control lysosomes. Sphinganine treatment causes the lysosome cholesterol content to increase about 5-fold while cholesteryl esters increase 40-fold compared with control cell lysosomes. These results are those expected if the inhibitors cause LDL cholesterol and cholesteryl esters to remain in the lysosome.

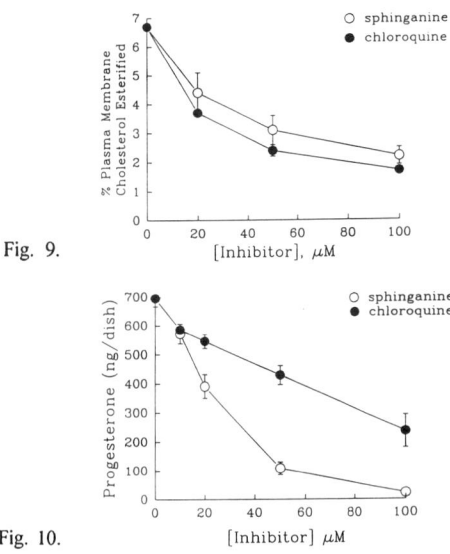

Fig. 9.

Fig. 10.

Fig. 9. **Effect of chloroquine and sphinganine on the esterification of plasma membrane cholesterol.** Dishes of MA-10 cells (60 X 15 mm) were incubated in 2 ml of 4°C assay medium containing 1 mCi [³H]-cholesterol (65.5 Ci/mmol). The dishes were incubated at 4°C for 2 h. After 2 h, the dishes were washed two times with 2 ml 37°C assay medium and then 4 ml of assay medium containing the indicated concentrations of either chloroquine (●) or sphinganine (O) was added. After 6 h at 37°C, cells were scraped from the dish and extracted. Radioactivity associated with cholesteryl esters and free cholesterol was quantitated and expressed as the percentage of total cell cholesterol radioactivity. Data shown are the average ± range of two independent experiments, triplicate dishes per experiment.

Fig. 10. **Effect of chloroquine and sphinganine on progesterone production by MA-10 cells.** Dishes of MA-10 cells (30 X 15 mm) were incubated in assay medium containing 1 mM dibutyryl-cAMP and the indicated concentrations of either chloroquine (●) or sphinganine (O). After 4 h at 37°C the medium was removed and the medium content of progesterone was quantitated by radioimmunoassay. Data shown are the average ± range of two independent experiments, triplicate joints per experiment.

Plasma membrane cholesterol can be internalized into the cell and either esterified (23-25) or in stimulated cells converted into steroid hormones (4). If the lysosome is an important transport intermediate between the plasma membrane and the cell, it would be predicted that these agents would disrupt these processes. The data of Figures 9 and 10 indicate that this is the case. In Figure 9 both chloroquine and sphinganine produce dose-dependent inhibition of plasma membrane cholesterol esterification. Chloroquine is most disruptive causing half-maximal inhibition at 30μm and inhibiting esterification to 25% of control at 100μm. Sphinganine causes half-maximal inhibition at 44μm and at 100μm inhibits plasma membrane cholesterol esterification to 33%. Figure 9 shows the effects of each inhibitor on progesterone synthesis. Both agents inhibit progesterone synthesis. Sphinganine is the more potent inhibitor resulting in half-maximal inhibition at 28μm and causing inhibition to 14% of control at 100μm. Chloroquine causes half-maximal inhibition of steroidogenesis at 50μm while inhibiting progesterone synthesis to 34% of control at 100μm.

These experiments would suggest that label applied to the plasma membrane would accumulate in the lysosomes of inhibitor-treated cells. In fact, lysosomes from cells incubated with chloroquine accumulated 1.5-1.9 more plasma membrane label than lysosomes from control cells. Sphinganine treatment resulted in 2.2-3.1-fold more intralysosomal plasma membrane label accumulation than that found in control lysosomes.

All the data of this series of experiments are consistent with the hypothesis that the lysosome acts as a transport intermediate between the plasma membrane and the cell interior. Since neither inhibitor drug is a specific inhibitor of lysosomes, the possibility remains that the drugs indeed inhibit lysosomes but that the effects on cholesterol transport are caused by effects on other organelles.

One organelle, the Golgi apparatus, is a known intermediate in protein transport and might be expected to be sensitive to these inhibitor drugs. To exclude Golgi inhibition underlying the inhibitor effect on plasma membrane internalization, we employed the relatively specific Golgi apparatus inhibitor, Brefeldin A (26). We identified concentrations of Brefeldin A that completely disrupted the Golgi apparatus. This treatment does not effect steroidogenesis and actually augments plasma membrane cholesterol esterification. From these experiments we conclude that the Golgi apparatus is not required for plasma membrane cholesterol internalization or for its subsequent esterification or utilization for steroidogenesis.

The possibility still remains that chloroquine and sphinganine inhibit the lysosome but that this inhibition does not underlie the drug effects on plasma membrane cholesterol utilization. In an attempt to visualize plasma membrane cholesterol internalization directly and relate this to the acidic vesicle lysosome pathway we performed the experiments of Figure 11. 3-[2,4-dinitroanilino] 3'-amino-n-methyl dipropylamine (DAMP) was used to identify acidic vesicles and lysosomes (27). A cholesterol analogue, N-(N-[6-[(7-nitrobenz-2 oxa 1,3 diazol-4-yl) amino]-caproyl) cholesterol (NBD-cholesterol) was used to visualize internalizing membrane. NBD-cholesterol clearly enters the cell in vesicular-like structures but these structures are discrete from the acidic vesicles and lysosomes. The impression that the fluorescence label is separate from the lysosome is reinforced by cell fractionalization experiments using NBD-cholesterol labeled cells. In these experiments fluorescence is found with the plasma membrane marker as expected. The peak of fluorescence is found at a density that overlaps the lysosomal peak but that is also clearly less dense than the lysosomes (22): The most straight forward interpretation of these data is that NBD-cholesterol enters the cell in vesicles but in vesicles that were different than the acidic vesicles or the lysosomes.

120

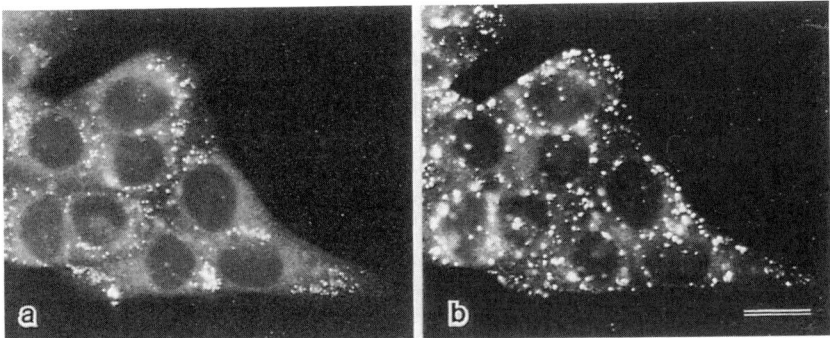

Fig. 11. **Internalized NBD-cholesterol and acid vesicle-lysosome concentrated DAMP are in different vesicles.** Cells were cell-surface labeled with NBD-cholesterol and loaded with DAMP. Cells observed using the fluorescein optical filter set demonstrated vesicles labeled with NBD-cholesterol (a), while the same cells observed using the rhodamine optical filter set illustrated vesicles labeled with anti-DNP antibody (b). The magnification bar shown is equal to 20 μm.

CONCLUSIONS

1. Most cholesterol destined to become a steroid hormone will pass through the plasma membrane.

2. cAMP increases the internalization rate of plasma membrane.

3. Plasma membrane steroidogenic cholesterol enters the cells in a vesicular compartment with many functional similarities to the acidic vesicle-lysosome.

4. The "cholesterol analogue?" NBD-cholesterol enters the cell in a vesicular compartment that is anatomically distinct from the acidic vesicles and lysosome.

REFERENCES

1. Wattenberg BW, Silbert DF. Sterol partitioning among intracellular membranes. J Biol Chem 1983;2-58:2284-89
2. Lange Y, Swaisgood MH, Ramos BV, Steck TL. Plasma membranes contain half the phospholipid and 90% of the cholesterol and sphingomyelin in cultured human fibroblasts. J Biol Chem 1989;264:3736-93
3. Hou JW, Collins DC, Schleicher RL. Sources of cholesterol for testosterone biosynthesis in murine Leydig cells. Endocrinology 1990;127:2047-55

4. Freeman DA, Ascoli M. Studies on the source of cholesterol used for steroid biosynthesis in cultured Leydig tumor cells. J Biol Chem 1982;257:14231-38

5. Gocze PM, Freeman DA. Plasma membrane cholesterol is used as steroidogenic substrate in Y-1 mouse adrenal tumor cells and normal sheep adrenal cells. Exp Cell Res 1993;209:21-25

6. Lange Y, Steck TL. Cholesterol-rich intracellular membranes: a precursor to the plasma membrane. J Biol Chem 1985;260:15592-97

7. Kaplan MR, Simoni RD. Transport of cholesterol from the endoplasmic reticulum to the plasma membrane J Cell Biol 1985;101:446-53

8. Freeman DA. Cyclic-AMP mediated modification of cholesterol traffic in Leydig tumor cells. J Biol Chem 1987;262:13061-8

9. Freeman DA. Plasma membrane cholesterol: removal and insertion into the membrane and utilization as substrate for steroidogenesis. Endocrinology 1989;124:2527-34

10. Nagy L, Freeman DA. Effect of cholesterol transport inhibitors on steroidogenesis and cholesterol transport in cultured Leydig cells. Endocrinology 1990;126:2267-76

11. Lange Y, Ramos BV. Analysis of the distribution of cholesterol in the intact cell. J Biol Chem 1983;25-8:15130-4

12. Harrison EH, Bernard DW, Scholm P, Quin DM, Rothblatt GH, Glich JM. Inhibitors of neutral cholesteryl ester hydrolase. J Lipid Res 1990;31:2187-93

13. Choi YS, Stocco DM, Freeman DA. Diethylumbelliferyl phosphate inhibits steroidogenesis by interfering with a long-lived factor acting between protein kinase A activation and induction of the steroidogenic acute regulatory StAR protein. Eur J Biochem 1995;234:680-5

14. Suzuki T, Dale GL. Biotinylated erythrocytes: in vitro survival and in vitro recovery. Blood 1987;70-:791-5

15. Marsh M, Schmidt S, Kern H, Herms E, Male P, Mellman I, Helenius A. Rapid analytical and preparative isolation of functional endosomes by free flow electrophoresis. J Cell Biol 1987;104:875-6

16. Griffiths G, Bach R, Marsh M. A quantitative analysis of the endocytic pathway in baby hamster kidney cells. J Cell Biol 1989;109:2703-20

17. Burgert HG, Thilo L. Internalization and recycling of plasma membrane glycoconjugates during pinocytosis in the macrophage cell line P 388D. Exp Cell Res 1983;144:127-42

18. Watts C, Marsh M. Endocytosis: what goes in and how. J Cell Sci 1992;103:1-8

19. Muller WA, Steinman RM, Cohn ZA. The membrane proteins of the vesicular system 1, analysis by a novel method of intralysosomal iodination. J Cell Biol 1980;86:292-303

20. Pentchev PG, Brady RO, Blanchette-Mackie EJ, Vanier MT, Carstea ED, Parker CC, Goldin E, Roff GF. The Niemann-Pick C lesion and its relationship to the intracellular distribution and utilization of LDL cholesterol. Biochem Biophys ACTA 1994;1225:235-43

21. Harmala AS, Porn MI, Slotte JP. Sphingosine inhibits sphingomyelinase-induced cholesteryl esterification in cultured fibroblasts. Biochem Biophys ACTA 1993;97-104

22. Porpaczy Z, Tomasek JJ, Freeman DA. Internalized plasma membrane cholesterol passes through an endosome compartment that is distinct from the acid-vesicler lysosome compartment. Exp Cell Res 1997;234:217-24

23. Tabas I, Rosoff WJ, Boykow GC. Acylcoenzyme A: cholesterol acyl transferase in macrophages utilizes a cellular pool of cholesterol oxidase-accessible cholesterol as substrate. J Biol Chem 1988;263:1266-72

24. Slotte JP, Bierman EL. Depletion of plasma-membrane sphingomyelin rapidly alters the distribution of cholesterol between plasma membranes and intracellular cholesterol pools in cultured fibroblasts. Biochem J 1988;250:653-8

25. Nagy L, Freeman DA. Cholesterol movement between the plasma membrane and the cholesteryl ester droplets of cultured Leydig tumor cells. Biochem J 1990;271:809-14

26. Klausner RD, Donaldson JG, Lippincott-Schwartz J. Brefeldin A: insights into the control of membrane traffic and organelles structure. J Cell Biol 1992;116:1071-80

27. Anderson RGW, Falck JR, Goldstein JL, Brown MS. Visualization of acidic organelles in intact cells by electron microscopy. Proc Natl Acad Sci USA 1984;81:4838-42

CHOLESTEROL TRAFFICKING IN CACO-2 CELLS

F. Jeffrey Field

University of Iowa College of Medicine, Department of Internal Medicine, Iowa City, Iowa 52242; e-mail f-jeffrey-field@uiowa.edu

KEY WORDS: CaCo-2 cells, cholesterol influx, triacylglycerol-rich lipoprotein assembly and secretion, acylcoenzyme A: cholesterol acyltransferase, caveolin, caveolae

---

ABSTRACT

In CaCo-2 cells, the transport of cholesterol from the plasma membrane to the endoplasmic reticulum occurs via transport vesicles, likely derived from specialized domains of the plasma membrane. Agents that inhibit p-glycoprotein activity interfere with the transport of plasma membrane cholesterol to the endoplasmic reticulum and the secretion of triacylglycerol-rich lipoproteins. The uptake of micellar cholesterol causes the "clustering" of plasma membrane cholesterol to specialized microdomains of the plasma membrane containing caveolin. It is likely that it is from these areas that plasma membrane cholesterol fluxes to the endoplasmic reticulum. Cholesterol derived from the plasma membrane is the major substrate for ACAT and triacylglycerol-rich lipoprotein cholesterol. The transport of newly-synthesized cholesterol from the endoplasmic reticulum to the plasma membrane is constitutive and independent of Golgi, microtubular function, new protein synthesis, or p-glycoprotein. CaCo-2 cells regulate the amount of newly-synthesized cholesterol arriving at the plasma membrane by altering the rate of cholesterol synthesis, not by altering the transport process. The pathways that transport cholesterol to and from the plasma membrane are distinct.

CONTENTS

INTRODUCTION

In man, cholesterol flux across the small intestinal cell contributes to the plasma cholesterol pool and to maintaining whole body cholesterol balance. Unlike any other cell within the body, the intestinal absorptive cell is bathed at its apical surface with dietary and biliary cholesterol. This constant influx of micellar cholesterol clearly adds to the total cholesterol pool of the enterocyte. In this regard, the intestinal cell is quite unique. Like other cells, enterocytes have the machinery to synthesize new cholesterol, and in fact, are quite efficient in doing so (1). In addition, the enterocyte has lipoprotein receptors and can procure cholesterol by the uptake of circulating lipoproteins (2). Unlike other cells, however, which rely almost entirely on lipoprotein cholesterol to supply their cholesterol needs, under normal physiological conditions, it is unlikely that the intestinal cell utilizes either newly-synthesized or lipoprotein cholesterol to meet its immediate sterol requirements. With the continuous influx of cholesterol at the apical membrane, cholesterol synthesis and LDL receptor expression are likely to be suppressed and indeed, there is experimental evidence to suggest that this is so (3,4). Since absorbed micellar cholesterol is likely the major source of cholesterol for the intestinal cell, its transport and utilization are of obvious importance.

It has been postulated that the major fate of absorbed cholesterol is its transport to the endoplasmic reticulum, the site of lipoprotein assembly (5). It is here that it enters a pool of cholesterol which is utilized as substrate by acylcoenzyme A:cholesterol acyltransferase (ACAT), an enzyme which catalyzes the esterification of cholesterol to a fatty acyl-CoA (5). This esterified cholesterol and the unesterified cholesterol which "escapes" ACAT are used to synthesize the core and surface components of a triacylglycerol-rich lipoprotein particle. The importance of ACAT in this process has been repeatedly shown. If ACAT activity is inhibited, micellar cholesterol is not esterified and secreted lipoprotein particles are deficient in cholesteryl esters (6,7). The proposed trafficking of micellar cholesterol to the endoplasmic reticulum makes good sense; after all, the intestinal absorptive cell is in the business to expeditiously transport all nutrients from the lumen into the blood and/or lymph. Although this makes good sense and seems quite straight-forward, the premise may not be entirely correct. For example, in a study by Klein and Rudel (8) investigating cholesterol absorption in nonhuman primates, 18 hours were required for lymph cholesterol specific activity to come into equilibrium with the specific activity of dietary cholesterol. Over this period, mass transport of cholesterol across the intestine remained essentially constant suggesting that dietary cholesterol was being diluted with endogenous cholesterol. These results suggest that the intestinal cell utilizes cholesterol from a source other than dietary to assemble and secrete a lipoprotein particle. Thus, to better understand the processes of cholesterol absorption and lipoprotein assembly and secretion, more information regarding intestinal cholesterol trafficking at the cellular level is necessary.

CaCo-2 cells have been used as a model for small intestinal absorptive cells for many years (9). They are derived from a human colon adenocarcinoma which, when grown to confluency in culture, spontaneously differentiate into cells that have morphological and biochemical characteristics of mature enterocytes (9). This cell line has been used extensively to investigate the regulation of cholesterol and lipoprotein metabolism (3,10, review), cholesterol absorption (11), fatty acid and triacylglycerol metabolism (12-14), and many other parameters of lipid metabolism (15-18). Although these cells are not perfect models and differences do exist between them and absorptive intestinal cells (10), they have provided investigators with a model to study specific metabolic processes at the cellular and molecular level. In this chapter, cholesterol trafficking within CaCo-2 cells and its role in lipoprotein synthesis and secretion will be discussed.

CHOLESTEROL TRANSPORT FROM THE PLASMA MEMBRANE TO THE ENDOPLASMIC RETICULUM

In one of our earlier reports (11), we made an observation that perplexed us. At the time, we were attempting to document the esterification of micellar cholesterol by ACAT. The experimental design was simple. CaCo-2 cells were incubated with labeled cholesterol solubilized in taurocholate micelles containing oleic acid to drive lipoprotein synthesis. Following the incubation, the amount of labeled cholesterol that was taken up and esterified was estimated (Figure 1).

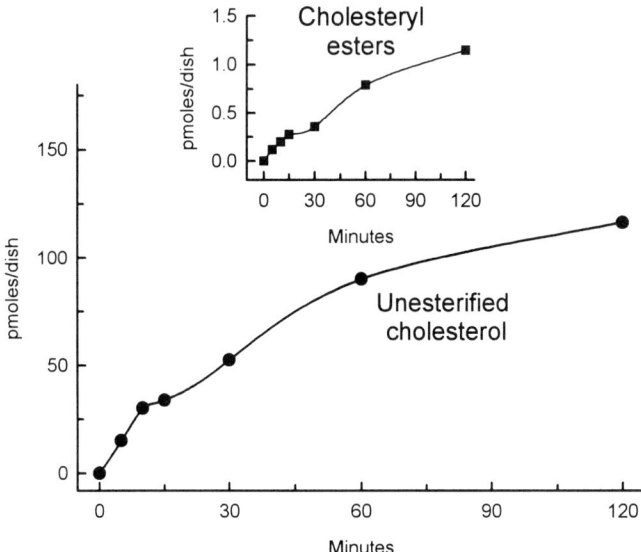

Figure 1:   Uptake and esterification of micellar cholesterol. Reprinted by permission from J. Lipid Res. 28, 1987. See page 1064 for more details.

As expected, cells readily took up the micellar cholesterol. The amount of cholesterol that was esterified, however, amounted to only 1% of the cell-associated labeled cholesterol. If that statement is turned around, it indicates that 99% of the labeled cholesterol taken up by the cell was not encountering ACAT. Moreover, no labeled cholesteryl esters were being secreted into the basolateral medium. In contrast, if cells were incubated with micelles containing labeled oleic acid, cellular cholesteryl esters were rapidly synthesized, and moreover, within minutes, these cellular labeled cholesteryl esters were secreted into the basolateral medium. One could argue that the labeled micellar cholesterol was only adsorbed to the apical cell surface and not actually taken up by the cell. This cannot be the case, however. Shortly after adding micellar cholesterol to CaCo-2 cells, cholesterol synthesis is decreased and cholesterol esterification is increased. This suggests that micellar cholesterol is entering metabolic pools of cholesterol resulting in the "expected" changes in cholesterol metabolism. At the time, this apparent discrepancy between the fate of labeled micellar cholesterol and fatty acid remained unexplained. It did suggest to us, however, that under conditions of fatty acid influx, the cell was utilizing cholesterol from an intracellular pool to assemble and secrete a lipoprotein particle. This observation was quite contrary to what we and others had expected.

In a very early study, it was suggested that the preferred substrate for ACAT was cholesterol derived not from the endoplasmic reticulum (the site of ACAT) but from cholesterol contained within the plasma membrane (19). Because no one knew at the time, how to transport plasma membrane cholesterol to the endoplasmic reticulum and ACAT, this observation was largely ignored. Now, there is ample evidence to suggest that cholesterol contained within the plasma membrane is shuttled from the plasma membrane to the endoplasmic reticulum (20-22). Moreover, as the earlier investigators had suspected, plasma membrane cholesterol is thought to be the major substrate for ACAT (21). From these results, we postulated that cholesterol trafficking from the plasma membrane to the endoplasmic reticulum was also occurring in intestinal cells and that during fatty acid flux, intestinal cells would recruit cholesterol derived from the plasma membrane to assemble a lipoprotein particle.

There is debate whether intestinal ACAT, by regulating the amount and availability of cholesterol and cholesteryl esters in the endoplasmic reticulum, plays an important role in lipoprotein assembly (23,24). Cholesterol and cholesteryl esters are required components of a triacylglcerol-rich lipoprotein particle. If either of these sterols were limiting at the site of lipoprotein assembly, lipoprotein synthesis and secretion may be disrupted. We postulated, therefore, that cholesterol trafficking from the plasma membrane to the endoplasmic reticulum is required for normal lipoprotein synthesis and secretion. In CaCo-2 cells, approximately 0.5-1.0% of plasma membrane cholesterol is esterified per hour (25). If one assumes that the esterification of cholesterol derived from the plasma membrane represents trafficking of cholesterol from the plasma membrane to the endoplasmic reticulum, the results would suggest that approximately 1% of plasma membrane cholesterol influxes to the endoplasmic reticulum every hour. This pathway in CaCo-2 cells is temperature-dependent as essentially no plasma membrane cholesterol arrives at the endoplasmic reticulum at temperatures of 4° or 15° C (25). New protein synthesis is also not required as cycloheximide has no effect on cholesterol transport to the endoplasmic reticulum. When we searched to find agents that had been used to disrupt cholesterol trafficking in other cell types, it became apparent that most of the agents were either amphiphiles or sterols. Many of these same agents that were used to interfere with cholesterol trafficking were also used to reverse the multidrug-resistant phenotype in cells having amplified expression of p-glycoprotein, a membrane-bound transport protein from a larger family of ATP'ase binding cassette proteins (26). This suggested, perhaps, that p-glycoprotein or a related transport protein may have a role in cholesterol trafficking. It has been shown that p-glycoprotein is abundant in small intestinal absorptive cells located in the upper villous (27,28). Its function, however, is unknown.

Progesterone, verapamil and trifluoperazine, agents that interact with p-glycoprotein, inhibit the arrival of plasma membrane cholesterol to the endoplasmic reticulum in CaCo-2 cells (Figure 2) (25).

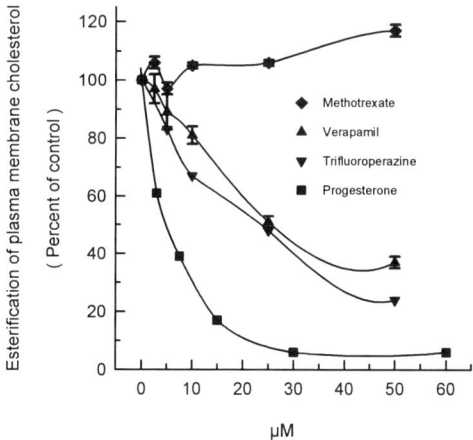

Figure 2: Effect of verapamil, trifluoperazine, progesterone and methotrexate on cholesterol trafficking from the plasma membrane to the endoplasmic reticulum. To label the plasma membrane cholesterol pool, cells were incubated for 90 min. at 4°C with trace amounts of [3][H]-cholesterol. The cells were warmed to 37°C in the presence or absence of increasing concentrations of the agents. The esterification of cholesterol derived from the plasma membrane was determined to estimate cholesterol trafficking from the plasma membrane to the endoplasmic reticulum. Reprinted by permission from J. Lipid Res. 36, 1995. See page 1535 for more details.

In contrast, methotrexate an antimetabolite that does not interact with p-glycoprotein, has no effect on this process. Moreover, these agents also inhibit the secretion of newly-synthesized apolipoprotein (apo) B and apoB mass (Figure 3). This strongly suggests that agents that interfere with cholesterol trafficking from the plasma membrane also interfere with the normal secretion of lipoprotein particles. Not only is the number of secreted lipoprotein particles decreased by these agents, but also affected is the amount of lipid that is carried in these particles. Progesterone, verapamil, and trifluoperazine decrease the secretion of newly-synthesized triacylglycerols and cholesteryl esters, as well as triacylglycerol mass suggesting that these agents interfere with the secretion of the entire triacylglycerol-rich lipoprotein particle. To further support the notion that p-glycoprotein is involved in cholesterol transport in CaCo-2 cells, other recognized inhibitors of p-glycoprotein function, such as amiodarone, colchicine, cyclosporine A, forskolin, and tamoxifen also interfere with both cholesterol trafficking from the plasma membrane to the endoplasmic reticulum and apoB secretion (25).

Although the data are very suggestive that in intestinal cells, p-glycoprotein is involved in cholesterol transport, the evidence remains circumstantial. Recently, in an animal model in which p-glycoprotein is eliminated, the mdr1a/mdr1b knock-out mouse, the transport function of the small intestine appears intact as the mice are phenotypically similar to their wild-type mates (29). Although these results do not completely exclude a role for another member of the p-glycoprotein family in intracellular cholesterol trafficking (nature has a way for making redundancies), more data will obviously be required to implicate this class of transport proteins in this process.

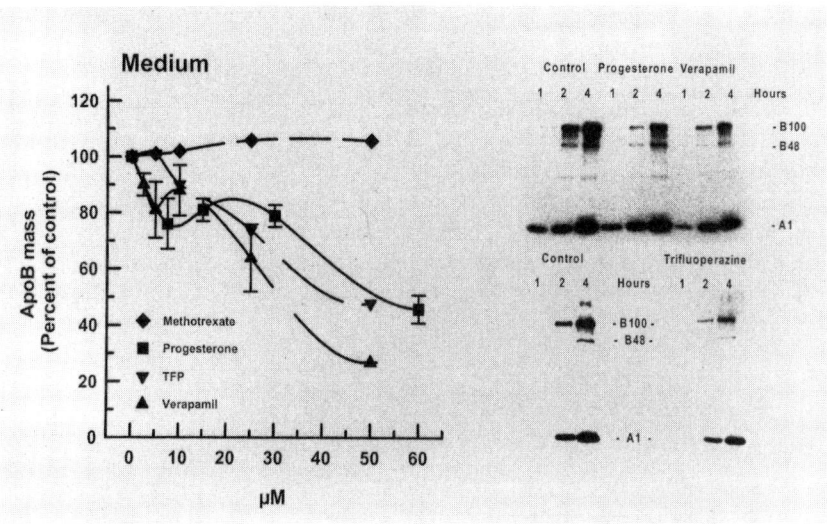

Figure 3: Effect of verapamil, trifluoperazine, and progesterone on: apoB mass secretion (left panel). Cells were incubated for 18 hr. with 1 mM oleic acid attached to albumin and in the presence or absence of increasing concentrations of the agents. The amount of apoB mass secreted into the basolateral medium was determined. Newly-synthesized apoB secretion (right panel). Cells were preincubated for 1 hr. with 50 μM of the agent to be tested. 1 mM oleic acid was added along with ^{35}S-methionine. At 1, 2, and 4 hr., the amount of labeled apoB100, apoB48, and apoAI secreted was determined following immunoprecipitation and polyacrylamide gel electrophoresis. Reprinted by permission from J. Lipid Res. 36, 1995. See page 1537 and 1538 for more details.

Progesterone, verapamil, and trifluoperazine decrease the arrival of plasma membrane cholesterol to the endoplasmic reticulum thereby causing a decrease in the formation of cholesteryl esters. Do these agents decrease lipoprotein secretion by limiting the amount of cholesteryl esters available for normal lipoprotein synthesis? The answer is no, at least, not acutely. The addition of an ACAT inhibitor to cells will prevent the esterification of plasma membrane cholesterol, or for that matter, cholesterol derived from any intracellular pool. During the acute influx of fatty acids, ACAT inhibition does not alter the number of lipoprotein particles being secreted by CaCo-2 cells (25). The intestinal cell, therefore, can assemble and secrete triacylglycerol-rich lipoproteins without esterification of plasma membrane cholesterol. The lipoprotein particles secreted by CaCo-2 cells incubated with an ACAT inhibitor, however, are deficient in cholesteryl esters and, as you might expect, so are lipoproteins secreted by cells incubated with progesterone, verapamil, or trifluoperazine (25).

The answer as to why triacylglycerol-rich lipoprotein secretion is decreased by agents that interfere with cholesterol transport is not clear. In CaCo-2 cells, transport vesicles are likely responsible for the trafficking of cholesterol from the plasma membrane to the endoplasmic reticulum (25). We would speculate that these transport vesicles are derived from specialized areas of the plasma membrane and agents which disrupt vesicular trafficking in general will result in the inhibition of metabolic processes which depend upon normal vesicular function, i.e., lipoprotein secretion.

ORIGINS OF TRIACYLGLYCEROL-RICH LIPOPROTEIN CHOLESTEROL

The mechanisms involved in the assembly and secretion of triacylglycerol-rich lipoprotein particles by the intestine are poorly understood. In response to an influx of fatty acid and the need to transport triacylglycerols, the intestinal cell secretes more apoB particles suggesting that more lipoproteins are being secreted to accommodate the increased flux of lipid (10). The intestinal cell must rapidly recruit or synthesize the necessary lipid components of a lipoprotein particle to ensure normal assembly and secretion. Although unesterified and esterified cholesterol are important components of the lipoprotein particle making up the surface and core, respectively, it is unclear where inside the cell this cholesterol is originating.

When CaCo-2 cells are driven to produce lipoprotein particles by adding oleic acid to the apical medium, more cholesterol derived from the plasma membrane influxes to the endoplasmic reticulum for esterification and more is being secreted into the basolateral medium (Figure 4) (30). Moreover, both unesterified and esterified cholesterol derived from the plasma membrane are secreted in triacylglycerol-rich lipoproteins. It is very clear, therefore, that in times of increased requirements for cholesterol in assembling the lipoprotein particle, the intestinal cell uses plasma membrane cholesterol to meet these immediate needs. There is also a good correlation between the ability of certain fatty acids to drive lipoprotein synthesis and their ability to increase cholesterol transport from the plasma membrane to the endoplasmic reticulum. For example, palmitic, arachidonic, and eicosapentaenoic acids, three fatty acids that do not increase the secretion of apoB by CaCo-2 cells, do not affect cholesterol trafficking. Whereas, oleic and linoleic acids, fatty acids that stimulate apoB secretion, cause more plasma membrane cholesterol to influx to the endoplasmic reticulum (30). Thus, it is apparent that the plasma membrane serves as a reservoir for cholesterol to be used for the assembly and secretion of triacylglycerol-rich lipoproteins by CaCo-2 cells.

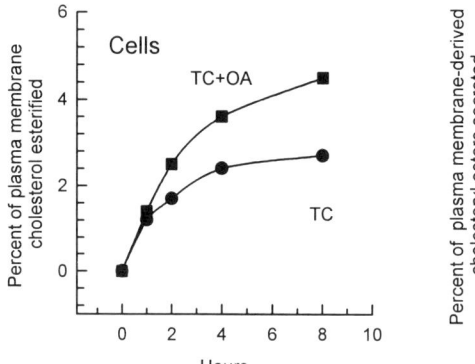 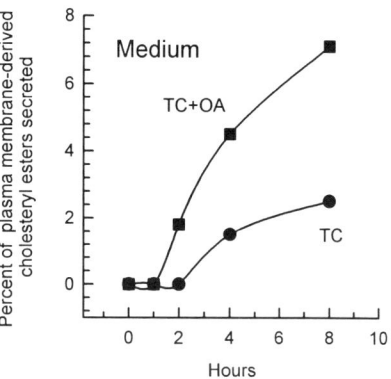

Figure 4: Effect of oleic acid on cholesterol trafficking from the plasma membrane to the endoplasmic reticulum. To label the plasma membrane cholesterol pool, cells were incubated for 90 min. at 4°C with trace amounts of ^{3}[H]-cholesterol. They were then warmed to 37° C in the presence or absence of 5 mM taurocholate (TC) or taurocholate and 0.5 mM oleic acid (TC+ OA). The esterification of cholesterol derived from the plasma membrane and its secretion into the basolateral medium were determined. Reprinted by permission from J. Lipid Res. 36, 1995. See page 2654 for more details.

It had been suggested that cholesterol taken up from micelles is the preferred substrate for intestinal ACAT (5). As mentioned earlier in this chapter, however, only 1% of cell-associated micellar cholesterol is esterified in CaCo-2 cells during a five hour period, a time by which lipids are being mobilized and secreted by the intestinal cell. In contrast, cholesterol derived from the plasma membrane is rapidly being esterified and secreted. So where is this micellar cholesterol and why is it not being shuttled to the endoplasmic reticulum? The answer to this question became clear when CaCo-2 cells (prelabeled with cholesterol) were incubated with micelles containing oleic acid and increasing concentrations of unlabeled cholesterol (Figure 5). Cholesterol that is being taken up from the micelle causes more labeled plasma membrane cholesterol to traffick to the endoplasmic reticulum for esterification and secretion. These results cannot be interpreted to mean that micellar cholesterol mixes and equilibrates with structural cholesterol of the plasma membrane. If this were the case, unlabeled micellar cholesterol would have diluted the labeled plasma membrane cholesterol pool resulting in less labeled cholesterol being esterified and secreted. This did not occur. The results suggest, instead, that cholesterol entering the apical membrane from micelles displaces cholesterol of the plasma membrane causing it to influx to the endoplasmic reticulum. The exact mechanism for this proposed displacement is unclear. Perhaps excess cholesterol entering the apical membrane causes a "flip-flop" of structural cholesterol within the bilayer so that it now can be efficiently transported; or, the absorbed excess micellar cholesterol

132

causes the lateral movement of structural cholesterol to specialized areas within the plasma membrane that are destined for influx. Either or both of these scenarios are possible. More information will be required, however, to delineate a specific mechanism. In somewhat similar experiments, Xu and Tabas (31), in macrophages, found that there was a particular threshold of cholesterol that the plasma membrane would tolerate. Once this threshold was reached, excess plasma membrane cholesterol influxed to the endoplasmic reticulum for esterification. In contrast to macrophages, in which the additional excess cholesterol is derived from the uptake and metabolism of lipoprotein cholesterol, in intestinal cells, excess cholesterol arrives directly into the apical membrane from dietary cholesterol solubilized in micelles.

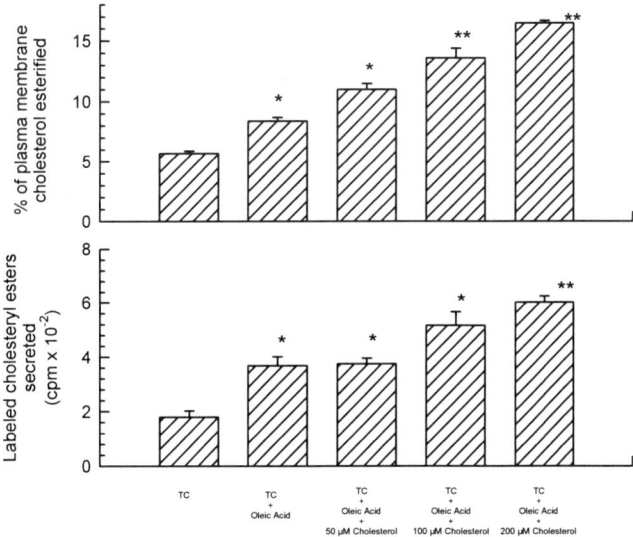

Figure 5: Effect of micellar cholesterol on the esterification and secretion of cholesterol derived from the plasma membrane. Cells were incubated for 90 min. at 4°C with trace amounts of ^{3}[H]-cholesterol. They were then warmed to 37°C in the presence of 5 mM taurocholate (TC), taurocholate and 0.5 mM oleic acid, or taurocholate, oleic acid, and increasing concentrations of unlabeled cholesterol. The esterification of cholesterol derived from the plasma membrane (top) and its secretion into the basolateral medium (bottom) were determined. Reprinted by permission from J. Lipid Res. 36, 1995. See page 2656 for more details.

It has been argued that during active lipoprotein assembly and secretion, the demand for more unesterified cholesterol to package the lipoprotein particle is partially met by an increase in cholesterol synthesis (32). A number of studies in liver, in fact, have shown that lipoprotein secretion can be disrupted by inhibiting

HMG-CoA reductase activity, suggesting that newly-synthesized cholesterol is important for the normal assembly and secretion of hepatic lipoproteins (33,34). Other data, however, would dispute that (35-38). In CaCo-2 cells, oleic acid, which is used to drive lipoprotein synthesis, does stimulate cholesterol synthesis (30). A small fraction of this newly-synthesized cholesterol is found in secreted triacylglycerol-rich lipoproteins. This cholesterol, which obviously originates in the endoplasmic reticulum, does not first travel to the plasma membrane prior to its secretion as part of a lipoprotein particle. Thus, there is a pathway in the intestinal cell by which the cell can utilize newly-synthesized cholesterol for lipoprotein assembly by directly shunting this cholesterol and its ester into the lumen of the endoplasmic reticulum. Most of the cholesterol that is synthesized during lipoprotein assembly, however, is transported to the plasma membrane. During fatty acid flux, one might have expected that the intestinal cell would redirect the normal trafficking of newly-synthesized cholesterol away from the plasma membrane and into the lumen of the endoplasmic reticulum, particularly if newly-synthesized cholesterol was critical for lipoprotein synthesis and secretion. This does not occur. In the next section, this observation will be expanded and discussed. Suffice it to say, during triacylglycerol-rich lipoprotein synthesis and secretion, and with this, an increased demand for unesterified cholesterol, the cell continues to deliver a similar fraction of newly-synthesized cholesterol to the plasma membrane. Moreover, in CaCo-2 cells, inhibiting the activity of HMG-CoA reductase does not interfere with the secretion of apoB mass suggesting that new cholesterol synthesis is not required for the assembly and secretion of lipoprotein particles (personal observations).

It can be concluded, therefore, that in CaCo-2 cells, plasma membrane cholesterol is the major source for triacylglycerol-rich lipoprotein cholesterol (30). This actually makes good sense. The intestinal cell must recruit cholesterol rapidly when it encounters a large flux of fatty acids. Since the plasma membrane contains 75% of the cell's cholesterol, the plasma membrane is an accessible and immediately available source from which to draw the required amount of cholesterol. Moreover, the cell maintains the cholesterol content of the plasma membrane within a very narrow range. By regulating the transport of cholesterol from this pool to the endoplasmic reticulum via transport vesicles, the cell can readily secure the necessary amount required to assemble the lipoprotein particle.

CHOLESTEROL TRANSPORT FROM THE ENDOPLASMIC RETICULUM TO THE PLASMA MEMBRANE

The trafficking of newly-synthesized cholesterol from its site of synthesis, the endoplasmic reticulum, to the plasma membrane is also important for maintaining cellular cholesterol homeostasis. As described above, the intestinal cell utilizes cholesterol derived from the plasma membrane to assemble a triacylglycerol-rich lipoprotein particle. During times of increased fatty acid flux, the intestinal cell draws from its cholesterol reserves contained within the plasma membrane to ensure normal lipoprotein synthesis and secretion. It could be argued then that during lipoprotein assembly and secretion, the transport of newly-synthesized cholesterol to the plasma membrane is essential for replacing the cholesterol that is being used for lipoprotein synthesis. This would ensure proper structure and function of the plasma membrane and also provide a continuous supply of accessible cholesterol. Moreover, if trafficking of cholesterol to the plasma membrane were a regulated process, the cell could control the amount of cholesterol reaching the plasma membrane by altering the efficiency of the transport pathway.

Information obtained from other cell types has suggested that cholesterol is transported from the endoplasmic reticulum to the plasma membrane in unique transport vesicles by a pathway that differs from the secretory pathway (39,40). The intestinal cell is different, however. This pathway may not be as important as in other cells because the intestinal cell can utilize cholesterol taken up from micelles to replenish the plasma membrane cholesterol pool. Thus, there are some important questions that need to be addressed specifically in intestinal cells. Is cholesterol trafficking from the endoplasmic reticulum to the plasma membrane a regulated process? Can the intestinal cell alter the efficiency of this pathway to meet its cholesterol requirements? Is this trafficking of cholesterol different from the plasma membrane cholesterol influx pathway? Is "p-glycoprotein" involved in the transport of cholesterol to the plasma membrane?

The transport of newly-synthesized cholesterol to the plasma membrane occurs rapidly in CaCo-2 cells with 50% of newly-synthesized cholesterol found on the plasma membrane by 30 minutes (Figure 6) (41). By 5 hours, 72% of the newly-synthesized cholesterol can be found on the cell's surface. The trafficking of cholesterol from the endoplasmic reticulum to the plasma membrane is not dependent upon normal Golgi function (Table 1). Disrupting Golgi with Brefeldin A or monensin decreases total cholesterol synthesis but does not alter the percent of newly-synthesized cholesterol reaching the plasma membrane. Similarly, colchicine and cycloheximide do not affect cholesterol transport to the plasma membrane suggesting that microtubular function or new protein synthesis is not required for this transport pathway. The transport of cholesterol from the endoplasmic reticulum to the plasma membrane in CaCo-2 cells, therefore, is not dissimilar to the transport of cholesterol to the plasma membrane that is described in other cells (39,40).

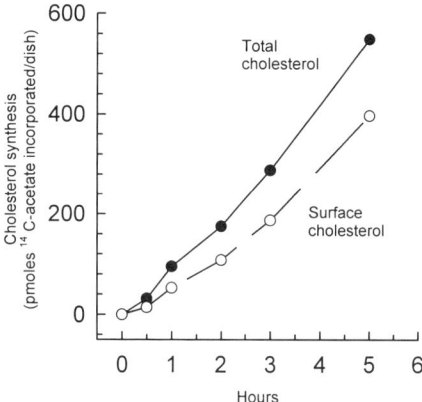

Figure 6: Trafficking of newly-synthesized cholesterol to the plasma membrane. Cells were incubated for up to 5 hr. with 60 μM [14][C]-acetate. At each time point, the cells were fixed and exposed to cholesterol oxidase. The radioactivity incorporated into cholesterol and cholestenone was determined to estimate total cholesterol synthesis. The amount of newly-synthesized cholesterol on the cell surface was calculated by dividing the radioactivity in cholestenone by the amount in total cholesterol. Reprinted by permission from J. Lipid Res. 39, 1998. See page 336 for more details.

Table 1. Effect of inhibition of Golgi function, microtubules and protein synthesis on the incorporation of ^{14}C-acetate into total and surface cholesterol

	Cholesterol							
	Total			Surface			Percent on surface		
	(pmoles ^{14}C-acetate incorporated/6 h)							
Control	707	±	17	561	±	17	79	±	1
Brefeldin A	338	±	22 [a]	255	±	18 [a]	76	±	1
Monensin	403	±	15 [a]	291	±	10 [a]	72	±	1
Colchicine	606	±	11 [a]	449	±	14 [a]	74	±	1
Cycloheximide	448	±	14 [a]	327	±	11 [a]	73	±	1

CaCo-2 cells grown on micropore membranes were incubated for 6 hours with buffer alone, 0.25 μg/ml brefeldin A, 10 μM monensin, 500 μM colchicine or 5 μM cycloheximide. To label newly-synthesized cholesterol, 60 μM ^{14}C-acetate (130 dpm/pmole) was added with the treatment at the beginning of the incubation. The amount of radioactivity incorporated into total cholesterol and cholesterol present on the cell surface was estimated using cholesterol oxidase. The values are the mean ± SEM of 4 filters. [a] $p < 0.05$ vs. control.

Following a fatty meal, the intestinal absorptive cell responds rapidly to changing requirements of cholesterol to ensure normal lipoprotein synthesis and secretion. It would seem reasonable to assume that an intestinal cell could divert newly-synthesized cholesterol that was normally destined for transport to the plasma membrane to a pool within the endoplasmic reticulum that could be used to assemble a lipoprotein particle. Although this sounds quite reasonable and practical, the intestinal cell does not seem to do this. Following acute changes occurring in cholesterol flux across the cell, the transport of cholesterol from the endoplasmic reticulum to the plasma membrane is unaffected (Table 2).

Table 2. Effect of decreasing the influx or increasing the efflux of plasma membrane cholesterol on the incorporation of ^{14}C-acetate into total and surface cholesterol

	Cholesterol							
	Total			Surface			Percent on surface		
	(pmoles ^{14}C-acetate incorporated /6 h)								
Control	438	±	13	344	±	13	78	±	1
Digitonin	601	±	23 [a]	460	±	17 [a]	77	±	1
Filipin	783	±	27 [a]	631	±	23 [a]	80	±	1
Taurocholate	381	±	11	308	±	10	81	±	1
Taurocholate + LPC	588	±	17 [b]	419	±	16 [b]	71	±	1

CaCo-2 cells grown on micropore membranes were incubated for 6 hours with buffer alone, 5 µg/ml digitonin or 5 µg/ml filipin. Another set of cells were incubated for 6 hours with 5 mM taurocholate or 5 mM taurocholate + 250 µM lysophosphatidylcholine (LPC). To label newly-synthesized cholesterol, 60 µM ^{14}C-acetate (130 dpm/pmole) was added with the treatment at the beginning of the incubation. Radioactivity incorporated into total and surface cholesterol was estimated using cholesterol oxidase. The values are the mean ± SEM of 6 filters. [a] $p < 0.05$ vs. control. [b] $p < 0.05$ vs. taurocholate.

For example, under experimental conditions in which cholesterol influx to the endoplasmic reticulum is markedly decreased by incubating cells with filipin, digitonin, or lysophosphatidylcholine, the cell perceives a cholesterol deficiency. Rates of cholesterol synthesis increase. The percent of newly-synthesized cholesterol reaching the plasma membrane, however, remains at approximately 75% suggesting that the cell is unable to alter the percent of cholesterol trafficking to the surface. Similarly, when cholesterol flux is increased by incubating cells with either cholesterol or 25-hydroxycholesterol, the cell perceives a cholesterol excess and cholesterol synthetic rates decrease (Table 3). Again, however, the percent of newly-synthesized cholesterol reaching the plasma membrane remains constant.

These results suggest that the intestinal cell does not regulate the transport process by which newly-synthesized cholesterol is shuttled to the plasma membrane. The transport process is constitutive. The cell regulates how much cholesterol is transported to the plasma membrane by altering its rate of cholesterol synthesis and not by altering the percent of newly-synthesized cholesterol transported there.

Table 3. Effect of micellar cholesterol on the incorporation of ^{14}C-acetate into total and surface cholesterol

	Cholesterol							
	Total			Surface			Percent on surface	
	(pmoles ^{14}C-acetate incorporated /6 h)							
5 mM Taurocholate + 50 µM oleate	762	±	68	527	±	43	69	±	1
+ 200 µM cholesterol	488	±	12 [a]	319	±	13 [a]	65	±	1
+ 2.5 µM 25-OH cholesterol	265	±	12 [a]	171	±	9 [a]	64	±	1

CaCo-2 cells grown on micropore membranes were incubated for 6 hours with 5 mM taurocholate + 50 µM oleate, 5 mM taurocholate + 50 µM oleate + 200 µM cholesterol or 5 mM taurocholate + 50 µM oleate + 2.5 µM 25-hydroxycholesterol. To label newly-synthesized cholesterol, 60 µM ^{14}C-acetate (130 dpm/pmole) was added with the treatment at the beginning of the incubation. Radioactivity incorporated into total and surface cholesterol was estimated using cholesterol oxidase. The values are the mean ± SEM of 6 filters. [a]p < 0.05 vs. control.

Although there is evidence to suggest that p-glycoprotein may be involved in the transport of cholesterol from the plasma membrane to the endoplasmic reticulum (42,43), p-glycoprotein is not involved in the transport of newly-synthesized sterols to the plasma membrane in CaCo-2 cells (41). Agents which inhibit p-glycoprotein and disrupt cholesterol trafficking from the plasma membrane to the endoplasmic reticulum cause an accumulation of sterol intermediates within CaCo-2 cells resulting in a decrease in the amount of total cholesterol being synthesized. These sterol intermediates, however, have no problem finding their way to the plasma membrane suggesting that the transport pathway of these sterols from the endoplasmic reticulum to the plasma membrane remains intact (41). By preventing these intermediate sterols from returning to the endoplasmic reticulum to complete the process of cholesterol synthesis, less cholesterol is being synthesized. Thus, p-glycoprotein inhibitors do not interfere with sterol transport from the endoplasmic reticulum to the plasma membrane, but instead, interfere with the return of cholesterol intermediates from the plasma membrane back to the endoplasmic reticulum for completion of cholesterol synthesis (41,42).

Caveolae/caveolin in intestinal cells?

The results in CaCo-2 cells described above suggest that when excess cholesterol transiently accumulates in the apical membrane following the uptake of micellar cholesterol, structural cholesterol of the plasma membrane is displaced to a particular microdomain within the membrane. It is this cholesterol that is destined for transport to the endoplasmic reticulum. We postulated, therefore, that in intestinal cells there were specialized areas within the plasma membrane that functioned to "cluster" excess cholesterol molecules directing them, likely by vesicles derived from these specialized areas of plasma membrane, to the endoplasmic reticulum. The plasma membrane of the intestinal cell is enriched in glycosylsphingolipids, sphingomyelin, and cholesterol (44). Domains of the plasma membrane that are enriched in these particular lipids are resistant to solubilization by nonionic detergents (45). In other cell types, these microdomains resistant to detergents are said to represent caveolae, flask-shaped invaginations of the plasma membrane which are proposed to be involved in potocytosis, signal transduction, and cholesterol efflux (46-52). Caveolin, a 21-kDa cholesterol-binding protein whose function is unknown, is found in these detergent-insoluble domains of the plasma membrane (53,54). It has been postulated that caveolin may have a role in forming dynamic domains within the membrane that are used for membrane trafficking or in organizing specific microdomains for the functions mentioned above. Moreover, following oxidation of surface cholesterol, caveolin has been shown to cycle to the endoplasmic reticulum and trans-Golgi (55,56). Because of its cholesterol-binding properties, its proposed role in organizing specialized microdomains within the plasma membrane, and its recognized cycling between the plasma membrane and endoplasmic reticulum, we postulated a possible role for caveolae/caveolin in cholesterol trafficking in CaCo-2 cells. There has been debate, however, as to whether intestinal cells even contain caveolin (57,58).

Anticipating difficulty in detecting a low abundance message in CaCo-2 cells, caveolin mRNA was first amplified by RT-PCR (Figure 7a). It is clear that CaCo-2 cells contain the expected PCR product. Moreover, RNA extracted from human intestinal biopsies and brushings also contain the RT-PCR product suggesting that mRNA for caveolin exists in human enterocytes as well. Northern blot hybridization of RNA extracted from CaCo-2 cells detects two transcripts of 3.0 and 0.8 kb size for caveolin mRNA (Figure 7b). It is apparent, therefore, that CaCo-2 cells and human small intestine express the gene for caveolin. In Figure 8, the results are shown of a Western analysis for caveolin protein in human intestinal brushings and in CaCo-2 cells. In human intestinal cells, caveolin exists primarily as a 600 kDa protein. In CaCo-2 cells, both a 600 and 21 kDa protein is recognized by the antibody. It is well-recognized that caveolin can exist as homooligomers within cells (59-61). Following the boiling of the samples, however, all the caveolin is detected as the 21 kDa protein. Moreover, using sucrose-density gradients to separate intracellular organelles, caveolin in CaCo-2 cells is found in detergent-resistant microdomains of the plasma membrane (Figure 9).

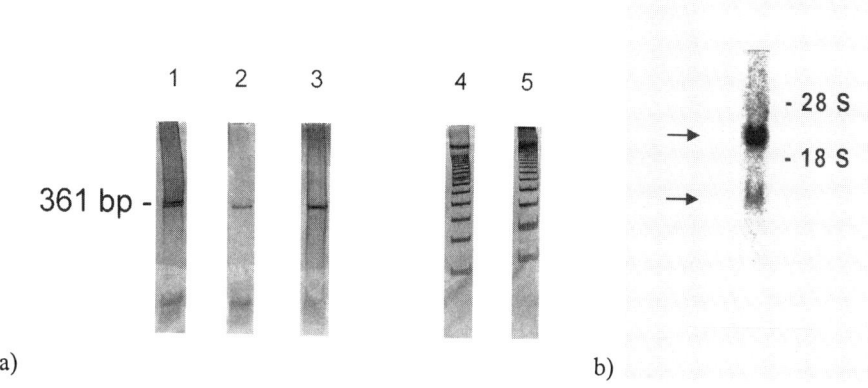

361 bp -

Figure 7:   a) Caveolin mRNA by RT-PCR.   Lane 1:   CaCo-2 cells; lane 2: human intestinal cells collected by biopsy; lane 3: human intestinal cells collected by brushing; lanes 4 and 5:   100 and 123 bp DNA ladder, respectively. b) Northern blot for caveolin mRNA from CaCo-2 cells.

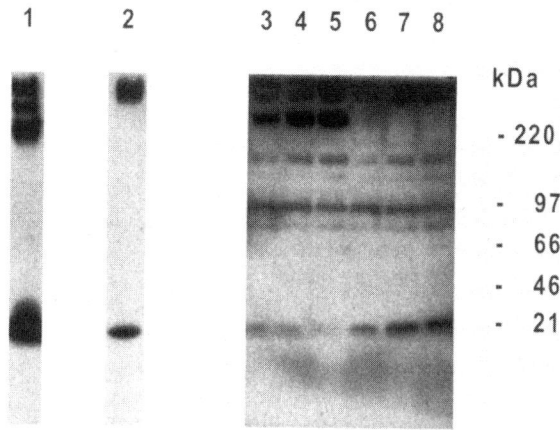

Figure 8:   Western analysis for caveolin protein in CaCo-2 cells and human intestine. Lane 1: human endothelial cells, room temperature; lane 2: human intestinal cells collected by brushings, heated at 100°C, 5 minutes; lanes 3-5: CaCo-2 cell homogenates containing 5, 10, or 30 μg protein, room temperature; lanes 6-8: CaCo-2 cell homogenates containing 5, 10, 30 μg protein, heated at 100° C, 5 minutes.

140

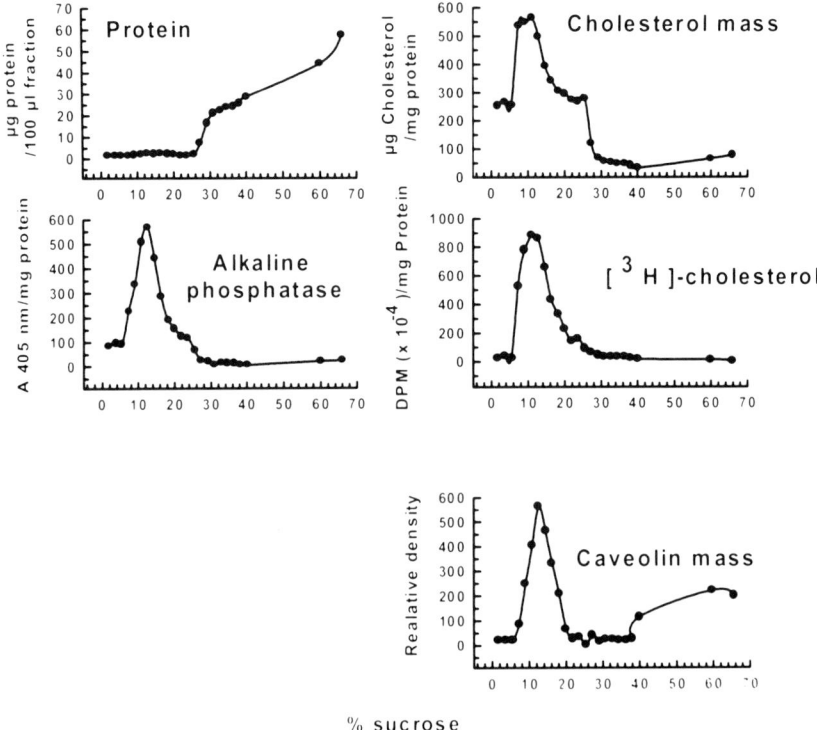

Figure 9: Detergent-resistant microdomains in CaCo-2 cells. CaCo-2 cells were prelabeled with cholesterol. The cells were freeze-thawed in buffer to prevent protein degradation and homogenized. The homogenates were further homogenized in 1% Triton X-100 and made to 40% sucrose. This mixture was overlaid with sucrose solution of 36-4% and centrifuged 18 hours at 118,610 x g. The fractions were collected and analyzed for protein, cholesterol mass, labeled cholesterol, and alkaline phosphatase activity. Insert shows caveolin mass in fractions.

As alluded to above, we had postulated that cholesterol transported from the plasma membrane to the endoplasmic reticulum originated from "specialized" domains within the plasma membrane. In studies with other cell types, the membrane-impermeable enzyme, cholesterol oxidase, has been used to selectively oxidize cholesterol contained within these detergent-resistant domains of the plasma membrane (51,55). To investigate whether an influx of cholesterol at the apical membrane would cause "clustering" of plasma membrane cholesterol to these regions of the membrane accessible to cholesterol oxidase, cells were equilibrated with labeled cholesterol and then incubated with micelles containing increasing concentrations of cholesterol (Figure 10).

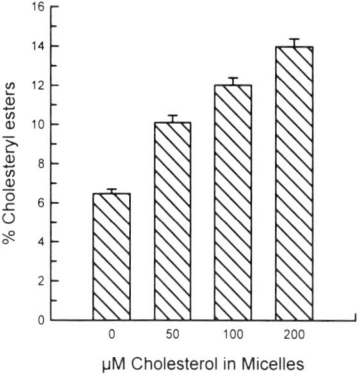
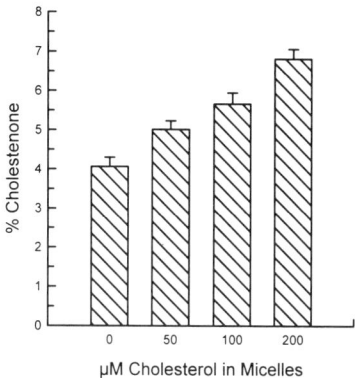

Figure 10: Effect of micellar cholesterol on the transport of plasma membrane cholesterol to the endoplasmic reticulum and cholesterol oxidase accessibility. CaCo-2 cells were prelabeled with cholesterol. They were then incubated at 37°C for 4 hours in medium containing 5 mM sodium taurocholate, 50 μM oleate and 0, 50, 100, 200 μM cholesterol. At the end of the incubation, cells were washed and incubated an additional hour at 37°C with 10 IU/ml cholesterol oxidase. The percent of cholesteryl esters and cholestenone in cells was determined.

Following the uptake of micellar cholesterol, the amount of cholesteryl esters synthesized from cholesterol derived from the plasma membrane increases with increasing concentrations of micellar cholesterol. Moreover, more plasma membrane cholesterol moves to the detergent-resistant domains now accessible to cholesterol oxidase. These results suggest that incoming micellar cholesterol does cause cholesterol of the plasma membrane to "cluster" in these microdomains; and, it is perhaps from these domains that cholesterol is transported to the endoplasmic reticulum.

When we investigated whether changes in cholesterol flux across the CaCo-2 cell would regulate the amount of caveolin mass or mRNA levels, the results were very clear. Caveolin mass nor gene expression were altered by changes in cholesterol flux. The half-life for caveolin has been estimated to be 10 ½ hours (56). With this long a half-life, it would seem unlikely that an intestinal cell would acutely regulate caveolin expression by altering its rate of synthesis. Two of the major participants in lipoprotein assembly and secretion in intestinal cells, apolipoprotein B and microsomal triglyceride transfer protein, are also not acutely regulated by changes in their synthetic rates or gene expression (62,63). The intestinal cell, therefore, must synthesize an abundance of these proteins to accommodate large influxes of fatty acids. If caveolin were to be involved in cholesterol transport in intestinal

142

cells, thereby linking it to lipoprotein assembly and secretion, it would not necessarily be expected that its mass or mRNA levels would change in response to fatty acid or cholesterol flux. The evidence that caveolae/caveolin play a role in cholesterol trafficking in intestinal cells remains circumstantial. Further investigation will be required to establish a definitive relationship.

CONCLUSIONS

1. Plasma membrane cholesterol serves as a reservoir for cholesterol to be used for the assembly and secretion of triacylglycerol-rich lipoproteins.

2. P-glycoprotein, or a related protein from the ATP'ase binding cassette family may have a role in cholesterol trafficking from the plasma membrane to the endoplasmic reticulum and triacylglycerol-rich lipoprotein secretion.

3. Micellar cholesterol causes "clustering" of plasma membrane cholesterol to specialized microdomains. It is this structural cholesterol within this microdomain that is transported to the endoplasmic reticulum for esterification and transport.

4. The transport of newly-synthesized cholesterol to the plasma membrane is constitutive and independent of Golgi, microtubular function, protein synthesis, and p-glycoprotein. The cell regulates how much cholesterol arrives at the plasma membrane by altering cholesterol synthesis, not by altering the transport process.

5. Human intestinal and CaCo-2 cells contain caveolin, a cholesterol-binding protein that could play a role in cholesterol trafficking.

REFERENCES

1. Field FJ, Kam N TP, Mathur, SN. Regulation of cholesterol metabolism in the intestine. Gastroenterol 1990;99:539-551.
2. Fong LG, Fujishima SE, Komaromy MC, Pak YK, Ellsworth JL, Cooper AD. Location and regulation of low-density lipoprotein receptors in intestinal epithelium. Am J Physiol 1995; 269:G60-72.
3. Field FJ, Shreves T, Fujiwara D, Murthy S, Albright E, Mathur S N. Regulation of gene expression and synthesis and degradation of 3-hydroxy-3-methylglutaryl coenzyme A reductase by micellar cholesterol in CaCo-2 cells. J Lipid Res 1991;32:1811-1821.
4. Field FJ, Fujiwara D, Born E, Chappell DA, Mathur SN. Regulation of LDL receptor expression by luminal sterol flux in CaCo-2 cells. Arterioscler Thromb 1993;3:729-737.
5. Suckling KE, Stange EF. Role of acyl-CoA:cholesterol acyltransferase in cellular cholesterol metabolism. J Lipid Res 1985;26:647-671.
6. Clark SB, Tercyak AM. Reduced cholesterol transmucosal transport in rats with inhibited mucosal acyl CoA: cholesterol acyltransferase and normal pancreatic function. J Lipid Res 1984;25:148-159.
7. Krause R, Anderson M, Bisgaier C L, Bocan T, Bousley R, DeHart P, Essenburg A, Hamelehle K, Homan R, Kieft K, McNalley W, Stanfield R, Newton RS. In vivo evidence that the lipid-regulating activity of the ACAT inhibitor CI-976 in rats is due to inhibition of both intestinal and liver ACAT. J Lipid Res 1993;34:279-294.
8. Klein RL, Rudel LL. Cholesterol absorption and transport in thoracic duct lymph lipoproteins of nonhuman primates. Effect of dietary cholesterol level. J Lipid Res 1983;24:343-357.
9. Pinto M, Robine-Leon S, Appay M-D, Kedinger M, Triadou N, Dussaulx E, Lacroix B, Simon-Assmann P, Haffen K, Fogh J, Zweibaum A. Enterocyte-like differentiation and polarization of the human colon carcinoma cell line Caco-2 in culture. Biol Cell 1983;47:323-330.
10. Field FJ, Mathur SN. Intestinal lipoprotein synthesis and secretion. Prog Lipid Res 1995;34:185-198.
11. Field FJ, Albright E, Mathur SN. Regulation of cholesterol esterification by micellar cholesterol in CaCo-2 cells. J Lipid Res 1987;28:1057-1066.
12. Trotter PJ, Storch J. Fatty acid uptake and metabolism in a human intestinal cell line (Caco-2): comparison of apical and basolateral incubation. J Lipid Res 1991;32:293-304.
13. Van Greevenbroek MMJ, Voorhout, WF, Erkelens, DW, vanMeer G, deBruin TWA. Palmitic acid and linoleic acid metabolism in Caco-2 cells: different triglyceride synthesis and lipoprotein secretion. J Lipid Res 1995;36:13-24.
14. Ranheim T, Gedde-Dahl A, Rustan AC, Drevon CA. Influence of eicosapentaenoic acid (20:5, n-3) on secretion of lipoproteins in CaCo-2 cells. J Lipid Res 1992;33:1281-1293.
15. Giannoni F, Field FJ, and Davidson NO. An improved reverse transcription-polymerase chain reaction method to study apolipoprotein gene expression in Caco-2 cells. J Lipid Res 1994;35:340-350.
16. Reisher SR, Hughes TE, Ordovas JM, Schaefer EJ, Feinstein SI. Increased expression of apolipoprotein genes accompanies differentiation in the intestinal cell line Caco-2. Proc Natl Acad Sci USA 1993;90:5757-5761.
17. Faust RA, Albers JJ. Regulated vectorial secretion of cholesteryl ester transfer protein (LTP-I) by the CaCo-2 model of human enterocyte epithelium. J Biol Chem 1988;263:8786-8789.
18. Rogler G, Herold G, Fahr C, Fahr M, Rogler D, Reimann FM, Stange EF. High-density lipoprotein 3 retroendocytosis: a new lipoprotein pathway in the enterocyte (CaCo-2). Gastroenterol 1992;103:469-480.
19. Mitropoulos KA, Venkatesan S, Synouri-Vrettakou S, Reeves BEA, Gallagher JJ. The role of plasma membranes in the transfer of non-esterified cholesterol to the Acyl-CoA: cholesterol acyltransferase substrate pool in liver microsomal fraction. Biochim Biophys Acta 1984;792:227-237.
20. Tabas I, Rosoff WJ, Boykow GC. Acyl coenzyme A:cholesterol acyl transferase in macrophages utilizes a cellular pool of cholesterol oxidase-accessible cholesterol as substrate. J Biol Chem 1988;263:1266-1272.
21. Lange Y, Strebel F, Steck TL. Role of the plasma membrane in cholesterol esterification in rat hepatoma cells. J Biol Chem 1993;268:13838-13843.

144

22. Lange Y. Cholesterol movement from plasma membrane to rough endoplasmic reticulium. J Biol Chem 1994;269:1-4.
23. Cianflone KM, Yasruel Z, Rodriguez MA, Vas D, Sniderman AD. Regulation of apoB secretion from HepG2 cells: evidence for a critical role for cholesteryl ester synthesis in the response to a fatty acid challenge. J Lipid Res 1990;31:2045-2055.
24. Dixon JL, Ginsberg HN. Regulation of hepatic secretion of apolipoprotein B-containing lipoproteins: information obtained from cultured liver cells. J Lipid Res 1993;34:167-179.
25. Field FJ, Born E, Chen H, Murthy S, Mathur SN. Esterification of plasma membrane cholesterol and triacylglycerol-rich lipoprotein secretion in CaCo-2 cells: possible role of p-glycoprotein. J Lipid Res 1995;36:1533-1543.
26. Gottesman MM, Pastan I. Biochemistry of multidrug resistance mediated by the multidrug transporter. Annu Rev Biochem 1993;62:385-427.
27. Fojo AT, Ueda K, Slamon DJ, Poplack DG, Gottesman MM, Pastan I. Expression of a multidrug-resistance gene in human tumors and tissues. Proc Natl Acad Sci USA 1987;84:265-269.
28. Thiebaut F, Tsuruo T, Hamada H, Gottesman MM, Pastan I, Willingham MC. Cellular localization of the multidrug-resistance gene product P-glycoprotein in normal human tissues. Proc Natl Acad Sci USA 1987;84:7735-7738.
29. Schinkel AH, Mayer U, Wagenaar E, Mol CAAM, vanDeemter L, Smit JJM, Van Der Valk MA, Voordouw AC, Spits H, van Tellingen O, Zulmans JMJM, Fibbe WE, Borst P. Normal viability and altered pharmacokinetics in mice lacking mdr1-type (drug-transporting) P-glycoproteins. Proc Nat Acad Sci USA 1997;94:4028-4033.
30. Field FJ, Born E, Mathur SN. Triacylglycerol-rich lipoprotein cholesterol is derived from the plasma membrane in Ca-Co2 cells. J Lipid Res 1995;36:2651-2660.
31. Xu XX, Tabas I. Lipoproteins activate acyl-coenzyme A:cholesterol acyltransferase in macrophages only after cellular cholesterol pools are expanded to a critical threshold level. J Biol Chem 1991;266:17040-17048.
32. Kam NTP, Albright E, Mathur SN, Field FJ. Effect of lovastatin on acyl-CoA:cholesterol O-acltransferase (ACAT) activity and the basolateral-membrane secretion of newly synthesized lipids by Caco-2 cells. Biochem J 1990;272:427-433.
33. Khan B, Wilcox HG, Heimberg M. Cholesterol is required for secretion of very-low-density lipoprotein by rat liver. Biochem J 1989;259:807-816.
34. Burnett JR, Wilcoc LJ, Telford DE, Kleinstiver SJ, Barrett PHR, Newton RS, Huff MW. Inhibition of HMG-CoA reductase by atorvastatin decreases both VLDL and LDL apolipoprotein B production in miniature pigs. Arterioscler Thromb Vasc Biol 1997;17:2589-2600.
35. David RA, Malone-McNeal M. Dietary cholesterol does not affect the synthesis of apolipoproteins B and E by rat hepatocytes. Biochem J 1985;227:29-35.
36. Sato R, Imanaka T, Takano T. The effect of HMG-CA reductase inhibitor (CS-514) on the synthesis and secretion of apolipoproteins B and A-I in the human hepatoblastoma HepG2. Biochim Biophys Acta 1990;1042:36-41.
37. Ribeiro A, Mangeney M, Loriette C, Thomas G, Pepin D, Janvier B, Chamaz J, Bereziat G. Effect of simvastatin on the synthesis and secretion of lipoproteins in relation to the metabolism of cholesterol. Biochim Biophys Acta 1991;1086:279-286.
38. Salam WH, Wilcox HG, Cagen LM, Heimberg M. Stimulation of hepatic cholesterol biosynthesis by fatty acids. Biochem J 1989;258:563-568.
39. DeGrella RF, Simoni RD. Intracellular transport of cholesterol to the plasma membrane. J Biol Chem 1982;257:14256-14262.
40. Urbani L, Simoni RD. Cholesterol and vesicular stomatitis virus G protein take separate routes from the endoplasmic reticulum to the plasma membrane. J Biol Chem 1990;265:1919-1923.
41. Field FJ, Born E, Murthy S, Mathur SN. Transport of cholesterol from the endoplasmic reticulum to the plasma membrane is constitutive in CaCo-2 cells and differs from the transport of plasma membrane cholesterol to the endoplasmic reticulum. J Lipid Res 1998;39:333-343.
42. Metherall JE, Li H, Waugh K. Role of multidrug resistance P-glycoproteins in cholesterol biosynthesis. J Biol Chem 1996; 271:2634-2640.
43. Debry P, Nash EA, Neklason DW, Metherall JE. Role of multidrug resistance P-glycoproteins in cholesterol esterification. J Biol Chem 1997;272:1026-1031.
44. Christiansen K, Carlsen J. Microvillus membrane vesicles from pig small intestine. Purity and lipid composition. Biochim Biophys Acta 1981;647: 188-195.

45. Brown DA, Rose JK.   Sorting of GPI-anchored proteins to glycolipid-enriched membrane subdomains during transport to the apical cell surface. Cell 1992;68: 533-544.

46. Anderson RG, Kamen BA, Rothberg KG, Lacey SW.   Potocytosis: sequestration and transport of small molecules by caveolae. Science 1992; 255: 410-411.

47. Anderson RG.   Caveolae: where incoming and outgoing messengers meet. Proc Nat Acad Sci USA. 1993.;90: 10909-10913.

48. Anderson RG.   Plasmalemmal caveolae and GPI-anchored membrane proteins. Curr Opin Cell Biol 1993; 5: 647-652.

49. Chang WJ, Ying YS, Rothberg KG, Hooper NM, Turner AJ, Gambliel HA, De Gunzburg J, Mumby SM, Gilman AG, Anderson RG.   Purification and characterization of smooth muscle cell caveolae. J Cell Biol 1994;126: 127-138.

50. Lisanti MP, Scherer PE, Tang Z, Sargiacomo M.   Caveolae, caveolin, and caveolin-rich membrane domains: a signalling hypothesis. Trends Cell Biol 1994;4: 231-235.

51   Fielding PE, Fielding CJ.   Plasma membrane caveolae mediate the efflux of cellular free cholesterol. Biochemistry 1995;34:14288-14292.

52. Liu J, Oh P, HornerT, Rogers RA, Schnitzer JE.   Organized endothelial cell surface signal transduction in caveolae distinct from glycosylphosphatidylinositol-anchored protein microdomains. J Biol Chem 1997;272: 7211-7222.

53. Lisanti MP, Tang ZL, Sargiacomo M.   Caveolin forms a hetero-oligomeric protein complex that interacts with an apical GPI-linked protein: implications for the biogenesis of caveolae. J Cell Biol 1993;123: 595-604.

54. Murata M, Peranen J, Schreiner R, Wieland F, Kurzchalia TV, Simons K. VIP21/caveolin is a cholesterol-binding protein. Proc Natl Acad Sci USA. 1995; 92: 10339-10343.

55. Smart EJ, Ying YS, Conrad PA, Anderson RG. Caveolin moves from caveolae to the Golgi apparatus in response to cholesterol oxidation. J Cell Biol 1994.;127: 1185-1197.

56. Conrad PA, Smart EJ, Ying YS, Anderson RG, Bloom GS.   Caveolin cycles between plasma membrane caveolae and the Golgi complex by microtubule-dependent and microtubule-independent steps. J Cell Biol 1995;131: 1421-1433.

57. Mayor S, Rothberg KG, Maxfield FR. Sequestration of GPI-anchored proteins in caveolae triggered by cross-linking. Science 1994.;264: 1948-1951.

58. Mirre C, Monlauzeur L, Garcia M, Delgrossi MH, Le Bivic A.   Detergent-resistant membrane microdomains from Caco-2 cells do not contain caveolin. Am J Physiol 1996;271:C887-894.

59. Monier S, Parton RG, Vogel F, Behlke J, Henske A, Kurzchalia TV. VIP21-caveolin, a membrane protein constituent of the caveolar coat, oligomerizes in vivo and in vitro. Mol Biol Cell 1995;6: 911-927.

60 . Monier S, Dietzen DJ, Hastings WR, Lublin DM, Kurzchalia TV.   Oligomerization of VIP21-caveolin in vitro is stabilized by long chain fatty acylation or cholesterol. FEBS Lett 1996; 388: 143-149.

61. Sargiacomo M, Scherer PE, Tang Z, Kubler E, Song KS, Sanders MC, Lisanti MP. Oligomeric structure of caveolin: implications for caveolae membrane organization. Proc Nat Acad Sci USA 1995;92: 9407-9411.

62. Murthy S, Albright E, Mathur SN, Davidson NO, and Field FJ 1992. Apolipoprotein B mRNA abundance is decreased by eicosapentaenoic acid in CaCo-2 cells. Effect on the synthesis and secretion of apolipoprotein B. Arterioscler Thromb 12: 691-700.

63. Mathur SN, Born E, Murthy S, Field FJ. 1997. Microsomal triglyceride transfer protein in CaCo-2 cells: characterization and regulation. J Lipid Res 38: 61-67.

EFFLUX AND PLASMA TRANSPORT OF BIOSYNTHETIC STEROLS

Jane Ellen Phillips and William J. Johnson

Department of Biochemistry, Allegheny University of the Health Sciences, 2900
Queen Lane, Philadelphia, PA 19129; e-mail: johnsonwj@auhs.edu

KEY WORDS: Caveolae, Desmosterol, Lipid-Rich Vesicle Fraction, Multidrug
Resistance Proteins, Sonic Hedgehog Protein, 7-Dehydrocholesterol, Smith-Lemli-
Opitz Syndrome, Sterol Carrier Protein, Triparanol, Unilamellar Vesicles.

ABSTRACT

The pathway of cholesterol biosynthesis has many steps, producing several
intermediates. Some of the intermediates, such as lanosterol, zymosterol
and desmosterol, are delivered to the plasma membrane, and are available
for efflux and entry into the pathway of reverse sterol transport. Our
studies with CHO and other extrahepatic cells suggest that these
intermediates constitute the majority of biosynthetic sterol released to
HDL; however, we see very little release from liver-derived cell lines,
which make mainly cholesterol. Interestingly, we find that one of the
intermediates, desmosterol, is released from cells three times more
efficiently than cholesterol to extracellular sterol acceptors. This release
may be an important mechanism in extrahepatic cells to prevent the
pathology that is associated with the build-up of intermediates in cells and
in tissues. In this chapter, we review what is known about biosynthetic
sterols and their transport within cells, and summarize our data on their
release to extracellular acceptors.

CONTENTS

INTRODUCTION

The majority of the body's sterol is thought to be provided through endogenous synthesis and not through the diet (1), with this synthesis occurring mainly in extrahepatic tissues. Therefore, the majority of sterol released in the pathway of reverse cholesterol transport is derived from this biosynthetic pool (1,2). During cholesterol biosynthesis mammalian cells in culture accumulate substantial amounts of late-stage biosynthetic intermediates such as lanosterol, desmosterol, zymosterol and 7-dehydrocholesterol (3) and these are found in the plasma membrane (4). In work from our laboratory, we see that during sterol synthesis in a variety of nonhepatic cells there is a significant accumulation of desmosterol or similarly structured sterols. In contrast, in liver-derived hepatoma cells, nearly all synthesized sterol consists of cholesterol, with very little accumulation of intermediates (5), indicating that hepatocytes may possess mechanisms that ensure the completion of cholesterol synthesis.

The excessive accumulation of sterol intermediates in cells can have serious consequences such as is seen with the Smith-Lemli-Opitz syndrome, where there is a deficiency of the sterol-Δ-7 reductase, resulting in the accumulation of 7-dehydrocholesterol and other sterol intermediates. This leads to several types of developmental defects, including mental retardation, failure to thrive and congenital heart anomalies (6). Similarly, administration of the drug Triparanol, an inhibitor of the sterol-Δ-24 reductase leads to the accumulation of desmosterol in the plasma and tissues. This accumulation is associated with developmental anomalies in utero, the development of irreversible cataracts and liver damage in adult humans, and aortic lesions with calcium deposition in egg-laying hens and in gerbils (7,8). A possible explanation for the developmental effects of sterol intermediates was provided by the recent demonstration that cholesterol associates covalently with at least some of the morphogenic proteins in the "hedgehog" family (9).

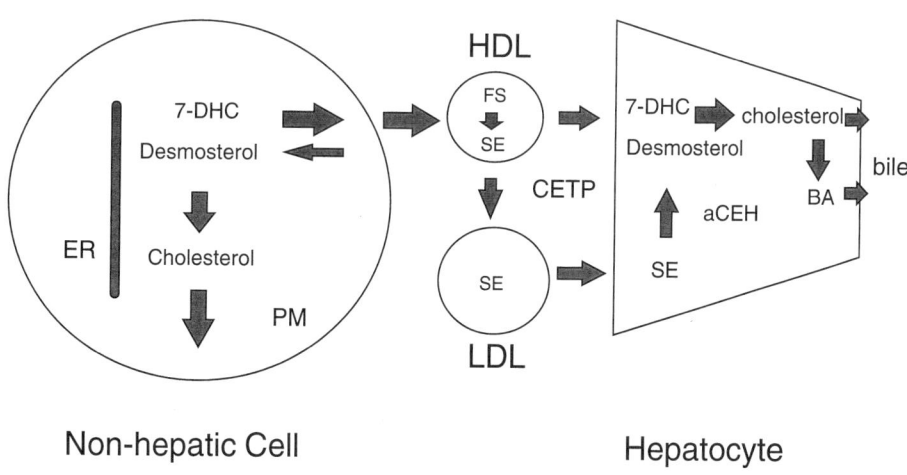

Figure 1. Proposed transport of biosynthetic cholesterol and sterol intermediates via the reverse-cholesterol-transport mechanism. *In many nonhepatic cells, it appears that both biosynthetic cholesterol and sterol intermediates reach the plasma membrane and are subject to HDL-mediated efflux and subsequent transport to the liver. Here hepatocytes may efficiently take up the sterols and convert them to cholesterol and/or bile acids. Abbreviations in this diagram are as follows: ER, endoplasmic reticulum; PM, plasma membrane; HDL, high density lipoprotein; LDL, low density lipoprotein; FS, free (unesterified) sterol; SE, steryl ester; 7-DHC, 7-dehydrocholesterol; aCEH, lysosomal acid cholesteryl ester hydrolase; BA, bile acid.*

We hypothesize that the rapid efflux of biosynthetic desmosterol and other intermediates from extrahepatic cells serves to limit the damage that these sterols otherwise would cause. This release would initiate transport to the liver, where uptake and efficient metabolism to cholesterol or bile acids could occur (Figure 1). Under normal conditions, these intermediates may constitute a large fraction of the sterol entering the pathway of reverse sterol transport. Investigation of this process may lead to a better understanding of why the intermediates appear to cause pathology and the role that they could play in heart disease.

STEROL BIOSYNTHESIS AND THE PRODUCTION OF INTERMEDIATES

In mammals, cholesterol is synthesized in a complex, multi-step process, and in addition, sterols cannot be oxidized completely to CO_2 and water. Associated with these constraints, tightly controlled mechanisms have evolved in animals to control the level of cholesterol in cells and tissues. Sterols can be acquired in two ways, first by taking up exogenously supplied and liver-derived cholesterol through a cell surface receptor known to bind circulating low density lipoproteins (LDL). This receptor cycles to the endosome whether ligand is bound or not, making LDL cholesterol uptake into the cell proportional to the concentration of receptors on its surface (10). Additionally cells acquire cholesterol through biosynthesis. Both LDL receptor levels, and the enzymes important in sterol synthesis are tightly regulated at the level of the gene, post-transcriptionally and post-translationally (11-13). However, if a cell has acquired too much sterol, it can rid itself of this excess by releasing cholesterol from its plasma membrane to circulating high density lipoprotein (HDL), which will transport it to either the liver or steroidogenic tissues. Hepatocytes can release cholesterol into bile or lipoproteins or convert it to bile acids, which are also secreted in the bile. Steroidogenic tissues can use cholesterol as a substrate for steroid synthesis. The transport to the liver is known as the reverse cholesterol transport (RCT) pathway and is important in preventing the over-accumulation of sterol in cells and in tissues (14).

Almost every cell in the body has the capacity to synthesize cholesterol. The synthetic pathway is complex with many later-stage branches (15). All of the carbons in the 27-carbon molecule are derived from acetate. The final structure contains a planar 4-ring nucleus with 2 methyl groups at carbons 18 and 19 (β-oriented) and a β-oriented hydroxyl group at carbon 3. This hydroxyl confers a degree of polarity to the sterol. The planar ring has an isooctyl side chain that is saturated, and this contributes to the molecule's hydrophobicity. The side chain, the hydroxyl group and the nuclear ring all contribute to cholesterol's ability to act as a structural component of membranes and its behavior within the phospholipid bilayer (15,16).

The first molecule produced in the pathway that can be defined as a sterol is lanosterol (4,4,14α-trimethyl-cholesta-8,24-dien-3β-ol), which although similar to cholesterol, contains three additional methyl groups (α-oriented), two at carbon 4 and one at carbon 14. Lanosterol also has a double bond in the side chain (Δ-24) while this bond is reduced in cholesterol. The double bond in the ring structure is found at carbon 8 (Δ-8) in lanosterol and not carbon 5 (Δ-5) as seen in cholesterol. However, cholesterol appears to be superior to lanosterol as a structural element in animal membranes. Evolution dictated further refinements: removal of the α-methyls, reduction of the side chain and transfer of the nuclear double bond. Beyond lanosterol, all of the enzymes in the biosynthetic pathway are independent of one another, and many of the steps can occur in different orders. Because the last three chemical modifications are not sequential, the number of cholesterol intermediates produced can be great (17). Due to small differences in polarity and stereochemistry, these sterols interact differently with surrounding molecules. Therefore they can behave quite differently in cells (17).

WHOLE-ANIMAL SITES OF STEROL SYNTHESIS

Sterol synthesis can occur in almost every cell in the body. The contribution of one tissue and the amounts of synthesis are different during the various stages of life (18) and under different dietary conditions (19). Adult animals can maintain consistent levels of sterol in tissues and plasma with a balance between what is absorbed, synthesized and excreted. The liver is the central organ involved in this control, with the ability to secrete and take up lipoproteins along with releasing cholesterol into bile or converting it to bile acids (1). It can also repress its own sterol synthesis when dietary cholesterol levels are high (2, 28). Exogenous dietary cholesterol is absorbed in the gut and transported to the liver, packaged initially in chylomicrons, which give rise in the circulation to chylomicron remnants. It can be repackaged and secreted into the circulation as very low density lipoproteins (VLDL). Hepatic tissue expresses LDL receptors which bind circulating VLDL and LDL. Levels of LDL receptors on the surface of the hepatocyte are controlled by cellular sterol concentrations at the level of the gene. If LDL levels are low, the liver decreases expression of receptors, which allow for the conversion of VLDL remnants to LDL in the plasma. When circulating LDL is high, transcription and translation of LDL receptor genes and RNA are upregulated, so receptor levels increase, and the liver takes up the LDL cholesterol. The sterol can then be converted to bile acids or secreted into bile. It is, however, not lost altogether since bile reabsorption in the terminal ileum is efficient, and much of the bile acids return to the liver via portal circulation. The sterols are once again packaged into chylomicrons and reabsorbed. It is possible to estimate absolute cholesterol turnover by measuring the rate of synthesis and absorption, or by measuring the levels of cholesterol and bile acid excreted in the feces (1).

Previously, it was assumed that the liver was also the primary site of sterol synthesis in the body (20) with as much as 97% produced in this organ (19). These data were collected using radiolabeled acetate precursor. It was later realized and reported by Dietschy and McGarry (21) that acetate labeling was not accurate because two assumptions were not always met, one being that the uptake of the precursor was not rate limiting between tissue types, and the other that the rate of conversion of the labeled acetate to acetyl Co-A would be high. Rates of acetate uptake were found to vary according to tissues and incorporation of the label into sterol was not consistent. Different tissues would dilute the acetyl Co-A pool to different degrees and would compartmentalize the acetate in different organelles. The compartments dictated what molecules would be produced (21). This altered the ability to calculate rates of synthesis due to varying initial specific activities. To circumvent these problems, researchers attempted to use the fatty acid octanoate. Conversion of these carbons to cholesterol was more efficient than the acetate (75% conversion versus 47%), but it too was compartmentalized in cells (21).

More recently D_2O and [^{3}H]water have been used to monitor synthetic activities. These measurements can be made more accurately because water can diffuse easily across cell membranes, and it equilibrates readily in cells labeling all pools to the same specific activity. Since cells will recycle these hydrogens in different oxidative reactions, the intracellular specific activity can be assumed to be identical to that of extracellular water, which can easily be measured in any body fluid sample (22). The precursor problems are eliminated because the labeled hydrogen is incorporated rapidly and uniformly into a co-factor important in sterol synthesis, NADPH, which donates hydrogen molecules during the reduction reactions. Hydrogen can also be obtained from the cellular water and from the acetyl Co-A pool formed in the presence of the labeled water. Because it is known approximately how many of these reactions are needed to synthesize a sterol molecule, it is possible to calculate a carbon-to-hydrogen ratio and determine synthetic rates in tissues (22). There are a few complications that must be considered. First, NADPH can be derived from the pentose phosphate pathway, and therefore be unlabeled, diluting the NADPH specific activity. It is also hard to determine if the labeled hydrogen is coming from the water or the acetyl Co-A (23). For accurate measurements, it is important to calculate the ^{3}H-to-carbon ratio experimentally. If this is impossible, it has been demonstrated that you can assume approximately 20-24 ^{3}H atoms per cholesterol molecule in most tissues that are actively synthesizing sterol (23). Using labeled H_2O, Andersen et al. (24) have shown that previous data on hepatic sterol synthesis greatly overestimates the contribution of the liver. Extrahepatic tissues were not as efficient at taking up the acetate, and there was also a large dilution of the specific activity of the precursor in some tissues. Both contributed to underestimating the levels of extrahepatic synthesis (24).

When animals are fed a low-cholesterol, low-fat diet, hepatic synthesis is active. However, when cholesterol is added to the diet, the rates of hepatic synthesis slow down while extrahepatic rates do not change (1). In some species such as rabbit and guinea pig, the levels of hepatic synthesis are inherently low, so suppression by a high-cholesterol diet does not allow for a great decrease. These animals exhibit high levels of circulating plasma cholesterol (25, 26). In many species including humans, if the diet is high in cholesterol, hepatic synthesis is suppressed. Supplementation of the diet with labeled cholesterol leads to a maximum steady-state specific activity (27, 28). Over time, any dilution represents extrahepatic sterol synthesis. In man, the specific activity of the circulating sterol never reaches more that 40% of what is fed, estimating that at least 60% of this sterol is provided by extrahepatic synthesis (27). Dietschy et al. estimate that under the conditions of dietary-suppressed hepatic sterol synthesis (a typical Western diet), the contribution of extrahepatic tissues to the biosynthetic sterol pool is 80% or greater in all of the species studied (1). In cynomolgus monkey fed a diet low in cholesterol, the liver contributed only 11.2% to total sterol synthesis (29). In squirrel monkeys, the liver contributes 40% of newly synthesized sterol while the skin, blood and colon contribute an additional 40% (26).

When considering the contribution of different tissues to the biosynthetic sterol pools it is important to keep in mind the size of the organ, its function and the species of animal under study. Using isotopic water, researchers have elucidated what tissues are the most active in synthesis in different animals, during different conditions. It appears in many species the intestine is a strong contributor (27, 1). If biliary circulation is diverted, there is an increase in the levels of sterol synthesis within the intestine (30). It is also interesting to note that the contribution of the intestine can easily be underestimated since its synthesized sterol is rapidly packaged into chylomicrons. Spady and Dietschy estimate that a large amount of this sterol is in circulation within one hour of introducing the labeled water (26).

Large tissues with low levels of sterol synthesis could still make major contributions, such as skeletal muscle and skin (30, 26). Additionally, it is important to consider the tissue's function and also its location. For example, skin produces a large amount of sterol, but these cells are continuously shed from the body and it is unknown whether much ends up in plasma circulation (26). When estimating sterol production levels in the whole animal, there are many variables between species, tissues and the tissue's functions.

STEROL INTERMEDIATES IN PLASMA CIRCULATION
Little is known about the plasma transport of the sterol intermediates. They are found in the serum of healthy individuals in very small quantities and the levels of some can be directly correlated with the activity of hepatic HMG-Co-A reductase (31). Patients who undergo ileal bypass have increased levels of hepatic sterol synthesis due to decreases in bile acid reuptake. Circulating methyl sterols (containing methyl groups at carbons 14 and 4 such as lanosterol, $\Delta8,2$-dimethylsterol, $\Delta8$-dimethylsterol and $\Delta8$-methostenol) are higher in these patients and it appears that their turnover is faster than the turnover of lipoprotein particles. They may be taken up by the liver without the required uptake of the lipoprotein (32). Isomeric dehydrocholesterols are also found in plasma (33). How they are transported and if they are metabolized to cholesterol, and also their contribution to different disease states are undefined. Ratios of circulating desmosterol to cholesterol are higher in patients with non-insulin dependent diabetes, due to increases in cholesterol synthesis (34). What is interesting is that some intermediates, such as desmosterol and squalene, are not reflective of hepatic biosynthesis. Serum levels of these molecules were also not affected by treatment with cholestryramine, a compound that inhibits the reuptake of sterols and bile acids in the lower intestine. Decreases in the reabsorption increases the levels of hepatic sterol synthesis. This lack of response could be because the majority of circulating desmosterol and squalene are not produced in the liver but instead in extrahepatic tissues.

INTRACELLULAR TRANSPORT OF BIOSYNTHETIC STEROLS

Intracellular sterol transport is partially defined in some areas and undelineated in others. Currently many theories exist implicating different proteins and cellular membrane structures that are important in moving sterols through the cell. The process however, is not that of passive diffusion since there is an extremely unequal distribution within the cell (35). It is believed that well over 90% of a cell's sterol is in the plasma membrane, with small amounts found in endoplasmic reticular membranes and the Golgi regions. The fate of exogenously-derived cholesterol (LDL cholesterol) and its transport are known to be different from the pathway followed by biosynthetic sterol. The mechanisms used to transport intracellular sterol to the plasma membrane do not appear to be the same as those from the plasma membrane to intracellular organelles (36, 37). Also, it appears that transport of intracellular sterol in cells differs depending on whether they are loaded with sterol or if they are have low levels and are actively synthesizing it (38).

Lipid-Rich Vesicle Fraction

Cholesterol and sterol intermediates are made at the endoplasmic reticulum and then transported to the plasma membrane. This movement is fast; cholesterol does not seem to accumulate in the interior of the cell, but instead is relocated to the plasma membrane soon after it is synthesized (39). Simoni and colleagues report the half-time of this transfer to be about 10-20 minutes, whereas Lange and her colleagues have reported this time to be one hour, later revising it to about 18 minutes (3, 35, 39, 40).

The transport can be stopped with energy poisons and this inhibition can be reversed by their removal (39). The movement of nascent sterols is also stopped by decreasing the cell's temperature to 15°C or below. The sterols then are trapped in a lipid-rich vesicle fraction, described by Kaplan and Simoni (39), which appears to have a very low density. While cells are kept at 15°C the newly synthesized sterol will build-up in the lipid-rich vesicle fraction and also in the endoplasmic reticulum, which precedes the vesicles, demonstrating that the synthesis of cholesterol is independent of its transport to the plasma membrane. It appears that the nascent sterols are unable to get to the plasma membrane until the cells are warmed to 37°C (40, 41). As this happens, the half-time of the appearance of the sterol in the lipid-rich vesicle fraction is 10-20 minutes, while the half-time of its disappearance from the vesicles is 5 minutes (41). These half-times support the lipid-rich vesicle fraction as an intermediate in the transport of newly synthesized sterols to the plasma membrane (41), and may imply that the rate-limiting step for transfer to the plasma membrane is either the formation of the vesicles, or their transport to, and not their association with, the plasma membrane.

The energy dependence and the presence of vesicles rich in sterols argue against passive diffusion of nascent sterols to the plasma membrane (41). Interestingly, the transport of the vesicles may not be mediated by the Golgi, since Golgi-disrupting agents do not stop their transport to the plasma membrane once the cell's temperature is increased (39, 42).

The Role of HDL Binding

Oram and colleagues have reported that the transport of newly synthesized cholesterol to the plasma membrane is dependent on intracellular signals generated when HDL binds to an extracellular receptor (43). This mechanism is most pronounced after cholesterol enrichment. Until this signal is induced, the newly synthesized cholesterol remains in intracellular domains. They report that a protein-to-protein interaction, (apo A-I and the proposed HDL receptor) causes the stimulation of intracellular protein kinase C in cholesterol-loaded cells. The kinase then phosphorylates proteins important in the trafficking of intracellular newly synthesized sterols. The sterols are brought to the cell surface where they are available for desorption to HDL. Oram proposes that while apo A-I will deplete a cell of intracellular cholesterol, the lipids in HDL are important for the removal of plasma membrane sterols. They speculate that this kinase-mediated transport pathway is already active in cells that are synthesizing cholesterol and no HDL stimulation is needed (43). The pathway exists to help rid cells of excess sterols by loading them into the plasma membrane. HDL upregulates the pathway, perhaps by stimulating vesicular transport (44)

Mendez has reported that if cells are loaded with exogenous cholesterol, the transport of newly synthesized sterols becomes Golgi dependent. Golgi disrupting agents inhibit mass cholesterol release to HDL by 50%, and sterol synthesis does not increase during this sterol loss. This argues for the presence of two pathways, since the lipid-rich vesicular transport appears to be independent of Golgi function. The second Golgi-mediated pathway could be upregulated when cells are overwhelmed by cholesterol (38).

This confinement of intracellular sterol pools in the absence of apo A-I does not seem to be absolute, however, as indicated by work showing that there is transport of newly synthesized sterols to the plasma membrane and efflux of these sterols from cells when they are exposed to extracellular unilamellar phospholipid vesicles, which lack the apoprotein necessary for receptor recognition (5).

Sterol Carrier Protein-2

A few proteins have been implicated in the intracellular transport pathway. Sterols are hydrophobic, and adverse to traveling through an aqueous environment. Proteins are thought to assist in this process and to help maintain the unequal sterol distribution within cells. Evidence exists that sterol carrier protein-2 (SCP-2, or nonspecific lipid transfer protein) plays an important role in sterol transport through the cells and in cholesterol synthesis. The mature protein is 13.2 kDa, but this must be processed from a larger precursor. It is believed that this processing occurs in peroxisomes and patients with Zellweger's syndrome, who lack peroxisomes, do not make functional SCP-2. The half times of transport of newly synthesized sterols to the plasma membrane are reduced in Zellweger cells compared to normal cells (45). Recent work has demonstrated that hepatic newly synthesized cholesterol is transported by two pathways, one, supporting Simoni's data, is rapid (10 minutes), Golgi and cytoskeleton independent. It also requires sterol carrier protein-2 (SCP-2). An additional pathway that takes a longer time and is dependent on the Golgi and microtubules is demonstrated in fibroblasts from patients with Zellweger's syndrome. These cells require a half-time of 20 minutes to transport newly synthesized sterol to the plasma membrane. Fibroblasts from normal patients can be forced to rely on the slower Golgi-dependent pathway when they are treated with SCP-2 antisense oligonucleotide. This suggests that cells use the rapid SCP-2 dependent pathway predominantly, but have a second alternative pathway. It has been suggested that *in vivo* the release of newly synthesized hepatic cholesterol destined for secretion in bile follows the rapid SCP-2 dependent pathway, since increases in bile acid secretion are seen when hepatocytes are transfected with, and overexpressing SCP-2 (46).

Baum et al. has shown that overexpression of SCP-2 in McA-RH 7777 cells increases the transport of newly synthesized sterol to the plasma membrane, as expected, but interestingly decreases cholesterol synthesis by 12%, while increasing cholesterol mass. The cholesterol levels are maintained in these cells by increasing the conversion of desmosterol to cholesterol. Baum proposes that the sterol intermediates, which are mainly found in the plasma membrane, represent a reserve pool for cholesterol synthesis. SCP-2 may increase the conversion of desmosterol to cholesterol, not through an upregulation of Δ-24 sterol reductase activity, but instead by increasing the internalization of desmosterol (47).

Caveolae
The cholesterol contained within the phospholipid bilayer of the plasma membrane appears to be present in both cholesterol-rich and cholesterol-poor regions or domains (48). One cholesterol-rich region of the membrane consists of flask-shaped invaginations called caveolae, which contain the structural protein caveolin. Caveolin cycles between the plasma membrane and intracellular locations (49). The exact function of caveolae is unknown, but they are diminished in number on the surface of cells that are starved of cholesterol. Recently, it has been reported that caveolae not only contain the majority of cholesterol that is released to extracellular acceptors, but also are a region where recently synthesized sterols are sequestered. (50). It has been suggested that caveolae could be instrumental in importing plasma membrane sterol to the ER (51). Smart et al. (52) recently demonstrated that the caveolin protein also participates in the reverse pathway, bringing nascent sterol to the plasma membrane from its site of synthesis. Caveolae are believed to be the region where the caveolin directs the newly synthesized sterol when it is delivered to the plasma membrane. From here it is free to move within the membrane in a lateral fashion (52). Consistent with the participation of caveolin in newly synthesized sterol transport, the lipid-rich vesicle fraction has the same buoyant density as intact caveolae. Additional information on the role of caveolae and caveolin in sterol transport appears in the chapters by Fielding and Fielding and by Smart.

TRANSPORT OF STEROL INTERMEDIATES FROM THE PLASMA MEMBRANE TO THE ER

It has been demonstrated that the multi-step process of cholesterol synthesis may not be completed when the molecule leaves the ER. Lange and Echevarria have shown that nascent lanosterol and zymosterol appear in the plasma membrane (4). It was thought that the final steps in synthesis could occur in areas other than the ER, and that perhaps late-stage enzymes such as the reductases that remove double bonds in the side chain and nucleus could be located in the plasma membrane. However, Lange later showed that some zymosterol returns to the ER where it is converted to cholesterol (35). The functional significance of this flux of sterol intermediates to the plasma membrane is unclear, but it could be due to indiscriminate sorting of the lipids into the vesicles at the ER (53).

It is thought that the transport of sterols to the plasma membrane from the ER is distinct from the internalization pathway. Field et al. (54) found that while they treated CaCo-2 cells with sphingomyelinase to activate the transport of cholesterol to the ER, the percent of newly synthesized sterols transported to the plasma membrane did not change. Furthermore, inhibiting the transport from the plasma membrane to the ER using progesterone, verapamil or nigericin did not alter the levels of newly synthesized sterols arriving in the membrane (54). Metherall, et al. (55) reports that in cells treated with progesterone, cholesterol biosynthesis slows down and sterol intermediates accumulate. Increased levels of progesterone can completely shut down the production of cholesterol and promote the production of lanosterol and a lanosterol-like intermediate. This might be explained by progesterone's ability to prevent the transport of cholesterol from the plasma membrane to the ER (56), and therefore prevent the transfer of the sterol intermediates back to the late-stage enzymes. Panini et al., have shown that desmosterol builds up in rat intestinal epithelial cells treated with progesterone while cholesterol levels decrease, but that progesterone has no effect on the activity of HMG Co-A reductase (57). Metherall proposes that progesterone causes a decrease in cholesterol synthesis and the build up of intermediates by interfering with the function of an isoform of the P-glycoprotein, MDR (mdr in mice), a family of multidrug resistance proteins, two of which are known to transport hydrophobic drugs from one side of the plasma membrane to the other and thereby facilitate their efflux. Isoforms are designated MDR1, 2 and 3. Expression of MDR1 and 3 increases in cells when they are challenged with drugs, and in the presence of ATP, these proteins can pump drugs out of the cellular interior; MDR2 however lacks drug efflux ability. This isoform is expressed in the canalicular membranes of hepatocytes and in rat and human mammary tumors (58). Its known function is to pump phospholipids across membrane bilayers from inner to outer leaflets. Also, it plays a role in the secretion of phospholipids, and subsequently cholesterol, into bile (59). Mice homozygous for a lesion in the mdr2 gene develop severe liver disease because they cannot incorporate cholesterol and phospholipids into bile (60). Metherall proposed that progesterone shuts down cholesterol biosynthesis by interfering with one of the MDR isoforms. Without the activity, cholesterol and the intermediates presumably cannot be transferred from the plasma membrane to the ER (61). Although there are intermediates evident in cells not treated with progesterone, the steroid appears to enhance their accumulation, possibly in the plasma membrane. Cells will not grow on progesterone in serum-free medium until cholesterol is added. This suggests that the return of the intermediates to the ER is a necessary step in the formation of biosynthetic cholesterol. The fact that non-steroidal inhibitors of MDR also decrease cholesterol biosynthesis points to a role of an isoform of MDR in sterol transport (61).

The release into the RCT pathway may be a major mechanism for cells to rid themselves of sterol intermediates, since the process of transport back to the ER is not always efficient. Lange reports that this transport is slow, with only 10% of plasma membrane derived zymosterol (cholest-8,24-diene-3β-ol) reaching the ER in one hour (35). Slotte and Bierman have shown that in quiescent human skin fibroblasts, 1.1% of [^{3}H]desmosterol that the cells incorporate into their plasma membranes is converted to [^{3}H]cholesterol per hour. The rate was slower in monkey arterial smooth muscle cells with 0.4% converted per hour (62). In contrast, in CHO cells, this process may be relatively efficient, as Metherall reports the disappearance of about 50% of the accumulated lanosterol within one hour of removing progesterone (55). Nonetheless, we have shown substantial efflux of desmosterol and other intermediates from CHO cells, when extracellular sterol acceptors are present during synthesis (5). In hepatocytes, it may also be a relatively efficient pathway, since we see little accumulation and release of the intermediates from liver and human hepatoma cells (5). Alternatively liver cells may have an efficient mechanism for completing cholesterol synthesis at the ER, so that significant amounts of the intermediates never reach the plasma membrane.

ENTRY OF BIOSYNTHETIC STEROLS INTO THE RCT PATHWAY

The efflux of biosynthetic cholesterol and sterol intermediates has been examined at the cellular level in the authors' laboratory (5). The initial objectives of the studies were to identify the major sterols accumulating in peripheral cells during typical labeling procedures (e.g., with ^{3}H-labeled acetate or mevalonate precursor), to quantify their efflux, and to compare the efflux promoted by different types of acceptors. Two critical aspects of the methodology were to be able to resolve the different biosynthetic sterols from each other and then to quantify them reliably. Towards these ends, reverse-phase HPLC was used for sterol separations, with the HPLC effluent directed into a flow-through liquid scintillation counter, which allowed for continuous monitoring and quantification of radioactivity in the various sterol peaks. As had been reported in earlier studies (63), we found that traditional thin-layer chromatography (TLC) was inadequate to resolve the various sterols that accumulated in most cells, as demonstrated by the fact that the newly synthesized "cholesterol" peak from TLC plates often could be resolved into multiple components by HPLC (5).

In HPLC comparisons of the synthesized sterols that accumulated in various cell types, two major trends were apparent (Figure 2): 1) in extrahepatic cells (CHO, smooth muscle, fibroblast), a complex sterol profile was the rule, consisting of cholesterol and several more polar intermediates. 2) In liver-derived cells (e.g., HepG2), a much simpler profile was obtained, with a major cholesterol peak and little of any other sterol accumulating. These results suggest that in liver cells, mechanisms exist to ensure the complete processing of nascent sterol to cholesterol, whereas in extrahepatic cells, it appears that either there are rate-limiting steps late in the pathway or there is diversion of the intermediates out of the endoplasmic reticulum so that synthesis is not completed. Based on retention times (RT) relative to cholesterol, some of the sterol intermediates accumulating in extrahepatic cells

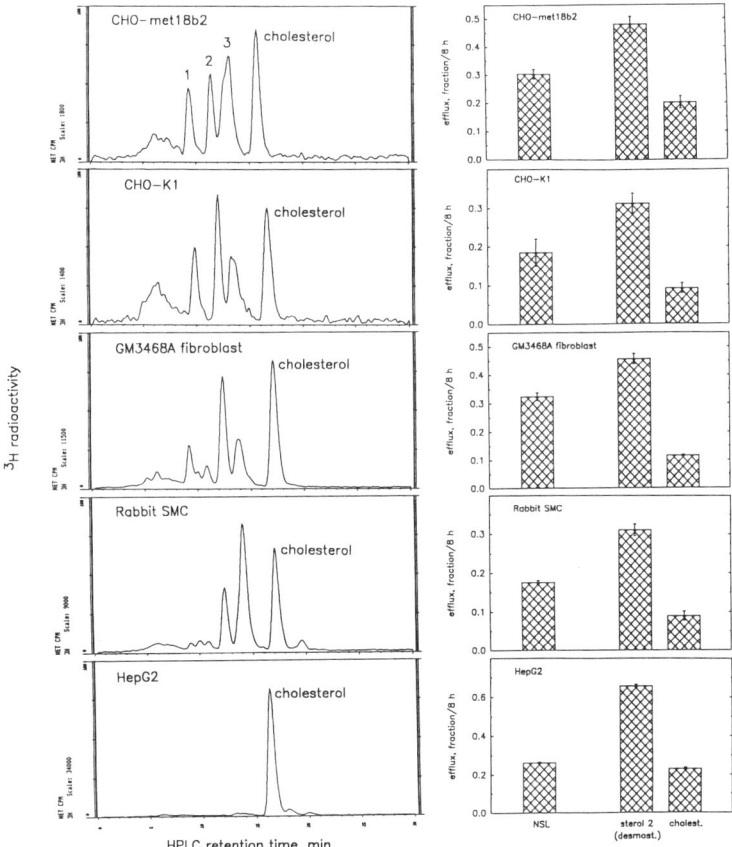

Figure 2. Production and efflux of newly synthesized sterols in a variety of cell types. *Cells were incubated in media containing [³H]acetate and either BSA (0.2%) or BSA plus HDL₃ (1 mg protein/ml), and then nonsaponifiable lipids were prepared from cells and media and analyzed by HPLC. The panels on the left show HPLC profiles of [³H]sterols obtained from cells incubated with BSA alone, which does not promote sterol efflux. Numbered peaks were identified tentatively as C-27 triene sterol (sterol 1), desmosterol (sterol 2), and 7-dehydrocholesterol plus lanosterol (sterol 3). The horizontal scale on the HPLC profiles is 0 to 30 minutes. The panels on the right show the fractional efflux of unfractionated nonsaponifiable lipid (NSL), cholesterol (cholest.), and desmosterol (demost.) when HDL₃ was present during the labeling period. (CHO, Chinese hamster ovary cells, SMC, smooth muscle cells). Reproduced with permission of the Journal of Biological Chemistry.*

were tentatively identified as desmosterol or zymosterol (RT = 0.7 relative to cholesterol), 7-dehydrocholesterol or lanosterol (RT = 0.8), and dihydrolanosterol (RT = 1.1). In CHO cells, the RT 0.7 peak was definitely identified as desmosterol by mass-spectral analysis (5).

Studies of the efflux of the biosynthetic sterols showed that all of them were subject to release that was at least as efficient as that of cholesterol (5). Interestingly, in all cell types studied the synthesized desmosterol underwent efflux that was approximately three times more efficient than that of cholesterol (Figure 2). Since approximately equal amounts of cholesterol and desmosterol accumulated in cells during sterol synthesis, this made desmosterol the major newly synthesized sterol released from cells in most experiments. The disparity in efflux between cholesterol and desmosterol was further explored using CHO cells as a representative extrahepatic cell model.

In a pulse-chase study (Figure 3), it was found that the intermediates that accumulated at low temperature (15°C), were converted only slowly to cholesterol during a chase at 37°C. For instance the disappearance of desmosterol from CHO cells during the chase occurred with a half-time of about 8 hours. Thus, once this intermediate leaves the cholesterol biosynthetic pathway, its return to the pathway is inefficient. In the same experiment, the half-time for efflux of the accumulated desmosterol with HDL_3 in the medium was approximately 4 hours. Thus, the more likely fate of the desmosterol appears to be efflux, rather than further metabolism in the cholesterol biosynthetic pathway.

The more rapid efflux of desmosterol persisted with different acceptors (HDL_3, phosphatidylcholine small unilamellar vesicles, artificial discoidal complexes of HDL apoproteins and phosphatidylcholine). Additionally, the enrichment of cells with cholesterol reduced synthesis, but did not eliminate the substantial accumulation of intermediates relative to cholesterol, and did not alter the efficient efflux of desmosterol in comparison to cholesterol (Table 1).

We conclude from these studies that the accumulation of biosynthetic intermediates during cholesterol biosynthesis is likely to be a common occurrence in extrahepatic cells, and that these intermediates are delivered along with cholesterol to the plasma membrane and there are available for efflux to HDL_3 and other acceptors. Desmosterol, or a diene sterol of similar structure, constitutes a major portion of the accumulating intermediates and is subject to much more rapid efflux than cholesterol. The newly synthesized cholesterol and desmosterol are available to both lipoprotein and protein-free acceptors, and this holds for both cholesterol-loaded and nonloaded cells. Thus, their efflux probably does not require interaction of acceptors with the HDL receptor that has been described by Oram and colleagues (43, 44). The rapid efflux of desmosterol and other intermediates may be part of a protective mechanism for moving these sterols efficiently to the liver, thereby preventing the pathology with which they are associated.

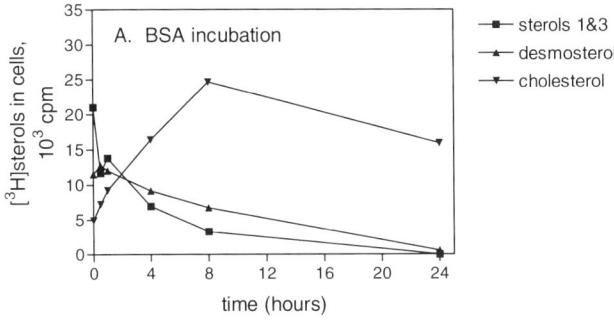

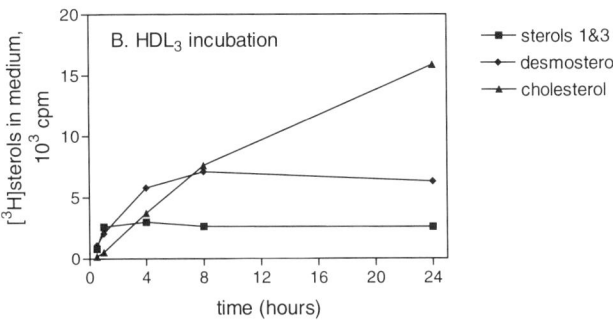

Figure 3. Pulse-chase analysis of biosynthetic sterol metabolism and efflux. CHO-met18b2 cells were incubated for 5 hours at 15°C with [³H]mevalonate (pulse), rinsed at 4°C, and then incubated for the indicated times at 37°C (chase) with either BSA alone or BSA plus HDL₃ (1 mg protein/ml). *Panel A.* [³H]sterols in the BSA-incubated cells during the chase. *Panel B.* [³H]sterols released to medium containing HDL₃ during the chase.

Table 1. Effects of cholesterol enrichment on efflux of biosynthetic sterols from CHO-K1 cells.

Treatment	*Cholesterol level µg/mg protein*	*Acceptor*	*Efflux, Fraction/24 hour*	
			Cholesterol	Desmosterol
Control	19	HDL₃	0.29 ± 0.04	0.62 ± 0.06
Enriched	38	HDL₃	0.19 ± 0.004 (66%)	0.47 ± 0.04 (76%)
Control	19	PC-SUV	0.062 ± 0.01	0.37 ± 0.04
Enriched	38	PC-SUV	0.038 ± 0.01 (61%)	0.17 ± 0.01 (46%)

Cells were enriched prior to labeling by exposure to LDL plus cholesterol-enriched liposomes in the presence of ACAT-inhibitor (Sandoz compound 58035). Simultaneous labeling and efflux were performed as described in Figure 2. The media contained either HDL₃ (1 mg protein/ml) or phosphatidylcholine small unilamellar vesicles (1 mg phospholipid/ml) to promote sterol efflux. The percentage values in parenthesis indicate the efflux of sterol from enriched cells relative to control cells.

We have begun to define the basis for the rapid efflux of desmosterol from cells, specifically addressing whether this is due to differences in the intracellular transport of demosterol and cholesterol, or alternatively to more rapid desorption of desmosterol from the plasma membrane. Comparisons of newly synthesized sterols in isolated plasma membranes vs. whole cells suggest similar efficiencies of delivery of cholesterol and desmosterol to the plasma membrane. Thus, the greater efflux of desmosterol from cells may be attributable to more efficient desorption from the cell surface, rather than more rapid delivery from the endoplasmic reticulum to the plasma membrane.

STEROL-PHOSPHOLIPID INTERACTIONS

Condensation of hydrocarbon acyl chains of membrane phospholipids and sphingolipids depends on two things, the acyl chain composition (64, 65) and the sterol structure (66-68). Sterols in a monolayer film have their hydroxyl group in the water while their skeleton and alkyl side chains are buried within the acyl groups of the phospholipids (66). There is no evidence of hydrogen bonding between the hydroxyl group and the phospholipids (69). Bloch proposes that lanosterol, the first sterol formed in the pathway of cholesterol synthesis later was refined to the cholesterol molecule because of its superior ability to condense phospholipids (16). Three regions of the molecule are required for this to occur: 1) a planar-nuclear ring, 2) a complete and branched side chain and 3) a β-oriented hydroxyl group on carbon number three (66). Lanosterol's α-oriented methyl groups interfere with condensation because they are bulky, although cholesterol's β-oriented methyl's (carbon 18 and 19) do not cause problems. Condensation increases as each α-methyl group is removed from lanosterol. This is probably because losing an axial-face methyl increases the opportunity for van der Waals interactions between the acyl chains and the planar ring (66) and the side chain (70).

The length of the side chain and the degree of branching are also important. Shorter side chains (less than eight carbons) decrease monolayer stability, whereas branching chains with eight to ten carbons increase stability. Lack of the side chain depletes the sterol's ability to condense the lipids (71). The reductive state of the side chain though appears to have no influence on condensation (72). This may prove important in the differences seen in the efflux properties of cholesterol and desmosterol.

PARTICIPATION OF STEROLS IN DEVELOPMENT AND METABOLIC REGULATION

How the accumulation of sterol intermediates leads to pathology in whole animals is unclear. Administration of the drug Triparanol, an inhibitor of the sterol-Δ-24 reductase, causes the accumulation of desmosterol in plasma and tissues. Short-term use of the drug has been associated with altered lipid metabolism. Studies on desmosterol esterification in J774 macrophages show that this sterol stimulates ACAT activity leading to an increase in stored desmosterol within the cell (73). This may explain the increase in aortic lesions in Triparanol-fed animals.

It appears, however, that sterol-associated pathology more often is caused by the participation of intermediates in processes that normally involve cholesterol, as may be the case with the Smith-Lemli-Opitz syndrome. Recently it was reported that developmental deformities also were associated with an increased level of desmosterol in the blood and tissues of a still-born infant (74). This suggests that a lack of function of sterol-Δ-24 reductase may also be an inherited disease, and that desmosterol may interfere with proper embryogenesis.

The disruption of morphogenesis by sterol intermediates may be linked to the embryonic morphogenic protein, sonic hedgehog (SHh). This protein is involved in creating segmental organizing centers in vertebrate animals, contributing to the formation of the central nervous system, vertebrae, eyes and the limbs. SHh undergoes autoproteolytic processing to yield a 19 kD amino terminal signaling region that is covalently linked to cholesterol. The protein/sterol linkage is necessary for the protein's cleavage and also for its localization (9). In mice lacking the SHh gene, physical deformities and mental retardation that mimic the Smith-Lemli-Opitz syndrome are evident. It has been shown that the severity of developmental defects in Smith-Lemli-Opitz syndrome is inversely correlated with the levels of serum cholesterol (75). Patients with less serum cholesterol usually have higher amounts of 7-dehydrocholesterol (76). This suggests that while the lack of cholesterol can cause defective SHh signaling, the presence of 7-dehydrocholesterol may also contribute. by covalently binding to the protein, preventing it from functioning properly (77).

CONCLUSIONS

Currently it is known that when cells are actively synthesizing cholesterol, they also make intermediate molecules that are very similar in structure. Minor differences in the intermediates make them behave quite differently from cholesterol. It has been established that these molecules are transported from the endoplasmic reticulum and are found in the plasma membrane, and their transport mechanism is thought to be vesicular. Why the intermediates appear in the transport vesicles may be due to indiscriminate sorting at the ER. Once in the plasma membrane, the intermediates, like cholesterol, are available for release into the pathway of reverse sterol transport. The pattern of their transport in the plasma is not well defined. It is not know which lipoprotein fraction they initially associate with. Many of the precursors have been found in an esterified form in plasma (31), but it is not established if they are acquired by HDL and transferred to LDL by the action of cholesteryl ester transfer protein (CETP). Somehow they may be delivered to the liver, which appears to have the ability to ensure that the majority of its sterol production is cholesterol. If this does happen, it may be an efficient process, since only small amounts of the intermediates are found in the plasma of healthy animals. Recent work has shown that sterol side chain structure affects the clearance of chylomicron-like emulsions from the plasma of rats. All sterols with altered side chains decrease the uptake of the particles when compared to cholesterol, except those that contained desmosterol, which were taken up more efficiently than cholesterol (78). It could be that the presence of this sterol increases the lipoprotein's uptake by the liver cell.

Additionally, the presence of desmosterol could alter its uptake in non-hepatic tissue. In this case, the molecule may be signaling the levels of sterol synthesis in surrounding cells.

Additional research needs to be conducted to understand how the body manages the intermediates. Cholesterol has begun to be recognized as a signaling as well as a structural molecule (79). To understand the intermediates' association with pathology in animals, it is important to ascertain how they differ from cholesterol as structural entities in a membrane. However, it could be more significant to establish how they alter cholesterol's signaling ability.

ACKNOWLEDGMENTS
The authors' research has been funded by NIH Program Project Grant HL22633, a grant from the Heinz Foundation (JEP) and a pre-doctoral grant from the Southeastern Pennsylvania affiliate of the American Heart Association (JEP).

REFERENCES
1. Dietschy JM, Turley SD, Spady DK. Role of liver in the maintenance of cholesterol and low density lipoprotein homeostasis in different animal species, including humans. *The Journal of Lipid Research* 1993; 34:1637-1657.

2. Turley SD, Spady DK, Dietschy JM. Identification of a metabolic difference accounting for the hyper-and hyporesponder phenotypes of cynomolgus monkey. *The Journal of Lipid Research* 1997; 38:1598-1611.

3. Lange Y, Matthies HJ. Transfer of cholesterol from its site of synthesis to the plasma membrane. *The Journal of Biological Chemistry*, 1984; 259(23):14624-14630.

4. Echevarria F, Norton RA, Nes WD, Lange, Y. Zymosterol is located in the plasma membrane of cultured human fibroblasts. *The Journal of Biological Chemistry*, 1990; 265(15):8484-8489.

5. Johnson WJ, Fisher RT, Phillips MC, Rothblat GH. Efflux of newly synthesized cholesterol and biosynthetic sterol intermediates from cells. *The Journal of Biological Chemistry*, 1995; 270(42):25037-25046.

6. Opitz JM, de la Cruz F. 1994. Cholesterol metabolism in the RSH/Smith-Lemli-Opitz syndrome: Summary of an NICHD conference. *The American Journal of Medical Genetics*, 1994; 50:326-338.

7. Laughlin RC, Carey TF. Cataracts in patients treated with Triparanol. *Journal of the American Medical Association*, 1962; 181(4):129-130.

8. Wong HYC, Vroman HE, and Mendez, HC. Atherogenic aspects of desmosterol metabolism caused by prolonged Triparanol administration. *Life Sciences,* 1966; 5:629-637.

9. Porter JA, Young, KE, Beachy PA. Cholesterol modification of hedgehog signaling proteins in animal development. *Science*, 1996; 274, October 11, 255-259.

10. Brown MS, Goldstein JL. A receptor mediated pathway for cholesterol homeostasis. *Science*, 1986; 232(4746):34-37.

11. Goldstein JL, Brown MS. Regulation of the mevalonate pathway [review]. *Nature*, 1990; 343(6257):425-430.

12. Simoni RD. Mutation in the lumenal part of the membrane domain of HMG-CoA reductase alters its regulated degradation. *Biochemical and Biophysical Research Communications*, 1995; 206(1):186-193.

13. Bradfute DL, Simoni RD. Non-sterol compounds that regulate cholesterogenesis. Analogues of farnesyl pyrophosphate reduce 3-hydroxy-3-methylglutaryl-coenzyme A reductase levels. *The Journal of Biological Chemistry*, 1994; 269(9):6645-6650.

14. Fruchart JC, DeGeteire C, Delfly B, Castro GR. Apolipoprotein A-I containing particles and reverse cholesterol transport: Evidence for connection between cholesterol efflux and atherosclerosis risk. *Atherosclerosis*, 1994; 110 Suppl:535-539.

15. Bloch KE. The biological synthesis of cholesterol [review]. *Science*, 1965; 150(692):19-28.

16. Bloch KE. Sterol structure and membrane function. *CRC Critical Reviews in Biochemistry*, 1973; 14:47-92.

17. Schroepfer GJ, Jr., Lutsky BN, Martin JA, Huntoon S, Fourcans B, Lee W-H, Vermillion J. Recent investigations on the nature of sterol intermediates in the biosynthesis of cholesterol. *Proceedings of the Royal Society of London, B*, 1972; 180:125-146.

18. Woollette LA. Origin of cholesterol in the fetal golden Syrian Hamster: Contribution of *de novo* sterol synthesis and maternal derived lipoprotein cholesterol. *The Journal of Lipid Research, 1996*; 37(6):1246-1257.

19. Dietschy JM, Wilson JD. Cholesterol synthesis in the squirrel monkey: Relative rates of synthesis in various tissues and mechanisms of control. *The Journal of Clinical Investigation*, 1968; 47:166-174.

20. Hotta S, Chaikoff IL. The role of the liver in the turnover of plasma cholesterol. *Archives of Biochemistry and Biophysics*, 1954; 56:28-37.

21. Dietschy JM, McGarry JD. Limitations of acetate as a substrate for measuring cholesterol synthesis in liver. *The Journal of Biochemistry*, 1974; 249(1):52-58.

22. Hellerstein MK. Methods for measurement of fatty acid and cholesterol metabolism. *Current Opinion in Lipidology*, 1995; 6:172-181.

23. Dietschy JM, Spady DK. Measurement of rates of cholesterol synthesis using tritiated water. *The Journal of Lipid Research*, 1984; 25:1469-1476.

24. Andersen JM, Dietschy JM. Absolute rates of cholesterol synthesis in extrahepatic tissues measured with [^{3}H]-labeled water and [^{14}C]-labeled substrates. *The Journal of Lipid Research*, 1979; 20:740-752.

25. Andersen JM, Turley SD, Dietschy JM. Relative rates of sterol synthesis in the liver and various extrahepatic tissues of normal and cholesterol-fed rabbits. *Biochimica et Biophysica Acta*, 1982; 711:421-430.

26. Spady DK, Dietschy JM. Sterol synthesis *in vivo* in 18 tissues of the squirrel monkey, guinea pig, rabbit, hamster and rat. *The Journal of Lipid Research*, 1983; 24:303-315.

27. Wilson JD. Biosynthetic origin of serum cholesterol in the squirrel monkey: Evidence for a contribution by the intestinal wall. *The Journal of Clinical Investigation*, 1968; 47:175-187.

28. Fujiwara T, Hirono H, Arakawa T. Idiopathic hypercholesterolemia: Demonstration of an impaired feedback control of cholesterol synthesis in vivo. *The Journal of Experimental Medicine*, 1965; 87:155.

29. Turley SD, Spady DK, Dietschy JM. Role of liver in the synthesis of cholesterol and the clearance of low density lipoproteins in the cynomolgus monkey. *The Journal of Lipid Research*, 1995; 36(1):67-79.

30. Dietschy JM. The role of bile salts in controlling the rate of intestinal cholesterogenesis. *The Journal of Clinical Investigation*, 1968; 47:286-300.

31. Björkhem I, Miettinen T, Reihnèr E, Ewerth S, Angelin B, Einarsson K. Correlation between serum levels of some cholesterol precursors and activity of HMG-CoA reductase in human liver. *The Journal of Lipid*

166

Research, 1987; 28(4):1137-1143.

32. Kolvisto PVI, Miettinen TA. Increased amounts of cholesterol precursors in lipoproteins after ileal exclusion. *Lipids*, 1988; 23(10):993-996.

33. Axelson M. Occurrence of isomeric dehydrocholesterols in human plasma. *The Journal of Lipid Research*, 1991; 32:1441-1448.

34. Stranberg TE, Aslomaa A, Vanhanen H, Miettininen TA. Associations of fasting blood glucose with cholesterol absorption and synthesis in nondiabetic middle-aged men. *Diabetes*, 1996; 45(6):755-761.

35. Lange Y, Echevarria F, Steck TL. Movement of zymosterol, a precursor of cholesterol, among three membranes in human fibroblasts. *The Journal of Biological Chemistry*, 1991; 266(32):21439-21443.

36. Johnson WJ, Phillips MC Rothblat GH. Lipoproteins and cellular cholesterol homeostasis. (Review). *Sub-Cellular Biochemistry*, 1997; 28:235-276.

37. Ikonen E. Molecular mechanisms of intracellular cholesterol transport. *Current Opinion in Lipidology*, 1997; 8(2):60-64.

38. Mendez AJ. Monensin and Brefeldin A inhibit high density lipoprotein-mediated cholesterol efflux from cholesterol enriched cells. *The Journal of Biological Chemistry*, 1995; 270(11):5891-5900.

39. Kaplan MR, Simoni RD. Transport of cholesterol from the endoplasmic reticulum to the plasma membrane. *The Journal of Cell Biology*, 1985; 101:446-453.

40. DeGrella RF, Simoni RD. Intracellular transport of cholesterol to the plasma membrane. *The Journal of Biological Chemistry*, 1982; 257(23):14256-14262.

41. Simoni R D. Intracellular transport of membrane lipids. *Progress in Clinical and Biological Research*, 1988; 282:29-41.

42. Urbani L, Simoni RD. Cholesterol and vesicular *stomatitis* virus G protein take separate routes from the endoplasmic reticulum to the plasma membrane. *The Journal of Biological Chemistry*, 1990; 265(4):1919-1923.

43. Aviram M, Bierman EL, Oram JF. High density lipoprotein stimulates sterol translocation between intracellular and plasma membrane pools in human monocyte-derived macrophages. *The Journal of Lipid Research*, 1989; 30:65-76.

44. Hokland BM, Slotte JP, Bierman EL, Oram JF. Cyclic AMP stimulates efflux of intracellular sterol from cholesterol-loaded cells. *The Journal of Biological Chemistry*, 1993; 268(34):25343-25349.

45. Puglielli L, Rigotti A, Greco AV, Santos MJ, Nervi F. Sterol carrier protein-2 is involved in cholesterol transfer from the endoplasmic reticulum to the plasma membrane in human fibroblasts. *The Journal of Biological Chemistry*, 1995; 270(32):18723-18726.

46. Puglielli L, Rigotti A, Amigo L, Nunez L, Greco AV, Santos MJ, Nervi F. Modulation of intrahepatic cholesterol trafficking: Evidence by *in vivo* antisense treatment for the involvement of sterol carrier protein-2 in newly synthesized cholesterol transport into rat bile. *Biochemical Journal*, 1996; 317(part 3):681-687.

47. Baum CL, Reschly EJ, Apurba KG, Groh ME, Schadick K. Sterol carrier protein-2 overexpression enhances sterol cycling and inhibits cholesterol ester synthesis and high density lipoprotein cholesterol secretion. *The Journal of Biological Chemistry*, 1997; 272(10):6490-6498.

48. Rothblat GH, Mahlberg FH, Johnson WJ, Phillips MC. Apolipoproteins, membrane cholesterol domains and the regulation of cholesterol efflux. *The Journal of Lipid Research*, 1992, 33(8):1091-1097.

49. Murata H, Peranen J, Schreiner R, Wieland F, Kurzchalia TV, Simons K. VIP21/caveolin is a cholesterol-binding protein. *Proceedings of the National Academy of Science of the United States of America*, 1995;

92(22):10339-10343.

50. Fielding PE, Fielding CJ. Plasma membrane caveolae mediate the efflux of cellular free cholesterol. *Biochemistry*, 1995; 34(44):14288-14292.

51. Smart EJ, Ying Y-u, Conrad PA, Anderson RGW. Caveolin moves from caveolae to the Golgi apparatus in response to cholesterol oxidation. *The Journal of Cell Biology*, 1994; 127:1185-1197.

52. Smart EJ, Ying Y, Donzell WC, Anderson RGW. A role for caveolin in transport of cholesterol from endoplasmic reticulum to plasma membrane. *The Journal of Biological Chemistry*, 1996; 271(46):29427-29435.

53. Liscum L, Dahl NK. Intracellular cholesterol transport. *The Journal of Lipid Research*, 1992; 33(9):1239-1254.

54. Field EJ, Born E, Murthy S, Mathur SN. Transport of cholesterol from the endoplasmic reticulum to the plasma membrane is constitutive in CaCo-2 cells and differs from the transport of plasma membrane cholesterol to the endoplasmic reticulum. *The Journal of Lipid Research*, 1998; 39:333-343.

55. Metherall JE, Waugh K, Li H. Progesterone inhibits cholesterol biosynthesis in cultured cells. *The Journal of Biological Chemistry*, 1996; 271(5):2627-2633.

56. Mazzone T, Krishna M, Lange Y. Progesterone blocks intracellular translocation of free cholesterol derived from cholesteryl ester in macrophages. *The Journal of Lipid Research*, 1995; 36(3):544-551.

57. Panini SR, Gupta A, Sexton RC, Parish EJ, Rudney H. Regulation of sterol biosynthesis and of 3 hydroxy-3-methylglutaryl-coenzyme A reductase activity in cultured cells by progesterone. *The Journal of Biological Chemistry*, 1987; 262(30):14435-14440.

58. Zhang F, Riley J, Gant TW. Intrinsic multidrug class 1 and 2 gene expression and localization in rat and human mammary tumors (Review). *Laboratory Investigation*, 1996; 75(3):413-426.

59. Higgins C. Flip-flop: The transmembrane translocation of lipids. *Cell*, 1994; 79:393-395.

60. Oude Elferink RP, Ottenhoff R, van Wijland M, Smit JJ, Schinkel AH, Groen AK. Regulation of biliary lipid secretion by mdr2 P-glycoprotein in the mouse. *The Journal of Clinical Investigation*, 1995; 95(1):31-38.

61. Metherall JE, Li H, Waugh K. Role of multidrug resistance P-glycoproteins in cholesterol biosynthesis. *The Journal of Biological Chemistry*, 1996; 271(5):2634-2640.

62. Slotte JP, Bierman EL. Movement of plasma-membrane sterols to the endoplasmic reticulum in cultured cells. *Biochemical Journal*, 1987; 248(1):237-242.

63. Burki E, Logel J, Sinensky M. Endogenous sterol synthesis is not required for regulation of 3-hydroxy-3-methylglutaryl coenzyme A reductase by low density lipoprotein. *The Journal of Lipid Research*, 1987; 28:1199-1205.

64. Smaby JM, Brockman HL, Brown RE. Cholesterol's interfacial interactions with sphingomyelins and phospatidylcholines: Hydrocarbon chain structure determines the magnitude of condensation. *Biochemistry*, 1994; 33:9135-9142.

65. Lund-Katz S, Laboda HM, McLean LR, Phillips MC. Influence of molecular packing and phospholipid type on rates of cholesterol exchange. *Biochemistry*, 1988; 24:3416-3423.

66. Demel R, Bruckdorfer KR VanDeenen LLM. Structural requirements of sterols for the interaction with lecithin at the air-water interface. *Biochimica and Biophysica Acta*, 1972; 255:311-320.

67. Evans RW. Aggregates of saturated phospholipids as the air-water interface. *Chemistry and Physics of Lipids*, 1995; 78(2):163-175.

168

68. Demel RA, De Kruyff B. The function of sterols in membranes. *Biochimica et Biophysica Acta*, 1976; 457:109-132.

69. Boggs JM. Lipid intramolecular hydrogen bonding influence on structural organization and membrane function. *Biochimica and Biophysica Acta*, 1987; 906(3):353-404.

70. Slotte JP. Effect of sterol structure on molecular interactions and lateral domain formation in monolayers containing dipalmitoyl phosphatidylcholine. *Biochimica and Biophysica Acta*, 1995; 1237:127-134.

71. Slotte JP, Junder M, Vilchèze C, and Bittman R. Effect of sterol side-chain on sterol-phosphatidylcholine interactions in monolayers and small unilamellar vesicles. *Biochimica et Biophysica Acta*, 1994; 1190:435-443.

72. Evans RW, Williams MA, Tinoco J. Surface areas of 1-palmitoyl phosphatidylcholines and their interactions with cholesterol. *Biochemical Journal*, 1987; 245:455-462.

73. Tabas I, Feinmark SJ, Beatini N. The reactivity of desmosterol and other shellfish-and xanthomatosis-associated sterols in the macrophage sterol esterification reaction. *The Journal of Clinical Investigation*, 1987; 84:1713-1721.

74. Clayton P, Mills K, Keeling J, FitzPatrick D. Desmosterolosis: A new inborn error of cholesterol biosynthesis. *The Lancet*, 1996; 348(9024): 404.

75. Tint GS, Salen G, Batta AK, Shefer S, Irons M, Elias ER, Morris CA, Hoganson G, Hughes-Benzie R. Correlation of severity and outcome with plasma sterol levels in variants of the Smith-Lemli-Opitz syndrome. *Journal of Pediatrics*, 1995; 127:82-87.

76. Tint GS, Seller M, Hughes-Benzie R, Batta AK, Shefer S, Genest D, Irons M, Elias E, Salen G. Markedly increased tissue concentration of 7-dehydrocholesterol combined with low levels of cholesterol are characteristic of the Smith-Lemli-Opitz syndrome. *The Journal of Lipid Research*, 1995; 36(1):89-95.

77. Kelley RI, Roessler E, Hennekam RCM, Feldman GL, Kosaki K, Jone MC, Palumbos JC, Muenke M. Holoprosencephaly in RSH/Smith-Lemli-Opitz syndrome: Does abnormal cholesterol metabolism affect the function of sonic hedgehog? *American Journal of Medical Genetics*, 1996; 66:478-484.

78. Martins IJ, Vilchèze C, Mortimer B-C, Bittman R, Redgrave, TG. Sterol side chain length and structure affect the clearance of chylomicron-like lipid emulsion in rats and mice. *The Journal of Lipid Research*, 1998; 39:302-312.

79. Jackson SM, Ericsson J, Edwards PA. Signaling molecules derived from the cholesterol biosynthetic pathway (Review). *Sub-Cellular Biochemistry*, 1997; 28:1-21.

INTRAMITOCHONDRIAL CHOLESTEROL TRANSFER IN STEROIDOGENIC CELLS

Douglas M. Stocco[1] and Jerome F. Strauss III[2]

[1]Department of Cell Biology and Biochemistry, Texas Tech University Health Sciences Center, Lubbock, Texas 79430. [2]Center for Research on Reproduction and Women's Health and Department of Obstetrics and Gynecology, University of Pennsylvania Medical Center, Philadelphia, Pennsylvania 19104.

KEY WORDS: steroidogenesis, mitochondria, P450 side-chain cleavage, steroidogenic acute regulatory protein (StAR), MLN64

ABSTRACT

The acute biosynthesis of steroid hormones in response to trophic hormone stimulation is controlled by transferring substrate cholesterol from the outer mitochondrial membrane to the inner mitochondrial membrane where it is converted into pregnenolone, the first steroid synthesized, by the cytochorme P450 side chain cleavage enzyme system. This regulation is cycloheximide sensitive and thus requires a protein factor whose role it is to mediate this transfer of cholesterol. One candidate for the regulatory protein is the mitochondrial phosphoprotein, the steroidogenic acute regulatory (StAR) protein. Cloning and sequencing of the StAR cDNA indicated that it was a novel protein. Transient transfections with the cDNA for the StAR protein resulted in increased steroid production in both MA-10 and COS-1 cells in the absence of stimulation. Mutations in the StAR gene have been shown to cause the potentially lethal disease, congenital lipoid adrenal hyperplasia, a condition in which cholesterol transfer to the P450scc is blocked. Recently, a protein with homology to a region in the C-terminal portion of the StAR protein has been cloned and shown to display some cholesterol transferring capacity. While it appears that StAR is the acute regulator of steroid biosynthesis, the mechanism which results in cholesterol transfer to the inner mitochondrial membrane is still unknown.

CONTENTS

170

Introduction

Steroid hormones make up a very important class of regulatory molecules which are synthesized mainly in the adrenal, the ovary, and the testis in response to steroidogenic stimuli. Adrenal glucocorticoids regulate carbohydrate metabolism and stress management and mineralocorticoids maintain salt balance. The estrogens and progestins, steroids synthesized in the ovary (and placenta), maintain secondary sex characteristics and are also essential for reproductive function. Lastly, testicular androgens are responsible for maintaining reproductive function and secondary sex characteristics in the male. While the steroid hormones can be distinguished from one another by their diverse physiological actions in the body, they do, however, share a common characteristic regardless of the tissue of origin. They are all synthesized from a common precursor substrate, namely cholesterol. Thus, the biosynthesis of all hormonal steroids in response to trophic hormone and other steroidogenic stimuli begins with the cleavage of the 27 carbon cholesterol molecule to form the first steroid synthesized, the 21 carbon containing molecule, pregnenolone. This reaction is catalyzed by the cytochrome P450 side chain cleavage enyme (P450*scc*), which is part of the cholesterol side chain cleavage enzyme system (CSCC) which is located on the matrix side of the inner mitochondrial membrane (1-4).

For many years it was the action of the P450*scc* enzyme in converting cholesterol to pregnenolone which was considered to be the rate-limiting step in steroidogenesis. However, it readily became clear that the activity of the P450scc enzyme was not rate-limiting in this process (5), and in order to initiate and sustain steroidogenesis, first, a constant supply of the substrate cholesterol for steroid biosynthesis must be available within the cell and second, a mechanism must exist for the delivery of this cholesterol to the site of cleavage in the inner mitochondrial membrane where the P450scc enzyme resides. The stores of cholesterol found inside steroidogenic cells may be supplied from serum in the form of high density lipoproteins (HDL) or low density lipoprotein (LDL) depending on the species and cell type in question (6,7), or, from what appears to be a less important source of cholesterol, *de novo* synthesis from acetate. Subsequently, given adequate intracellular cholesterol supplies, two separate but equally important processes must occur. First, mobilization of cholesterol from cellular stores such as lipid droplets or other cellular membranes to the outer mitochondrial membrane and second, the transfer of this cholesterol from the outer to the inner mitochondrial membrane (8-10). The factors and processes responsible for the mobilization of cholesterol to the outer mitochondrial membrane are thought to involve changes in cellullar architecture and putative transport proteins but their mechanisms of action are not well understood. The processes involved in intracellular cholesterol movement are the subjects of other chapters in this text and will not be discussed in any detail here. Suffice it to state that a steady supply of cholesterol to the outer mitochondrial membrane is an absolute requirement to maintain maximum steroidogenesis in response to stimulation of steroidogenic cells.

While the activity of the P450*scc* was considered as the rate limiting step in steroidogenesis in early studies (11-13), a number of observations indicated that the *true* rate limiting step effected by hormone stimulation was the delivery of cholesterol to the inner mitochondrial membrane and to the P450*scc* (12, 14-17). Proof of this was provided when hydroxylated analogs of cholesterol such as 22R-hydroxycholesterol, 20α-hydroxycholesterol or 25-hydroxycholesterol, all of which can readily diffuse across the mitochondrial membranes to the P450*scc*, were placed on steroidogenic cells. Given these substrates, high levels of steroids could be formed in the absence of hormone stimulation of the cells (18-21). These observations indicated that the P450*scc* was fully active and that it was the lack of availability of cholesterol for cleavage which prevented the production of pregnenolone. With these observations came the understanding that the major barrier to be overcome in the translocation of cholesterol to the P450*scc* was the aqueous space between the outer and inner mitochondrial membranes through which the hydrophobic cholesterol must pass. Since the aqueous diffusion of cholesterol is very slow (22-24), and could not provide sufficient substrate to account for the rapid and large increase in steroid production observed, it followed that stimulation of steroidogenesis required a mechanism which rapidly transported this steroid precursor across this barrier.

In summary, the overall production of steroids is controlled by events which facilitate the transport of cholesterol from lipid droplets and other cellular stores, first to the mitochondrial outer membrane and second, its subsequent translocation across the aqueous intermembrane space of the mitochondria to the inner membrane. While both processes are necessary to ensure maximal rates of steroid production in response to stimulation, it is the second process which is readily accepted as being the rate-limiting step in hormone regulated steroidogenesis.

The Acute Regulation of Steroidogenesis

The molecular events regulating the rapid production of steroids in response to trophic hormone stimulation have been the subject of intense investigation for over four decades. Regardless of the type of steroidogenic cell studied, the acute response to trophic hormone stimulation usually share many of the same characteristics. The steroidogenic responses are dose-dependent, have the same temporal relationship with the onset of steroid production and are sensitive to protein synthesis inhibitors. For these and other reasons, it is highly likely that the mechanism(s) involved in this acute regulation is similar in most steroidogenically active cell types.

Early studies on the regulation of steroid biosynthesis were performed in the adrenal gland where it had been observed that adrenocorticotropic hormone (ACTH) could stimulate the biosynthesis of steroids *in vitro* (11,25-27). This *in vitro* system was the first such model developed and was employed in subsequent studies to study this phenomenon. One of the first and most fundamental observations concerning steroidogenesis was that acute steroid production in response to hormone stimulation had an absolute requirement for the synthesis of new proteins. The first of such studies were performed by Ferguson (28,29) who demonstrated that the acute stimulation of corticoid synthesis in adrenal glands by ACTH was sensitive to the protein synthesis inhibitor puromycin. At approximately the same time, Garren and co-workers also demonstrated that steroidogenesis in adrenal tissue was dependent upon the ACTH stimulated synthesis of new proteins (30-33). Importantly, these studies also indicated the hormonally controlled step was distal to cholesterol ester hydrolysis but proximal to its side chain cleavage ie. at the delivery of cholesterol to the P450*scc* enzyme (32). Further, they concluded that the half-life of the putative stimulating factor was very short thus giving rise to two adjectives commonly used to describe the putative protein regulator, *cycloheximide sensitive* and *highly labile* (30). Many similar studies confirmed the need for *de novo* protein synthesis in the hormone regulated, acute production of steroids (12,16,33-40). Importantly, Simpson and Boyd, (1) determined that the cycloheximide sensitive step was located in the mitochondria, but, just as importantly, it was also noted by Arthur and Boyd (41), that protein synthesis inhibitors had no effect on the activity of the P450*scc* itself. These observations were quickly followed by studies which demonstrated that inhibition of protein synthesis had no effect on the increased delivery of cellular cholesterol to the outer mitochondrial membrane, but that the delivery of this substrate from the outer membrane to the inner mitochondrial membrane was completely inhibited by cycloheximide (16,42). Furthermore, it has been demonstrated that while cholesterol could be delivered to a so called pre-steroidogenic pool in the presence of cycloheximide, pregnenolone production did not occur until the inhibitor was removed and the cells subsequently stimulated with hormone (40). As a result of these many studies, the precise site of the cycloheximide inhibited regulation had been pinpointed, namely to the transfer of cholesterol to the P450*scc* enzyme in the inner mitochondrial membrane. The observation that *de novo* protein synthesis was indispensable for the acute production of steroids in response to hormone stimulation has also been made more recently in several different steroidogenic tissues (43-48).

In summary, many observations resulted in the overall characterization of the acute regulation of steroidogenesis. Essentially, these characteristics indicated that the acute production of steroids was dependent upon a hormone stimulated, rapidly synthesized, cycloheximide sensitive and highly labile protein whose function was to mediate the transfer of cholesterol from the outer mitochondrial membrane to the inner mitochondrial membrane and the P450scc enzyme. The effort to identify and characterize this acute regulatory protein(s) has been ongoing since the early observations of Ferguson and Garren and their colleagues. Several candidates have emerged from these efforts. A listing of these proteins and the data supporting their candidacies has been collectively reviewed (10), as has the characteristics for individual candidates (49). Only the observations made for one of these candidates, the Steroidogenic Acute Regulatory (StAR) protein will be summarizd here.

The Steroidogenic Acute Regulatory Protein

This review will summarize some of the studies performed on a protein which has been proposed as the acute regulator of cholesterol transfer to the inner mitochondrial membrane and hence, steroid biosynthesis. This protein was initially described by Orme-Johnson and colleagues as an ACTH-induced 30 kDa phosphoprotein in hormone-treated rat and mouse adrenocortical cells, and as an LH-induced protein in rat corpus luteum cells and mouse Leydig cells (43-46,50-53). This series of studies indicated that a close relationship between the appearance of the 30 kDa proteins and steroid hormone biosynthesis existed in several different steroidogenic tissues and that the synthesis of these proteins, as was steroidogenesis, was sensitive to cycloheximide. Proteins probably identical to those described by Orme-Johnson have been characterized in hormone stimulated MA-10 mouse Leydig tumor cells by Stocco and colleagues (47,48,54-58). During the course of studies in both laboratories, these proteins were found to be localized to the mitochondria and consisted of several forms of a newly synthesized 30 kDa protein. In addition to the 30 kDa mitochondrial proteins, a 37 kDa precursor form of these proteins was also detected, a common observation with mitochondrial proteins (45,47). Since the initial observation of these proteins, there have been a number of studies in which correlations between the synthesis of steroids and the synthesis of the 30 kDa proteins have been made (43-46,48,50-52,56-58). However, in spite of the many positive correlations noted, a direct cause-and-effect relationship between 30 kDa protein expression and steroidogenesis was lacking, and it became increasingly clear that it would be necessary to clone the 30 kDa proteins to unequivocally prove its function in steroidogenesis. The cloning of the cDNA for the 30 kDa protein was successfully accomplished in 1994 (59). When compared with other sequences in the data base both the nucleic acid sequence and protein sequence were found to be unique indicating the 30 kDa protein represented a novel protein. Most importantly, transient transfection experiments demonstrated that expression of the cDNA-derived protein in MA-10 mouse Leydig tumor cells resulted in a significant increase in steroid production in the absence of hormone stimulation. In addition, transient transfection of COS-1 cells with the cDNA for the 37 kDa protein resulted in a several fold increase in the conversion of cholesterol to pregnenolone (10,60,61). These results substantiated and extended the previous correlative studies and indicated a direct role for the 30kDa proteins in hormone-regulated steroid production. As a result of these observations, the protein was named the Steroidogenic Acute Regulatory (StAR) protein (59). To date, while the mechanism which StAR uses to transfer cholesterol to the inner mitochondrial membrane and the P450scc is not very well understood, it appears to be the best candidate protein for the acute regulator of steroidogenesis as proposed by Ferguson and Garren and others.

Consequences of a Mutated StAR gene

Congenital lipoid adrenal hyperplasia (lipoid CAH) is a lethal condition resulting from an almost complete inability of the newborn to synthesize steroids. This condition is manifested by the presence of large adrenals containing very high levels of cholesterol and cholesterol esters and also by an increased amount of lipid accumulation in testicular Leydig cells. This disease was originally thought to be due to a mutation of P450*scc* (20,22 desmolase) enzyme activity (62), and this belief had persisted until relatively recent times (62-65). However, the gene for this enzyme (66) has been shown to be normal in this disease and it was deduced that the defect lies upstream of P450*scc* at the point of cholesterol delivery to the enzyme. In studies designed to determine if StAR may be involved in lipoid CAH, Lin et al, (61) prepared StAR cDNA by RT-PCR using RNA isolated from testicular tissue of patients with lipoid CAH and identified nonsense mutations in the sequence. These mutations, were confirmed in the genomic DNA, and, importantly, while expression of the normal human StAR protein in COS-1 cells resulted in an 8-fold increase in steroid production, expression of the StAR cDNA from these patients indicated the proteins were completely inactive in promoting steroidogenesis. In addition to the first reports on StAR mutations causing lipoid CAH, many additional examples of mutations in StAR resulting in this disease are being reported (67-71), and perhaps it is not as rare as previously thought.

That mutations in the STAR gene resulted in lipoid CAH produced compelling evidence for the essential role of this protein in the acute regulation of steroidogenesis. Therefore, an obvious strategy was to produce a knockout of the StAR gene in an animal system with the goal of having a model system to study this protein. Caron et al used targeted disruption of the StAR gene in mice to successfully produce StAR knockout mice (72). Initial observations of these mice have indicated that regardless of genotype, all mice have female external genitalia, as seen in the human condition. Following birth, all animals failed to grow normally and death within a short period of time occured, presumably as a result of adrenocortical insufficiency. This was confirmed by the observation that serum levels of corticosterone and aldosterone were depressed while levels of ACTH and CRH were elevated. These observations indicated an impairment in the production of adrenal steroids with an accompanying loss of feedback regulation at the level of the hypothalamus or pituitary. Inspection of the adrenal gland revealed a normal medulla but an abnormal cortex, having a disrupted fascicular zone. Specific staining procedures revealed elevated lipid deposits in the adrenal cortex region of the StAR knockout mouse. While the StAR knockout mice were all phenotypically sex reversed, the testes of these animals appeared normal upon gross inspection. However, once again specific staining indicated the presence of elevated levels of lipid within this organ. In contrast, the ovaries of the StAR knockout mice were essentially indistinguishable from wild type animals, a similar situation as found with human StAR mutations (67-69).

Intramitochondrial Cholesterol Transfer

StAR would appear to have a critical function in the acutely regulated transfer of cholesterol from the outer mitochondrial membrane to the inner mitochondrial membrane. A model was earlier proposed whereby StAR may mediate cholesterol transfer to the P450*scc* and has appeared in a previous review (10). It was proposed that as the precursor protein was being imported into the mitochondrial inner compartment, contact sites between the inner and outer membranes were formed. With the formation of contact sites, the water barrier was removed and cholesterol was transfered from the outer to the inner mitochondrial membrane and hence, was available to the P450*scc* for pregnenolone synthesis (9,45,47). After processing, the membranes separated and no further cholesterol transfer could occur without additional synthesis and processing of StAR precursor proteins. Since the half life of the precursors of the 30 kDa mitochondrial proteins have been shown to be very short (45), this would explain the observation that steroidogenesis decays very quickly in the absence of new protein synthesis.

However, a recent report by Arakane et al has indicated that a revision of this model is required (73). It has been shown that N-terminal truncations of the StAR protein which remove as many as 62 amino acids have no inhibitory effect on steroid production in COS-1 cells transfected with the cDNAs containing the truncations. Western analysis and immunostaining for StAR protein indicated that the truncated StAR protein was not imported into the mitochondria and therefore, it appears that import of the StAR protein is not required for cholesterol transfer. Conversely, truncation of the C-terminus by 10 amino acids resulted in a decrease in steroid production of 50%, while a 28 amino acid truncation resulted in a complete loss of steroid production (73) indicating that the C-terminal region of the StAR protein is important in cholesterol transfer. A further indication of the importance of the C-terminal region of the StAR protein in cholesterol transfer can be seen in exciting findings recently reported by Watari et al (74). In this study, these investigators report the steroidogenic properties of a protein known as MLN64, which has significant homology to the C-terminal region of StAR. This protein was originally described as a gene product of unknown function which was highly expressed in specific breast tumors (75,76). Importantly, expression of the MLN64 protein in COS-1 cells resulted in a two-fold increase in steroid production, and removal of N-terminal sequences resulted in a further increase. Proteins having similar homologies to the C-terminal region of StAR and MLN64 were also found in *Caenorhabditis elegans* and were proposed to be as much as 600 million years old. Perhaps these proteins represent cellular proteins whose function was to aid in the trafficking of sterols within the cell. The relationship between StAR and MLN64 as well as the role of MLN64 in the cell remain to be determined and hopefully useful information concerning sterol movement in the cell will be obtained.

Phosphorylation of StAR appears to influence its steroidogenic activity. Two consensus protein kinase A phosphorylation sites are present in all StAR sequences determined to date (77). Cyclic AMP analogues rapidly induce phosphorylation of these sites. Mutation of one of these sites, serine 194 in the murine sequence or serine 195 in the human sequence, to an alanine residue causes a reduction in the steroidogenic activity of the mutant protein by more than 50%. In contrast, mutating this residue to an aspartic acid, which mimics the charge effect of phosphorylation, causes a modest increase in steroidogenic activity. They also provide a mechanism by which trophic stimulation can acutely increase steroidogenesis through the co- or post-translational modification of StAR.

While the mechanism of action of the StAR protein is still unknown, it is becoming increasingly clear that cholesterol transfer requires that it interact with as yet unknown proteins and/or other factors on the outside of the outer mitochondrial membrane and produce alterations which result in cholesterol transfer. This model could still incorporate the formation of contact sites between the two membranes and the idea of specific protein-protein interactions forming hydrophobic channels through which cholesterol can move also remains a possibility. Therefore, much remains to be determined concerning the mechanism whereby StAR can effect cholesterol transfer to the inner mitochondrial membrane. In this regard, the identification of the components with which StAR interacts on the outer mitochondrial membrane becomes of critical importance in understanding its mechanism of action.

The finding that expression of StAR can directly increase steroid output would suggest a direct effect of StAR expression on cholesterol transport. In a recent study, Cherradi et al (78), demonstrated that treatment of bovine adrenal glomerulosa cells with Ca^{+2} resulted in a cycloheximide sensitive stimulation of mitochondrial cholesterol transfer from the outer membrane to the contact sites and inner membrane. The significance of these studies was later demonstrated when it was shown that concomitant with the increase in cholesterol transfer was a cycloheximide sensitive increase in the synthesis of the StAR protein (79). This finding was further corroborated in a study in which Ca^{+2} stimulated cholesterol transfer, steroidogenesis and StAR synthesis were all inhibited by atrial natriuretic peptide in bovine adrenal cells (80). Lastly, we have recently demonstrated that incubation of mitochondria isolated from unstimulated MA-10 cells with COS-1 lysates

176

containing wild type StAR resulted in significant increases in both cholesterol transfer from outer to inner mitochondrial membranes and pregnenelone biosynthesis (81) and that recombinant human StAR protein stimulates pregnenolone synthesis by isolated bovine mitochondria and the conversion of ^{3}H-cholesterol into ^{3}H-pregnenolone (82). It is also intriguing that mitochondrial contact sites in bovine adrenocortical cells and MA-10 cells contain the first two enzymes in the steroidogenic pathway, P450*scc* and 3β HSD (78,79,83). Thus, it is tempting to speculate that the interaction of StAR with the mitochondria may cause the formation of a protein complex consisting of P450*scc* and 3β HSD, the enzymes required for the first two steps in steroidogenesis. Thus, cholesterol could quickly be converted to progesterone as speculated in an earlier study (78).

In summary, the demonstrated characteristics of the StAR protein make it the most attractive candidate for the long-sought hormone stimulated protein factor responsible for acutely regulating the transfer of cholesterol from the outer to the inner mitochondrial membrane, and thus, acutely regulating steroid hormone biosynthesis. However, the mechanism of action of StAR in transferring cholesterol to the inner mitochondrial membrane remains to be determined and as such should prove to be one of the most interesting questions for the future as the picture of the acute regulation of steroidogenesis continues to unfold.

Ackowledgments: the authors would like to acknowledge the support of NIH grants HD 17481 (to DMS) and HD06274 (to JFS).

References

1.   Simpson ER, Boyd GS. The cholesterol side-chain cleavage system of the adrenal cortex: a mixed function oxidase. Biochem Biophys Res Commun 1966; 24: 10-17
2.   Simpson ER, Boyd GS. The cholesterol side-chain cleavage system of the adrenal cortex. Eur J Biochem 1967; 2: 275-285
3.   Yago N, Ichii S. Submitochondrial distribution of components of the steroid 11-β-hydroxylase and cholesterol side chain-cleavage enzyme systems in hog adrenal cortex. J Biochem 1969; 65: 215-224
4.   Churchill PF, Kimura T. Topological studies of cytochromes P450scc and P45011β in bovine adrenocortical inner mitochondrial membranes. J Biol Chem 1979; 254: 10443-10448
5.   Hanukoglu I, Hanukoglu Z. Stoichiometry of mitochondrials, cytochromes P450, adrenodoxin and adrenodoxin reductase in adrenal cortex and corpus luteum. Eur J Biochem 1986; 157: 27-31
6.   Kovanen PT, Goldstein JL, Chappell DA, Brown MS. Regulation of low density lipoprotein receptors by adrenocorticotropin in the adrenal gland of mice and rats in vivo. J Biol Chem 1980; 255: 5591-5598
7.   Gwynne JT, Mahaffee DD. Rat adrenal uptake and metabolism of high density lipoprotein cholesteryl ester. J Biol Chem 1989; 264: 8141-8150
8.   Liscum L, Dahl NK. Intracellular cholesterol transport. J Lipid Res 1992; 33: 1239-1254
9.   Jefcoate CR, McNamara BC, Artemenko I, Yamazaki T. Regulation of cholesterol movement to mitochondrial cytochrome P450scc in steroid hormone synthesis. J Steroid Biochem Mol Biol 1992; 43: 751-767

10. Stocco DM, Clark BJ. Regulation of the acute production of steroids in steroidogenic cells. Endocr Rev 1996; 17: 221-244

11. Stone D, Hechter O. Studies on ACTH action in perfused bovine adrenals: Site of action of ACTH in corticosteroidogenesis. Arch Biochem Biophys 1954; 51: 457-469

12. Karaboyas GC, Koritz SB. Identity of the site of action of cAMP and ACTH in corticosteroidogenesis in rat adrenal and beef adrenal cortex slices. Biochemistry 1965; 4: 462-468

13. Garren LD, Gill GN, Masui H, Walton GM. On the mechanism of action of ACTH. Rec Prog Horm Res 1971; 27: 433-478

14. Crivello JF, Jefcoate CR. Intracellular movement of cholesterol in rat adrenal cells. J Biol Chem 1980; 255: 8144-8151

15. Mori M, Marsh JM. The site of luteinizing hormone stimulation of steroidogenesis in mitochondria of the rat corpus luteum. J Biol Chem 1982; 257: 6178-6183

16. Privalle CT, Crivello J, Jefcoate CR. Regulation of intramitochondrial cholesterol transfer to side-chain cleavage cytochrome P450scc in rat adrenal gland. Proc Natl Acad Sci USA 1983; 80: 702-706

17. Jefcoate CR, DiBartolomeos MJ, Williams CA, McNamara BC. ACTH regulation of cholesterol movement in isolated adrenal cells. J Steroid Biochem 1987; 27: 721-729

18. Lambeth JD, Kitchen SE, Farooqui AA, Tuckey R, Kamin H. Cytochrome P-450scc substrate interactions: studies of binding and catalytic activity using hydroxycholesterols. J Biol Chem 1982; 257: 1876-1884

19. Tuckey TC, Stevenson PM. Properties of bovine luteal cytochrome P-450scc incorporated into artificial phospholipid vesicles. Int J Biochem 1984; 16: 479-503

20. Tuckey RC, Atkinson HC. Pregnenolone synthesis from cholesterol and hydroxycholesterols by mitochondria from ovaries following the stimulation of immature rats with pregnant mare's serum gonadotropin and human choriogonadotropin. Eur J Biochem 1989; 186: 255-259

21. Tuckey RC. Cholesterol side-chain cleavage by mitochondria from the human placenta. Studies using hydroxycholesterols as substrates. J Steroid Biochem Mol Biol 1992; 42: 883-890

22. Phillips MC, Johnson WJ, Rothblat GH. Mechanisms and consequences of cellular cholesterol exchange and transfer. Biochim Biophys Acta 1987; 906: 223-276

23. Schroeder F, Jefferson JR, Kier AB, Knittel J, Scallen TJ, Wood WG, Hapala I. Membrane cholesterol dynamics: cholesterol domains and kinetic pools. Proc Soc Exptl Biol Med 1991; 196: 235-252

24. Rennert H, Chang YJ, Strauss III JF. Intracellular cholesterol dynamics in steroidogenic cells: a contemporary view. In: "The Ovary" (E.Y., Adashi, P.C.K., Leung, Eds.), 1993; pp. 147-164. Raven Press, New York.

25. Hechter O, Zaffaroni A, Jacobsen RP, Levy H, Jeanloz RW, SchenkerV, Pincus G. The nature and the biogenesis of the adrenal secretory product. Rec Prog Horm Res 1951; 6: 215-246

26. Haynes R, Savard K, Dorfman RI. An action of ACTH on adrenal slices. Science 1952; 116: 690-691

27. Saffran M, Grad B, Bayliss MJ. Production of corticoids by rat adrenals in vitro. Endocrinology 1952; 50: 639-643

28. Ferguson JJ. Puromycin and adrenal responsiveness to adrenocorticotropic hormone. Biochim Biophys Acta 1962; 57: 616-617

29. Ferguson JJ. Protein synthesis and adrenocorticotropin responsiveness. J Biol Chem 1963; 238: 2754-2759

30. Garren LD, Ney RL, Davis WW. Studies on the role of protein synthesis in the regulation of corticosterone production by ACTH in vivo. Proc Natl Acad Sci USA 1965; 53: 1443-1450

31. Garren LD, Davis WW, Crocco RM. Puromycin analogs: action of adrenocorticotropic hormone and the role of glycogen. Science 1966; 152: 1386-1388

32. Davis WW, Garren LD On the mechanism of action of adrenocorticotropic hormone. The inhibitory site of cycloheximide in the pathway of steroid biosynthesis. J Biol Chem 1968; 243: 5153-5157

33. Garren LD The mechanism of action of adrenocorticotropic hormone. In "Vitamins and Hormones" (R.S. Harris, I.G. Wool, J.A. Loraine, Eds.) 1968; 26: pp.119-145. Academic Press, New York

34. Cooke BA, Janszen FHA, Clotscher WF, van der Molen HJ. Effect of protein-synthesis inhibitors on testosterone production in rat testis interstitial tissue and Leydig-cell preparations. Biochem J 1975; 150: 413-418

35. Paul DP, Gallant S, Orme-Johnson NR, Orme-Johnson WH, Brownie AC. Temperature dependence of cholesterol binding to cytochrome P-450scc of the rat adrenal. Effect of adrenocorticotropic hormone and cycloheximide. J Biol Chem 1976; 251: 7120-7126

36. Farese RV, Prudente WJ. On the requirement for protein synthesis during corticotropin-induced stimulation of cholesterol side chain cleavage in rat adrenal mitochondrial and solubilized desmolase preparations. Biochim Biophys Acta 1977; 496: 567-570

37. Crivello JF, Jefcoate CR. Mechanism of corticotropin action in rat adrenal cells. 1. Effect of inhibitors of protein synthesis and of microfilament formation on corticosterone synthesis. Biochim Biophys Acta 1978; 542: 315-329

38. Toaff ME, Strauss III JF, Flickinger GL, Shattil SJ. Relationship of cholesterol supply to luteal mitochondrial steroid synthesis. J Biol Chem 1979; 254: 3977-3982

39. Solano AR, Neger R, Podesta EJ. Rat adrenal cycloheximide-sensitive factors and phospholipids in the control of acute steroidogenesis. J Steroid Biochem 1984; 21: 111-116

40. Stevens VL, Xu T, Lambeth JD. Cholesterol trafficking in steroidogenic cells: reversible cycloheximide-dependent accumulation of cholesterol in a pre-steroidogenic pool. Eur J Biochem 1993; 216: 557-563

41. Arthur JR, Boyd GS. The effect of inhibitors of protein synthesis on cholesterol side-chain cleavage in the mitochondria of luteinized rat ovaries. Eur J Biochem 1976; 49: 117-127

42. Ohno Y, Yanagibashi K, Yonezawa Y, Ishiwatari S, Matsuba M. A possible role of "steroidogenic factor" in the corticoidogenic response to ACTH; Effect of ACTH, cycloheximide and aminoglutethimide on the content of

cholesterol in the outer and inner mitochondrial membrane of rat adrenal cortex. Endocrinology (Jpn) 1983; 30: 335-338

43. Krueger RJ, Orme-Johnson NR. Acute adrenocorticotropic hormone stimulation of adrenal corticosteroidogenesis. J Biol Chem 1983; 258: 10159-10167

44. Epstein LF, Orme-Johnson NR Acute action of luteinizing hormone on mouse Leydig cells: Accumulation of mitochondrial phosphoproteins and stimulation of testosterone biosynthesis. Mol Cell Endocrinol 1991a; 81: 113-126

45. Epstein LF, Orme-Johnson NR. Regulation of steroid hormone biosynthesis: Identification of precursors of a phosphoprotein targeted to the mitochondrion in stimulated rat adrenal cortex cells. J Biol Chem 1991b; 266: 19739-19745

46. Pon LA, Orme-Johnson NR Acute stimulation of corpus luteum cells by gonadotropin or adenosine 3'5'-monophosphate causes accumulation of a phosphoprotein concurrent with acceleration of steroid synthesis. Endocrinology 1988; 123: 1942-1948

47. Stocco DM, Sodeman TC. The 30-kDa mitochondrial proteins induced by hormone stimulation in MA-10 mouse Leydig tumor cells are processed from larger precursors. J Biol Chem 1991; 266: 19731-19738

48. Stocco DM, Chen W. Presence of identical mitochondrial proteins in unstimulated constitutive steroid-producing R2C rat Leydig tumor and stimulated nonconstitutive steroid-producing MA-10 mouse Leydig tumor cells. Endocrinology 1991; 128: 1918-1926

49. Papadopoulos V. Peripheral-type benzodiazepine/diazepam binding inhibitor receptor: biological role in steroidogenic cell function. Endocr Rev 1993; 14: 222-240

50. Pon LA, Orme-Johnson NR. Acute stimulation of steroidogenesis in corpus luteum and adrenal cortex by peptide hormones: Rapid induction of a similar protein in both tissues. J Biol Chem 1986; 261: 6594-6599

51. Pon LA, Hartigan JA, Orme-Johnson NR. Acute ACTH regulation of adrenal corticosteroid biosynthesis: rapid accumulation of a phosphoprotein. J Biol Chem 1986a; 261: 13309-13316

52. Pon LA, Epstein LF, Orme-Johnson NR. Acute cAMP stimulation in Leydig cells: Rapid accumulation of a protein similar to that detected in adrenal cortex and corpus luteum. Endocr Res 1986b; 12: 429-446

53. Alberta JA, Epstein LF, Pon LA, Orme-Johnson NR. Mitochondrial localization of a phosphoprotein that rapidly accumulates in adrenal cortex cells exposed to adrenocorticotropic hormone or to cAMP. J Biol Chem 1989; 264: 2368-2372

54. Stocco DM, Kilgore MW. Induction of mitochondrial proteins in MA-10 Leydig tumour cells with human choriogonadotropin. Biochem J 1988; 249: 95-103

55. Stocco DM, Chaudhary LR. Evidence for the functional coupling of cAMP in MA-10 mouse Leydig tumor cells. Cell Signal 1990; 2: 161-170

56. Stocco DM. Further evidence that the mitochondrial proteins induced by hormone stimulation in MA-10 mouse Leydig tumor cells are involved in the acute regulation of steroidogenesis. J Steroid Biochem Mol Biol 1992; 43: 319-333

180

57. Stocco DM, Ascoli M. The use of genetic manipulation of MA-10 Leydig tumor cells to demonstrate the role of mitochondrial proteins in the acute regulation of steroidogenesis. Endocrinology 1993; 132: 959-967

58. Stocco DM, King S, Clark BJ. Differential effects of dimethylsulfoxide on steroidogenesis in mouse MA-10 and rat R2C Leydig tumor cells. Endocrinology 1995; 136: 2993-2999

59. Clark BJ, Wells J, King SR, Stocco DM. The purification, cloning, and expression of a novel LH-induced mitochondrial protein in MA-10 mouse Leydig tumor cells: characterization of the steroidogenic acute regulatory protein (StAR). J. Biol. Chem. 1994; 269: 28314-28322

60. Sugawara T, Holt JA, Driscoll D, Strauss III JF, Lin D, Miller WL, Patterson D, Clancy KP, Hart IM, Clark BJ, Stocco DM. Human steroidogenic acute regulatory protein: functional activity in COS-1 cells, tissue-specific expression, and mapping of the gene to 8p11.2 and a pseudogene to chromosome 13. Proc Natl Acad Sci USA 1995a; 92: 4778-4782

61. Lin D, Sugawara T, Strauss III JF, Clark BJ, Stocco DM, Saenger P, Rogol, A, Miller WL. Role of steroidogenic acute regulatory protein in adrenal and gonadal steroidogenesis. Science 1995; 267: 1828-1831

62. Degenhart HJ, Visser HKA, Boon H, O'Doherty NJ. Evidence for deficient 20α-cholesterol-hydroxylase activity in adrenal tissue of a patient with lipoid adrenal hyperplasia. Acta Endocrinol 1972; 71: 512-518

63. Hauffa PT, Miller WL, Grumbach MM, Conte FA, Kaplan SL. Congenital adrenal hyperplasia due to deficient cholesterol side-chain cleavage activity (20,22-desmolase) in a patient treated for 18 years. Clin Endocrinol 1985; 23: 481-493

64. Matteson KJ, Chung BC, Urdea MS, Miller WL. Study of cholesterol side-chain cleavage (20,22 desmolase) deficiency causing congenital lipoid adrenal hyperplasia using bovine-sequence P450scc oligodeoxyribonucleotide probes. Endocrinology 1986; 118: 1296-1305

65. Muller J, Torsson A, Neilsen MD, Petersen KE, Christoffesen J, Skakkeboek NE. Gonadal development and growth in 46, XX and 46 XY individuals with P450scc deficiency (congenital lipoid adrenal hyperplasia). Horm Res 1991; 36: 203-208

66. Lin D, Gitelman SE, Saenger P, Miller WL. Normal genes for the cholesterol side chain cleavage enzyme, P450scc, in congenital lipoid adrenal hyperplasia. J Clin Invest 1991; 88: 1955-1962

67. Bose HS, Sugawara T, Strauss III JF, Miller WL. The pathophysiology and genetics of congenital lipoid adrenal hyperplasia. N Engl J Med 1996; 335: 1870-1878

68. Bose HS, Pescovitz OH, Miller WL. Spontaneous feminization in a 46,XX female patient with congenital lipoid adrenal hyperplasia due to a homozygous frameshift mutation in the steroidogenic acute regulatory protein. J Clin Endocrinol Metab 1997; 82: 1511-1515

69. Fujieda K, Tajima T, Nakae J, Sageshima S, Tachibana K, Suwa S, Sugawara T, Strauss III JF. Spontaneous puberty in 46,XX subjects with congenital lipoid adrenal hyperplasia. J Clin Invest 1997; 99: 1265-1271

70. Nakae J, Tajima T, Sugawara T, Arakane F, Hanaki K, Hotsubo T, Igarashi N, Igarashi Y, Ishii T, Koda N, Kondo T, Kohno H, Nakagawa Y,

Tachibana K, Takeshima Y, Tsubouchi K, Strauss III JF, Fujieda K. Analysis of the steroidogenic acute regulatory protein (StAR) gene in Japanese patients with congenital lipoid adrenal hyperplasia. Human Mol Gen 1997; 6, 571-576

71. Okuyama E, Nishi N, Onishi S, Itoh S, Ishii Y, Miyanake H, Fujita K, Ichikawa Y. A novel splicing junction mutation in the gene for the steroidogenic acute regulatory protein causes congenital lipoid adrenal hyperplasia. J Clin Endocrinol Metab 1997; 82: 2337-2342

72. Caron KM, Soo SC, Wetsel WC, Stocco DM, Clark BJ, Parker KL. Targeted disruption of the mouse gene encoding steroidogenic acute regulatory protein provides insights into congenital lipoid adrenal hyperplasia. Proc Natl Acad Sci USA 1997b; 94, 11540-11545.

73. Arakane F, Sugawara T, Nishino H, Liu Z, Holt JA, Pain D, Stocco DM, Miller WL, Strauss JF. Steroidogenic acute regulatory protein (StAR) retains activity in the absence of its mitochondrial import sequence: Implications for the mechanism of StAR action. Proc Natl Acad Sci 1996; 93: 13731-13736

74. Watari H, Arakane F, Moog-Lutz C, Kallen CB, Tomasetto C. Gerton GL, Rio M, Baker ME, Strauss III JF. MLN64 contains a domain with homology to the steroidogenic acute regulatory protein (StAR) that stimulates steroidogenesis. Proc Natl Acad Sci 1997; 94: 8462-8467

75. Bieche I, Tomasetto C, Regnier CH, Moog-Lutz C, Rio MC, Lidereau R. Two distinct amplified regions at 17q11-q21 involved in human primary breast cancer. Cancer Res 1996; 56: 3886-3890

76. Moog-Lutz C, Tomasetto C, Regnier CH, Wendling C, Lutz Y, Muller D, Chenard MP, Basset P, Rio MC. MLN64 exhibits homology with the steroidogenic acute regulatory protein (StAR) and is overexpressed in human breast carcinomas. Intl J Canc 1997; 71: 183-191

77. Arakane F, King SR, Du Y, Kallen CB, Walsh LP, Watari H, Stocco DM, Strauss III JF. Phosphorylation of steroidogenic acute regulatory protein (StAR) modulates its steroidogenic avtivity. J Biol Chem 1997; 272: 32656-32662

78. Cherradi N, Rossier MF, Vallotton MB, Capponi AM. Calcium stimulates intramitochondrial cholesterol transfer and StAR protein in bovine adrenal glomerulosa cells. J Biol Chem 1996; 271: 25971-25975

79. Cherradi N, Rossier MF, Vallotton MB, Timberg R, Friedberg I, Orly J, Wang XJ, Stocco DM, Capponi AM. (1997). Submitochondrial distribution of three key steroidogenic proteins (steroidogenic acute regulatory protein, P450 side-chain cleavage and 3β-hydroxysteroid dehydrogenase isomerase enzymes) upon stimulation by intracellular calcium in adrenal glomerulosa cells J Biol Chem 272, 7899-7907

80. Cherradi N, Brandenburger Y, Rossier MF, Vallotton MB, Stocco DM Capponi AM. Atrial natriuretic peptide inhibits calcium-induced steroidogenic acute regulatory protein gene transcription in adrenal glomerulosa cells. Molec Endocrinol 1998; 12: 962-972

81. Wang XJ, Liu Z, Eimerl S, Weiss AM, Orly J, Stocco DM. Effect of truncated forms of the steroidogenic acute regulatory (StAR) protein on intramitochondrial cholesterol transfer. Endocrinology (in press).

182

82.   Arakane F, Kallen CB, Watari H, Foster JA, Sepuri NBV, Pain D, Staybrook SE, Lewis M, Gerton GL, Strauss III JF. The mechanism of action of steroidogenic acute regulatory protein (StAR) StAR acts on the outside of mitochondria to stimulate steroidogenesis. J Biol Chem 1998; 273: 16339-16345

83.   Cherradi N, Defaye G, Chambaz EM. Characterization of the 3β-hydroxysteroid dehydrogenase activity associated with bovine adrenocortical mitochondria. Endocrinology 1994; 134: 1358-1364

MECHANISMS AND CONSEQUENCES OF CHOLESTEROL LOADING IN MACROPHAGES

Ira Tabas

Departments of Medicine and Anatomy & Cell Biology, Columbia University, New York, New York 10032; e-mail: iat1@columbia.edu

KEY WORDS: atherosclerosis, foam cells, lipoproteins, acyl-CoA:cholesterol acyltransferase

ABSTRACT

A prominent and functionally important event in atherogenesis is the accumulation of cholesterol by subendothelial macrophages, a process known as foam cell formation. Macrophage foam cells form by the internalization and metabolism of "atherogenic" lipoproteins that accumulate on the subendothelial matrix. Among the most potent inducers of foam cells are aggregated LDL and chylomicron remnants. Cholesterol molecules from these lipoproteins are esterified to fatty acids by the enzyme acyl-CoA:cholesterol acyltransferase (ACAT); the major form of regulation of this key reaction appears to be through changes in cholesterol trafficking to ACAT. Importantly, the accumulation of cholesterol has profound effects on macrophage biology, including changes in the synthesis of specific proteins and in phospholipid metabolism. These effects may give insight into the mechanisms by which macrophage foam cells affect atherogenesis.

CONTENTS

INTRODUCTION

ATHEROGENIC LIPOPROTEINS THOUGHT TO BE INVOLVED IN
MACROPHAGE FOAM CELL FORMATION
    General Considerations
    Modified Forms of LDL
    Aggregated Lipoproteins
    Chylomicron Remnants

INTRACELLULAR CHOLESTEROL TRAFFICKING AND METABOLISM
DURING MACROPHAGE FOAM CELL FORMATION
    Overall Pathways
    Intracellular Location of ACAT in Macrophages
    Regulation of the ACAT Pathway in Macrophages

EFFECTS OF CHOLESTEROL LOADING ON MACROPHAGE BIOLOGY
    Molecules Increased or Decreased by Cholesterol Loading of Macrophages
    Phospholipid Metabolism in Macrophages Loaded with Excess Free Cholesterol

CONCLUSIONS

INTRODUCTION

Atherosclerotic vascular disease, which is the major underlying etiology of myocardial infarction, stroke, peripheral vascular disease, and aortic aneurysms, is the leading cause of death in many areas of the world, including the United States (1). It is therefore imperative that we gain a thorough understanding of the basic cellular and molecular processes that are involved in the initiation and propagation of atherogenesis. In this light, results from a large number of studies suggest that the key initiating event in atherogenesis is the retention of "atherogenic" lipoproteins in the subendothelial matrix of "susceptible" regions of the arterial tree (2). Biological responses to the retained lipoproteins, which may be modified by oxidation and aggregation, then triggers a series of events that eventually lead to the development of the occlusive and potentially rupture-prone atheroma (2).

One the earliest and most important responses to retained lipoproteins is the entry of blood-borne monocytes into the subendothelial space (3-5). A likely mechanism is that retained and modified lipoproteins induce the overlying endothelium to secrete cytokines, such as monocyte chemotactic protein-1, and adhesion molecules, such as vascular cell adhesion molecule-1, that attract and cause the adhesion of monocytes (6). The monocytes then diapedese through the endothelial layer and differentiate into macrophages, perhaps after one or more rounds of cell division, under the influence of endothelial-derived factors, notably macrophage colony-stimulatory factor (M-CSF) (7). Over a period of time, a rather unique cellular event occurs in these macrophages, namely, the accumulation of very large amounts of intracellular cholesteryl fatty acid ester (CE) (3-5). The CE accumulates in cytoplasmic membrane-bound droplets that appear "foamy" when viewed by microscopy, giving rise to the name "foam cell".

Recent evidence suggests that macrophage foam cells play important roles both in atherogenesis (8) and in the clinical progression of atherosclerotic lesions (9). The study of foam cell biology, therefore, may lead to a clearer understanding of both of these processes. This chapter will review interactions between macrophages and atherogenic lipoproteins, the intracellular trafficking and metabolism of lipoprotein-derived cholesterol in macrophages, and consequences of cellular cholesterol loading on the biology and atherogenicity of macrophages. More detailed information on intracellular cholesterol trafficking and metabolism in other cell types as well as on cholesterol efflux from cells can be found in other chapters in this book. Finally, it should be noted that CE accumulation also occurs in smooth muscle cells, particularly in advanced atherosclerotic lesions. Little is known, however, about the mechanisms or consequences of smooth muscle cell foam cell formation, and therefore this chapter will focus on the macrophage foam cell.

ATHEROGENIC LIPOPROTEINS THOUGHT TO BE INVOLVED IN MACROPHAGE FOAM CELL FORMATION

General Considerations

Exogenous cholesterol, not endogenously synthesized cholesterol, is the source of cholesterol for foam cell formation. *De novo* synthesis of cholesterol functions to maintain cellular cholesterol levels in the absence of an exogenous source, and key enzymes in the cholesterol biosynthetic pathway are down-regulated when an exogenous source of cholesterol is present (10). Plasma lipoproteins are the major forms of exogenous cholesterol that lead to foam cell formation, although other sources of cholesterol, such as cholesterol-rich platelet fragments (11), have been shown to cause cholesterol loading of macrophages in cell culture. As described elsewhere in this book, lipoproteins are emulsion-like particles of varying densities that consist of a hydrophobic core containing the neutral lipids CE and triglycerides surrounded by an outer phospholipid monolayer containing a small amount of free cholesterol (FC) and protein (apolipoproteins). The bulk of the cholesterol (*e.g.*, ~80% in LDL) is carried as CE in the core of the particle.

Modified Forms of LDL

Which lipoproteins are responsible for macrophage foam cell formation? Although fasting plasma levels of LDL are strongly correlated with the extent of atherosclerosis, native plasma LDL usually, but not always (12), has been shown to be a poor inducer of foam cell formation in cultured macrophages (13,14). This finding can be explained by a combination of down-regulation of LDL receptors (13) and, as described below, specific characteristics of LDL-cholesterol metabolism in macrophages (15). How, therefore, can we account for the epidemiological data linking LDL to atherogenesis and thus to foam cell formation? One possible resolution to this apparent paradox is that LDL is modified to an "atherogenic" form in the subendothelium (16). As an *in-vitro* model to explore this concept, LDL modified by acetylation has been widely studied. Native LDL is recognized by the LDL receptor by virtue of specific receptor-binding domains of the apolipoprotein B-100 moiety of LDL (17). Acetyl-LDL, due to acetylation of key lysine residues on apolipoprotein B-100, is not recognized by the LDL receptor but rather by another class of receptors, the scavenger receptors (18,19). The class A scavenger receptors are widely expressed on differentiated macrophages and mediate the constitutive uptake of acetyl-LDL, leading to massive CE accumulation in most cultured macrophage models (13,19). Although acetylation of LDL does not occur *in vivo*, oxidation of LDL is a physiologically plausible process that renders LDL a ligand for scavenger receptors and other cell-surface receptors on macrophages that are not down-regulated by cellular cholesterol loading (20). Indeed, immunohistochemical studies have documented the presence of oxidized LDL in atherosclerotic lesions, and a wide variety of cell culture studies have demonstrated potentially atherogenic effects of oxidized LDL, such as induction of atherogenic endothelial cell responses and smooth muscle cell proliferation (20). A widely held misconception, however, is that LDL oxidized by most standard *in-vitro* methods induces foam cell formation (*i.e.*, marked CE accumulation) in cultured macrophages. In fact, quite a few studies have shown that while macrophages incubated with oxidized LDL internalize the lipoprotein extensively, the cells accumulate mostly FC (21-24). The mechanism may be related to reduced CE content of the lipoprotein as a result of oxidation (22) or to incomplete lysosomal degradation of oxidized LDL components, which in turn may lead to decreased export of FC from lysosomes (25-27). Additional *in-vitro* modifications of oxidized LDL, such as by FC enrichment (28), are able to convert this lipoprotein into one that can cause substantial CE accumulation in cultured macrophages, but it is still not certain whether these additional modifications of oxidized LDL occur *in vivo*. Recently, investigators have shown that class A scavenger receptor knockout mice have reduced, but by no means absent, foam cell lesions (29). Whereas proponents of the oxidation hypothesis might argue that the presence of foam cells in these knockout mice implicates a role for other oxidized LDL receptors in foam cell formation *in vivo*, an equally plausible interpretation is that a significant portion of foam cell formation involves other types of atherogenic lipoproteins altogether.

Aggregated Lipoproteins

In this light, aggregated lipoproteins, which are not necessarily recognized by class A scavenger receptors, have been shown to exist in both early and late atherosclerotic lesions and are very potent inducers of foam cell formation in cultured macrophages (30-33). Freeze-fracture images of the subendothelium (34) as well as analyses of LDL isolated from atherosclerotic lesions (35,36) indicate that a large proportion of lesional LDL is in the form of aggregates and fused particles averaging approximately 100 nm in diameter. Although the mechanism of LDL aggregation is not known, physiological plausible hypotheses include extensive oxidation (37), hydrolysis by mast cell-derived proteases (38), and hydrolysis by a macrophage- and endothelial-derived secretory sphingomyelinase

(39-42). Importantly, Frank and colleagues (34) have shown that subendothelial LDL aggregation occurs within two hours after intravenous injection of LDL into normocholesterolemic rabbits (*i.e.*, without any evidence of atherosclerosis), which is almost certainly too early for either extensive oxidation or for processes that depend upon cells (*e.g.*, mast cells) that enter the arterial wall only after lesion initiation. In any case, in contrast to the situation with oxidized LDL, cultured macrophages incubated with aggregated LDL formed by a variety of methods *in vitro* accumulate massive amounts of CE.

Chylomicron Remnants

Post-prandial chylomicron remnants are a class of lipoproteins other than LDL that are epidemiologically associated with atherosclerosis, exist in atherosclerotic lesions, and are potent inducers of foam cell formation in cultured macrophages (43-45). Chylomicrons are very buoyant, lipid-rich lipoproteins formed in intestinal epithelial cells during the absorption of dietary fat (45). After entry into the circulation, the triglyceride component of the core of chylomicrons are rapidly hydrolyzed by the enzyme lipoprotein lipase bound to the lumenal surface of the endothelium (45). The remaining particle, called a "chylomicron remnant", may either be rapidly cleared by the liver or remain in the circulation for an extended period of time, depending upon specific metabolic and genetic factors that differ among individuals (45). If the chylomicron remnants remain in the circulation, they can become enriched in CE, enter the arterial wall, and initiate lesion development. A widely studied model of CE-rich chylomicron remnants is a lipoprotein, called β-VLDL, found at very high levels in the plasma of cholesterol-fed rabbits (46). As mentioned above and discussed in detail below, both chylomicron remnants and β-VLDL lead to marked accumulation of CE when incubated with cultured macrophages (47).

INTRACELLULAR CHOLESTEROL TRAFFICKING AND METABOLISM DURING MACROPHAGE FOAM CELL FORMATION

Overall Pathways

In the simplest scenario, based upon experiments in which cultured macrophages are incubated with atherogenic lipoproteins dissolved in tissue culture medium, CE accumulation occurs as follows (48): the lipoproteins interact with specific cell-surface receptors, leading to endocytosis of the lipoproteins and delivery to lysosomes. Under certain circumstances, lipoprotein-CE may enter cells in the absence of whole particle internalization by a process known as "selective uptake" (49). Lipoprotein-CE is hydrolyzed in lysosomes by an acidic CE hydrolase (50), and this newly liberated FC, together with the small proportion of FC that was part of the original lipoprotein (above), is eventually delivered to the intracellular cholesterol esterifying enzyme, acyl-CoA:cholesterol acyltransferase (ACAT) (see chapter by Chang & Chang). The resulting newly esterified cholesterol is sequestered in membrane-bound inclusions in the cytoplasm. These CE stores can be rehydrolyzed to FC when cholesterol efflux is induced, such as by incubation of foam cells with a cholesterol acceptor like HDL (51). Rehydrolysis is catalyzed by a neutral CE hydrolase that, in the case of murine macrophages, appears to be cAMP-regulatable hormone-sensitive lipase (52,53).

Much of the trafficking of FC to ACAT in macrophages involves lysosome-to-plasma membrane transport followed by plasma membrane-to-ACAT transport (48), though recent evidence in fibroblasts and CHO cells suggests that a portion of lysosomal FC may be transported to ACAT without traveling through the plasma membrane (54,55). The exact mechanism of lysosome-to-plasma membrane transport is not known, but the newly cloned NPC1 protein appears to play a critical role, because mutations in this protein result in the accumulation of cholesterol in lysosomes (56,57) (see chapter by Neufeld & Pentchev). Plasma membrane-to-ACAT transport is an energy-dependent process that most likely involves vesicular trafficking (58,59). It is this second transport pathway, which we have called the "distal" pathway (58), that appears to be subject to regulation (see below).

Intracellular Location of ACAT in Macrophages

A key issue in our understanding of cholesterol delivery to ACAT is the precise intracellular location(s) of ACAT. Subcellular fractionation studies using rat liver suggested that a prominent site of ACAT is the rough endoplasmic reticulum (ER) (60), and a major part of the immunofluorescence pattern using an anti-N-terminal ACAT antibody in a melanoma cell line was a reticular pattern that overlapped with a fluorescent dye that is known to stain the ER (61). A recent immunofluorescence study in macrophages using another anti-N-terminal ACAT antibody also showed a prominent reticular pattern that partly overlapped with two proteins markers of the ER, protein disulfide isomerase (PDI) and ribophorin (62). ACAT staining was also seen, however, in a perinuclear region that did not co-localize with these two ER markers (62). Definitive identification of this non-overlapping site is pending. In addition, 10-20% of ACAT on macrophages in suspension, but not on macrophages attached to tissue culture dishes, was accessible to cell-surface biotinylation (62). Therefore, at least in macrophages, cholesterol trafficking to ACAT may involve two or more destinations.

Regulation of the ACAT Pathway in Macrophages

The regulation of the ACAT reaction represents another key issue in foam cell formation. The increase in ACAT activity seen when macrophages and other cell types are incubated with "atherogenic" lipoproteins does not involve transcriptional or translational induction of the ACAT enzyme (63). Rather, cholesterol transport to the enzyme, leading to post-translational enzymatic activation, appears to be the major form of regulation (63). ACAT is activated by newly delivered cholesterol by both allosteric mechanisms and by substrate provision (*i.e.*, under basal conditions, ACAT is not saturated with cholesterol) (64). Studies in macrophages have revealed several interesting regulatory aspects of cholesterol delivery to ACAT involving the plasma membrane-to-ACAT pathway (see above). First, cholesterol transport to ACAT from the plasma membrane appears to be triggered only after a threshold amount of FC has been delivered to the cell (65). According to this model, lipoprotein-derived FC is transported to the plasma membrane (see above) and mixes with the cholesterol in that site. When the cholesterol content of the plasma membrane reaches a critical level, a mixture of lipoprotein-derived and cellular cholesterol is transported to ACAT (65). This mechanism allows ACAT to optimally "sample" and thus regulate the overall state of cholesterol in the cell, especially in the plasma membrane, rather than simply monitoring cholesterol flux through the lysosomes. Whether cholesterol levels above threshold actually induce a vesicular transport pathway (see above) or simply raise the gradient of cholesterol in constitutively produced vesicles is not known. What is known is that the sphingomyelin content of the plasma membrane, probably by influencing the "capacity" of the membrane for FC, can affect the actual threshold level: increased sphingomyelin raises the threshold, and decreased sphingomyelin lowers it (66).

A second possible component of the regulation of the cholesterol esterification pathway in macrophages and also CHO cells was initially revealed in studies using inhibitors of protein synthesis. The basal level of ACAT-mediated esterification (*i.e.*, esterification in the absence of exogenous cholesterol) was found to be increased by treatment of these cells with cycloheximide or puromycin (67,68). One hypothesis to explain these data is that under basal conditions, cells possess a short-lived protein that functions to inhibit some step in the cholesterol esterification pathway. Since ACAT itself is constitutively expressed (see above), this mechanism may prevent cells from depleting essential FC in membranes. According to this model, the putative inhibitor would have to be destroyed or disabled when cells are faced with a significant FC load (*e.g.*, during lipoprotein uptake). Recent data provided some support for this idea by showing that treatment of cells with cysteine protease inhibitors, which were shown not to be direct inhibitors of the ACAT enzyme, inhibited lipoprotein-induced, but not sphingomyelinase- or hydroxysterol-induced, cholesterol esterification activity in macrophages (69). Since only lipoprotein-induced ACAT activity was affected, the protease inhibitors may specifically influence a step related to lipoprotein-induced cholesterol transport to ACAT (see above). Definitive proof of the "protein inhibitor" hypothesis must await the isolation of one or more proteins that have inhibitory effects on the cholesterol esterification pathway.

A third potential area of regulation of cholesterol esterification in macrophages is related to the way these cells initially interact with atherogenic lipoproteins. For example, LDL and β-VLDL (above) have widely different potencies in stimulating cholesterol esterification in macrophages even under conditions of similar cholesterol internalization (15,47). Immunofluorescence and immunogold electron microscopy studies revealed that although both of these lipoproteins initially bound to cell-surface LDL receptors, they were internalized by different pathways: LDL was rapidly internalized and degraded in lysosomes, whereas a portion of β-VLDL resided on the surface of macrophages for a prolonged period of time in deep, wide invaginations called STEMs (surface tubules for entry in to macrophages) (15,70,71). Importantly, alterations in β-VLDL that caused the particles to be taken up more rapidly (*i.e.*, as occurs in the LDL pathway) resulted in a decrease in the stimulation of cholesterol esterification (70). Furthermore, acetyl-LDL and aggregated LDL (above), which are also potent inducers of cholesterol esterification, were found to engage in prolonged association with surface invaginations of both mouse and human macrophages (72-74). The mechanism whereby prolonged interaction with the cell surface leads to an increased stimulation of the cholesterol esterification pathway needs to be defined at a molecular level. Nonetheless, this issue may be quite relevant to foam cell formation in the subendothelium of developing lesions *in vivo*, where most of the lipoproteins are not free in solution but rather are very tightly bound to extracellular matrix (75), often in an aggregated form (34-36). In contrast to typical cell-culture experiments in which internalization of monomeric lipoprotein particles in solution predominates, developing foam cells *in vivo* may be expected to engage in prolonged cell-surface contact with avidly retained lipoproteins.

EFFECTS OF CHOLESTEROL LOADING ON MACROPHAGE BIOLOGY

Molecules Increased or Decreased by Cholesterol Loading of Macrophages

The above sections have emphasized mechanisms of foam cell formation. Another important area of foam cell biology is the consequences of cholesterol loading on the physiologic and molecular events in these cells. In addition to the well-studied regulatory effects of cholesterol in many cell types, such as down-regulation of the LDL receptor and cholesterol biosynthetic enzymes (17), studies with macrophages have revealed novel effects that are only beginning to be understood at the molecular level. Some of these include increased synthesis and secretion of apolipoprotein E, which may be important in cellular cholesterol efflux (76); increased secretion of products of the lipoxygenase pathway, which may influence smooth muscle cell migration during atherogenesis (77); increased synthesis of α-enolase, an enzyme that inhibits neutral CE hydrolysis (78); increased expression of a receptor for an atherogenic lipoprotein called lipoprotein(a) (79); increased expression of plasminogen activator, which promotes plasmin-mediated processes such as degradation of fibrin and activation of transforming growth factor-β and matrix metalloproteinases (80) increased expression and secretion of interleukin-8, a chemokine that appears to promote monocyte migration into developing atherosclerotic lesions (81); increased expression of a 150-kD protein called vigilin, whose function is yet to be elucidated (82,83); and decreased expression of fucosyltransferases, which may result in decreased expression of cell-surface selectin-binding molecules (84).

Phospholipid Metabolism in Macrophages Loaded with Excess Free Cholesterol

Consequences of *excessive* FC loading of macrophages may be particularly relevant to the foam cell of advanced atherosclerotic lesions, because macrophages in advanced atherosclerotic lesions have been noted to accumulate large amounts of FC (85-89). Although the mechanism is not known, one might speculate that defects in cholesterol transport to ACAT, in the ACAT enzyme itself, or in cholesterol efflux might contribute to this event, as might excessive neutral CE hydrolase activity. Whatever the mechanism, it is well established that excess FC in cellular membranes, particularly the plasma membrane, leads to membrane protein dysfunction and cytotoxicity (90). One possible defense mechanism might be an increase in phospholipid biosynthesis, which could serve to "buffer" or sequester the excess FC (91). Indeed, advanced lesional foam cells have been observed to have intracellular membrane whorls and increased phospholipid biosynthesis (91).

In support of this idea, studies with cultured macrophages have revealed that FC loading, such as by incubation with acetyl-LDL plus an ACAT inhibitor, directly leads to an increase in PC biosynthesis (92). The mechanism involves the post-translational activation of CTP:phosphocholine cytidylyltransferase (CT), a rate-limiting enzyme in the PC biosynthetic pathway (92,93) Protein dephosphorylation, perhaps of CT itself, appears necessary for this FC-mediated activation process (93). Interestingly, agents that block FC exit from lysosomes do not inhibit the activation of CT (93), suggesting that the initial signaling event may originate in the lysosome. In this light, much of the FC that accumulates in advanced lesional foam cells is lysosomal (88,89). Further cell-culture studies demonstrated that blunting of the PC biosynthesis response in FC-loaded macrophages leads to rapid cellular necrosis (94), consistent with the idea that the induction of CT and PC biosynthesis by FC loading is adaptive. Macrophage foam cell necrosis, leading to the development of the lipid, or necrotic, core, is an important event in advanced atherosclerotic lesions because it may promote plaque rupture and acute clinical events (90). Whether or not FC-mediated cytotoxicity, perhaps related to an eventual blunting of the above-described PC response *in vivo*, plays a role in the development of the necrotic core remains to be investigated.

CONCLUSIONS

The macrophage foam cell is a prominent and important component of the atherosclerotic lesion, playing roles in both lesion initiation and lesion progression. Foam cell biology as it pertains to both of these roles can be understood only by a thorough analysis of how macrophages interact with and internalize atherogenic lipoproteins and how they metabolize lipoprotein-derived cholesterol. Two key points regarding future research should be emphasized. First, much of our current knowledge of these foam cell processes come from studying cultured macrophages, often permanent cell lines, interacting with monomeric lipoproteins dissolved in tissue culture medium. In fact, macrophage subtypes are known to differ in important ways, and lesional macrophages may possess important differences from those studied in the laboratory. Likewise, the form of lipoprotein that interacts with macrophages in lesions is almost certainly different from those investigated in most cell culture studies. Therefore, studies examining the interaction of macrophages with lipoproteins retained and aggregated on three-dimensional extracellular matrices will be an important goal of future research.

Second, the oft-stated conclusion that foam cells promote atherogenesis may require careful examination. Clearly, macrophage foam cells can secrete molecules, such as oxidants, growth factors, inflammatory cytokines, and metalloproteinases, that may promote lesion development and plaque breakdown. But the ability of macrophages to scavenge potentially harmful molecules, including oxidized lipids, may be beneficial, such is often the case in other types of inflammatory and infectious lesions. Thus, as we identify specific molecules related to foam cell biology, and as we increasingly use *in-vivo* systems, such as transgenic and knockout mice, to study these molecules, experimental strategies must specifically address this critical issue. Only through such studies will we be able to use our knowledge of foam cell biology to rationally design anti-atherogenic therapeutic interventions.

REFERENCES

1. Braunwald, E. 1997. Cardiovascular medicine at the turn of the millennium: triumphs, concerns, and opportunities. *N. Engl. J. Med.* 337:1360-1369.
2. Williams, K. J. and I. Tabas. 1995. The response-to-retention hypothesis of early atherogenesis. *Arterioscler. Thromb. Vasc. Biol.* 15:551-561.
3. Schaffner, T., K. Taylor, E. Bartucci, K. Fischer-Dzoga, J. Beeson, S. Glagov, and R Wissler. 1980. Arterial foam cells with distinctive immunomorphologic and histochemical features of macrophages. *Am. J. Pathol.* 100:57-73.
4. Gerrity, R. G. 1991. The role of the monocyte in atherogenesis. I. Transition of blood-borne monocytes into foam cells in fatty lesions. *Am. J. Pathol.* 103:181-190.
5. Faggioto, A., R. Ross, and L. Harker. 1984. Studies of hypercholesterolemia in the nonhuman primate. I. Changes that lead to fatty streak formation. *Arteriosclerosis* 4:323-340.
6. Ross, R. 1995. Cell biology of atherosclerosis. *Annual Review of Physiology* 57:791-804.
7. Antonov, A. S., D. H. Munn, F. D. Kolodgie, R. Virmani, and R. G. Gerrity. 1997. Aortic endothelial cells regulate proliferation of human monocytes in vitro via a mechanism synergistic with macrophage colony-stimulating factor. Convergence at the cyclin E/p27(Kip1) regulatory checkpoint. *J. Clin. Invest.* 99:2867-2876.
8. Smith, J. D., E. Trogan, M. Ginsberg, C. Grigaux, J. Tian, and M. Miyata. 1995. Decreased atherosclerosis in mice deficient in both macrophage colony-stimulating factor (*op*) and apolipoprotein E. *Proc. Natl. Acad. Sci. USA* 92:8264-8268.
9. Libby, P. and S. K. Clinton. 1993. The role of macrophages in atherogenesis. *Curr. Opin. Lipidol.* 4:355-363.
10. Goldstein, J. L. and M. S. Brown. 1990. Regulation of the mevalonate pathway. *Nature* 343:425-430.
11. Curtiss, L. K., A. S. Black, Y. Takagi, and E. F. Plow. 1987. New mechanism for foam cell generation in atherosclerotic lesions. *J. Clin. Invest.* 80:367-373.
12. Tabas, I., D. A. Weiland, and A. R. Tall. 1985. Unmodified low density lipoprotein causes cholesteryl ester accumulation in J774 macrophages. *Proc. Natl. Acad. Sci. USA* 82:416-420.
13. Brown, M. S. and J. L. Goldstein. 1983. Lipoprotein metabolism in the macrophage: Implications for cholesterol deposition in atherosclerosis. *Annu. Rev. Biochem.* 52:223-261.
14. Steinberg, D., S. Parthasarathy, T. E. Carew, J. C. Khoo, and J. L. Witztum. 1989. Beyond cholesterol: modifications of low-density lipoprotein that increase its atherogenicity. *N. Engl. J. Med.* 320:915-924.
15. Tabas, I., S. Lim, X. Xu, and F. R. Maxfield. 1990. Endocytosed β-VLDL and LDL are delivered to different intracellular vesicles in mouse peritoneal macrophages. *J. Cell Biol.* 111:929-940
16. Tabas, I., G. C. Boykow, and A. R. Tall. 1987. Foam cell-forming J774 macrophages have markedly elevated acyl coenzyme A:cholesterol acyl transferase activity compared with mouse peritoneal

macrophages in the presence of low density lipoproteins (LDL) despite similar LDL receptor activity. *J. Clin. Invest.* 79:418-426.

17. Brown, M. S. and J. L. Goldstein. 1986. A receptor-mediated pathway for cholesterol homeostasis. *Science* 232:32-47.

18. Goldstein, J. L., Y. K. Ho, S. K. Basu, and M. S. Brown. 1979. Binding site on macrophages that mediates uptake and degradation of acetylated low density lipoprotein producing massive cholesterol deposition. *Proc. Natl. Acad. Sci. USA* 76:333-337.

19. Krieger, M. and J. Herz. 1994. Structures and functions of multiligand lipoprotein receptors: macrophage scavenger receptors and LDL receptor-related protein (LRP). *Annu. Rev. Biochem.* 63:601-637.

20. Steinberg, D. 1997. Low density lipoprotein oxidation and its pathobiological significance. *J. Biol. Chem.* 272:20963-20966.

21. Roma, P., A. L. Catapano, S. M. Bertulli, L. Varesi, R. Fumagalli, and F. Bernini. 1990. Oxidized LDL increase free cholesterol and fail to stimulate cholesterol esterification in murine macrophages. *Biochem. Biophys. Research Comm.* 171:123-131.

22. Ryu, B. H., F. W. Mao, P. Lou, R. L. Gutman, and P. Greenspan. 1995. Cholesteryl ester accumulation in macrophages treated with oxidized low density lipoprotein. *Biosci. Biotech. Biochem.* 59:1619-1622.

23. Klinkner, A. M., C. R. Waites, W. D. Kerns, and P. J. Bugelski. 1995. Evidence of foam cell and cholesterol crystal formation in macrophages incubated with oxidized LDL by fluorescence and electron microscopy. *J. Histochem. Cytochem.* 43:1071-1078.

24. Musanti, R. and G. Ghiselli. 1993. Interaction of oxidized HDLs with J774-A1 macrophages causes intracellular accumulation of unesterified cholesterol. *Arterioscler.Thromb.* 13:1334-1345.

25. Lougheed, M., H. F. Zhang, and U. P. Steinbrecher. 1991. Oxidized low density lipoprotein is resistant to cathepsins and accumulates within macrophages. *J. Biol. Chem.* 266:14519-14525.

26. Hoppe, G., J. O'Neill, and H. F. Hoff. 1994. Inactivation of lysosomal proteases by oxidized low density lipoprotein is partially responsible for its poor degradation by mouse peritoneal macrophages. *J. Clin. Invest.* 94:1506-1512.

27. Maor, I., H. Mandel, and M. Aviram. 1995. Macrophage uptake of oxidized LDL inhibits lysosomal sphingomyelinase, thus causing the accumulation of unesterified cholesterol-sphingomyelin-rich particles in the lysosomes. A possible role for 7-ketocholesterol. *Arterioscler. Thromb. Vasc. Biol.* 15:1378-1387.

28. Greenspan, P., H. Yu, F. Mao, and R. L. Gutman. 1997. Cholesterol deposition in macrophages: foam cell formation mediated by cholesterol-enriched oxidized low density lipoprotein. *J. Lip. l Res.* 38:101-109.

29. Suzuki, H., Y. Kurihara, M. Takeya, N. Kamada, M. Kataoka, K. Jishage, O. Ueda, H. Sakaguchi, T. Higashi, T. Suzuki, Y. Takashima, Y. Kawabe, O. Cynshi, Y. Wada, M. Honda, H. Kurihara, H. Aburatani, T. Doi, A. Matsumoto, S. Azuma, T. Noda, Y. Toyoda, H. Itakura, Y. Yazaki, T. Kodama, and et al. 1997. A role for macrophage scavenger receptors in atherosclerosis and susceptibility to infection. *Nature* 386:292-296.

30. Hoff, H. F., J. O'Neill, J. M. Pepin, and T. B. Cole. 1990. Macrophage uptake of cholesterol-containing particles derived from LDL and isolated from atherosclerotic lesions. *Eur. Heart J* 11:105-115.

31. Khoo, J. C., E. Miller, P. McLoughlin, and D. Steinberg. 1988. Enhanced macrophage uptake of low density lipoprotein after self-aggregation. *Arteriosclerosis* 8:348-358.

32. Suits, A. G., A. Chait, M. Aviram, and J. W. Heinecke. 1989. Phagocytosis of aggregated lipoprotein by macrophages: low density lipoprotein receptor-dependent foam-cell formation. *Proc. Natl. Acad. Sci. USA* 86:2713-2717.

33. Xu, X. and I. Tabas. 1991. Sphingomyelinase enhances low density lipoprotein uptake and ability to induce cholesteryl ester accumulation in macrophages. *J. Biol. Chem.* 266:24849-24858.

34. Nievelstein, P. F. E. M., A. M. Fogelman, G. Mottino, and J. S. Frank. 1991. Lipid accumulation in rabbit aortic intima 2 hours after bolus infusion of low density lipoprotein. *Arteriosclerosis and Thrombosis* 11:1795-1805.

35. Hoff, H. F. and R. E. Morton. 1985. Lipoproteins containing apo B extracted from human aortas: structure and function. *Ann. N. Y. Acad. Sci.* 454:183-194.

36. Guyton, J. R. and K. F. Klemp. 1996. Development of the lipid-rich core in human atherosclerosis. *Arterioscler. Thromb. Vasc. Biol.* 16:4-11.

37. Hoff, H. F., J. O'Neill, G. M. Chisolm,III, T. B. Cole, O. Quehenberger, H. Esterbauer, and G. Jürgens. 1989. Modification of low density lipoprotein with 4-hydroxynonenal induces uptake by macrophages. *Arteriosclerosis* 9:538-549.

38. Kovanen, P. T. 1991. Mast cell granule-mediated uptake of low density lipoproteins by macrophages: a novel mechanism leading to the formation of foam cells. *Ann. Med.* 23:551-559.

194

39. Schissel, S. L., J. Tweedie-Hardman, J. H. Rapp, G. Graham, K. J. Williams, and I. Tabas. 1996. Rabbit aorta and human atherosclerotic lesions hydolyze the sphingomyelin of retained low-density lipoprotein. Proposed role for arterial-wall sphingomyelinase in subendothelial retention and aggregation of atherogenic lipoproteins. *J. Clin. Invest.* 98:1455-1464.

40. Schissel, S. L., E. H. Schuchman, K. J. Williams, and I. Tabas. 1996. Zn^{2+}-stimulated sphingomyelinase is secreted by many cell types and is a product of the acid sphingomyelinase gene. *J. Biol. Chem.* 271:18431-18436.

41. Schissel, S. L., X. C. Jiang, J. Tweedie-Hardman, T. S. Jeong, E. H. Camejo, J. Najib, J. H. Rapp, K. J. Williams, and I. Tabas. 1998. Secretory sphingomyelinase, a product of the acid sphingomyelinase gene, can hydrolyze atherogenic lipoproteins at neutral pH. Implications for atherosclerotic lesion development. *J. Biol. Chem.* 273:2738-2746.

42. Marathe, S., S. L. Schissel, M. J. Yellin, N. Beatini, R. Mintzer, K. J. Williams, and I. Tabas. 1998. Human vascular endothelial cells are a rich and regulatable source of secretory sphingomyelinase. Implications for early atherogenesis and ceramide-mediated cell signaling. *J. Biol. Chem.* 273:4081-4088.

43. Zilversmit, D. B. 1979. Atherogenesis: a postprandial phenomenon. *Circulation* 60:473-485.

44. Goldstein, J. L., Y. K. Ho, M. S. Brown, T. L. Innerarity, and R. W. Mahley. 1980. Cholesteryl ester accumulation in macrophages resulting from receptor-mediated uptake and degradation of hypercholesterolemic canine β-very low density lipoproteins. *J. Biol. Chem.* 255:1839-1848.

45. Havel, R. J. 1995. Chylomicron remnants: hepatic receptors and metabolism. *Curr. Opin. Lipidol.* 6:312-316.

46. Salzman, N. H. and F. R. Maxfield. 1989. Fusion accessibility of endocytic compartments along the recycling and lysosomal endocytic pathways in intact cells. *J. Cell Biol.* 109:2097-2104.

47. Mahley, R. W., T. L. Innerarity, M. S. Brown, Y. K. Ho, and J. L. Goldstein. 1980. Cholesterol ester synthesis in macrophages: stimulation by β-very low density lipoproteins from cholesterol-fed animals of several species. *J. Lipid Res.* 21:970-980.

48. Tabas, I. 1995. The stimulation of the cholesterol esterification pathway by atherogenic lipoproteins in macrophages. *Curr. Opin. Lipidol.* 6:260-268.

49. Rinninger, F., M. Brundert, S. Jackle, T. Kaiser, and H. Greten. 1995. Selective uptake of low-density lipoprotein-associated cholesteryl esters by human fibroblasts, human HepG2 hepatoma cells and J774 macrophages in culture. *Biochim. Biophys. Acta* 1255:141-153.

50. Anderson, R. A. and G. N. Sando. 1991. Cloning and expression of cDNA encoding human lysosomal acid lipase/cholesteryl ester hydrolase. Similarities to gastric and lingual lipases. *J. Biol. Chem.* 266:22479-22484.

51. Brown, M. S., Y. K. Ho, and J. L. Goldstein. 1980. The cholesteryl ester cycle in macrophage foam cells: Continual hydrolysis and re-esterification of cytoplasmic cholesteryl esters. *J. Biol. Chem.* 255:9344-9352.

52. Small, C. A., J. A. Goodacre, and S. J. Yeaman. 1989. Hormone-sensitive lipase is responsible for the neutral cholesterol ester hydrolase activity in macrophages. *FEBS Letters* 247:205-208.

53. Khoo, J. C., K. Reue, D. Steinberg, and M. C. Schotz. 1993. Expression of hormone-sensitive lipase mRNA in macrophages. *J. Lipid Res.* 34:1969-1974.

54. Neufeld, E. B., A. M. Cooney, J. Pitha, E. A. Dawidowicz, N. K. Dwyer, P. G. Pentchev, and E. J. Blanchette-Mackie. 1996. Intracellular trafficking of cholesterol monitored with a cyclodextran. *J. Biol. Chem.* 271:21604-21613.

55. Underwood, K. W., N. L. Jacobs, A. Howley, and L. Liscum. 1998. Evidence for a cholesterol transport pathway from lysosomes to the endoplasmic reticulum that is independent of the plasma membrane. *J. Biol. Chem.* 273:4266-4274.

56. Carstea, E. D., J. A. Morris, K. G. Coleman, S. K. Loftus, D. Zhang, C. Cummings, J. Gu, M. A. Rosenfeld, W. J. Pavan, D. B. Krizman, J. Nagle, M. H. Polymeropoulos, S. L. Sturley, Y. A. Ioannou, M. E. Higgins, M. Comly, A. Cooney, A. Brown, C. R. Kaneski, E. J. Blanchette-Mackie, N. K. Dwyer, E. B. Neufeld, T. Y. Chang, L. Liscum, D. A. Tagle, and et al. 1997. Niemann-Pick C1 disease gene: homology to mediators of cholesterol homeostasis. *Science* 277:228-231.

57. Loftus, S. K., J. A. Morris, E. D. Carstea, J. Z. Gu, C. Cummings, A. Brown, J. Ellison, K. Ohno, M. A. Rosenfeld, D. A. Tagle, P. G. Pentchev, and W. J. Pavan. 1997. Murine model of Niemann-Pick C disease: mutation in a cholesterol homeostasis gene. *Science* 277:232-235.

58. Skiba, P. J., X. Zha, S. L. Schissel, F. R. Maxfield, and I. Tabas. 1996. The distal pathway of lipoprotein-induced cholesterol esterification, but not sphingomyelinase-induced cholesterol esterification, is energy-dependent. *J. Biol. Chem.* 271:13392-13400.

59. Lange, Y., J. Ye, and F. Strebel. 1995. Movement of 25-hydroxycholesterol from the plasma membrane to the rough endoplasmic reticulum in cultured hepatoma cells. *J. Lipid Res.* 36:1092-1097.

60. Balasubramanian, S., S. Venkatesen, K. A. Mitropoulos, and T. J. Peters. 1978. The submicrosomal localization of acyl-coenzyme A:cholesterol acyltransferase and its substrate and of cholesteryl esters in rat liver. *Biochem. J.* 174:863-872.

61. Chang, C. C. Y., J. Chen, M. A. Thomas, D. Cheng, V. A. Del Priore, R. S. Newton, M. E. Pape, and T. Chang. 1995. Regulation and immunolocalization of acyl-coenzyme A:cholesterol acyltransferase in mammalian cells as studied with specific antibodies. *J. Biol. Chem.* 270:29532-29540.

62. Khelef, N., X. Buton, N. Beatini, H. Wang, V. Meiner, T. Y. Chang, R. V. Farese, Jr., F. R. Maxfield, and I. Tabas. 1998. Immunolocalization of ACAT in macrophages. *J. Biol. Chem.*

63. Chang, T. Y., C. C. Y. Chang, and D. Cheng. 1997. Acyl-coenzyme A:cholesterol acyltransferase. *Annu. Rev. Biochem.* 66:613-638.

64. Cheng, D., C. C. Y. Chang, X. Qu, and T. Chang. 1995. Activation of acyl-coenzyme A:cholesterol acyltransferase by cholesterol or by oxysterol in a cell-free system. *J. Biol. Chem.* 270:685-695.

65. Xu, X. and I. Tabas. 1991. Lipoproteins activate acyl coenzyme A:cholesterol acyl transferase in macrophages only after cellular cholesterol pools are expanded to a critical threshold level. *J. Biol. Chem.* 266:17040-17048.

66. Okwu, A. K., X. Xu, Y. Shiratori, and I. Tabas. 1994. Regulation of the threshold for lipoprotein-induced acyl-CoA:cholesterol *O*-acyltransferase stimulation in macrophages by cellular sphingomyelin content. *J. Lipid Res.* 35:644-655.

67. Chang, C. C. Y., G. M. Doolittle, and T. Y. Chang. 1986. Cycloheximide sensitivity in regulation of acyl coenzyme A: cholesterol acyltransferase activity in Chinese hamster ovary cells. I. Effect of exogenous sterols. *Biochemistry* 25:1693-1699.

68. Tabas, I. and G. C. Boykow. 1987. Protein synthesis inhibition in mouse peritoneal macrophages results in increased acyl coenzyme A:cholesterol acyl transferase activity and cholesteryl ester accumulation in the presence of native low density lipoprotein. *J. Biol. Chem.* 262:12175-12181.

69. Schissel, S. L., N. Beatini, X. Zha, F. R. Maxfield, and I. Tabas. 1995. Effect and cellular site of action of cysteine protease inhibitors on the cholesterol esterification pathway in macrophages and Chinese hamster ovary cells. *Biochemistry* 34:10463-10473.

70. Tabas, I., J. N. Myers, T. L. Innerarity, X. Xu, K. Arnold, J. Boyles, and F. R. Maxfield. 1991. The influence of particle size and multiple apoprotein E-receptor interactions on the endocytic targeting of β-VLDL in mouse peritoneal macrophages. *J. Cell Biol.* 115:1547-1560.

71. Myers, J. N., I. Tabas, N. L. Jones, and F. R. Maxfield. 1993. β-VLDL is sequestered in surface-connected tubules in mouse peritoneal macrophages. *J. Cell Biol.* 123:1389-1402.

72. Zha, X., I. Tabas, P. L. Leopold, N. L. Jones, and F. R. Maxfield. 1996. Evidence for prolonged cell-surface contact of acetyl-LDL before entry into macrophages. *Arterioscler. Thromb. Vasc. Biol.* 90:1421-1431.

73. Kruth, H. S., S. I. Skarlatos, K. Lilly, J. Chang, and I. Irim. 1995. Sequestration of acetylated LDL and cholesterol crystals by human monocyte-derived macrophages. *J. Cell Biol.* 129:133-145.

74. Zhang, W., P. M. Gaynor, and H. S. Kruth. 1997. Aggregated low density lipoprotein induces and enters surface-connected compartments of human monocyte-macrophages. Uptake occurs independently of the low density lipoprotein receptor. *J. Biol. Chem.* 272:31700-31706.

75. Smith, E. B., I. B. Massie, and K. M. Alexander. 1976. The release of an immobilized lipoprotein fraction from atherosclerotic lesions by incubation with plasmin. *Atherosclerosis* 25:71-84.

76. Basu, S. K., Y. K. Ho, M. S. Brown, D. W. Bilheimer, R. G. W. Anderson, and J. L. Goldstein. 1982. Biochemical and genetic studies of the apolipoprotein E secreted by mouse macrophages and human monocytes. *J. Biol. Chem.* 257:9788-9795.

77. Mathur, S. N. and F. J. Field. 1987. Effect of cholesterol enrichment on 12-hydroxyeicosatetraenoic acid metabolism by mouse peritoneal macrophages. *J. Lipid Res.* 28:1166-1176.

78. Shand, J. H. and D. W. West. 1995. Inhibition of neutral cholesteryl ester hydrolase by the glycolytic enzyme enolase. Is this a secondary function of enolase? *Lipids* 30:763-770.

79. Bottalico, L. A., G. A. Keesler, G. M. Fless, and I. Tabas. 1993. Cholesterol loading of macrophages leads to marked enhancement of native lipoprotein(a) and apoprotein(a) internalization and degradation. *J. Biol. Chem.* 268:8569-8573.

80. Falcone, D. J., T. A. McCaffrey, A. Haimovitz-Friedman, J. Vergilio, and A. C. Nicholson. 1993. Macrophage and foam cell release of matrix-bound growth factors. Role of plasminogen activation. *J. Biol. Chem.* 268:11951-11958.

81. Wang, N., I. Tabas, R. Winchester, S. Ravalli, L. E. Rabbani, and A. Tall. 1996. Interleukin-8 is induced by cholesterol loading of macrophages and expressed by macrophage foam cells in human atheroma. *J. Biol. Chem.* 271:8837-8842.

82. McKnight, G. L., J. Reasoner, T. Gilbert, K. O. Sundquist, B. Hokland, P. A. McKernan, J. Champagne, C. J. Johnson, M. C. Bailey, R. Holly, and et al. 1992. Cloning and expression of a cellular high density lipoprotein-binding protein that is up-regulated by cholesterol loading of cells. *J. Biol. Chem.* 267:12131-12141.

83. Schmidt, C., B. Henkel, E. Poschl, H. Zorbas, W. G. Purschke, T. R. Gloe, and P. K. Muller. 1992. Complete cDNA sequence of chicken vigilin, a novel protein with amplified and evolutionary conserved domains. *Eur. J. Biochem.* 206:625-634.

84. Cullen, P., S. Mohr, B. Brennhausen, A. Cignarella, and G. Assmann. 1997. Downregulation of the selectin ligand-producing fucosyltransferases Fuc-TIV and Fuc-TVII during foam cell formation in monocyte-derived macrophages. *Arterioscler. Thromb. Vasc. Biol.* 17:1591-1598.

85. Lundberg, B. 1985. Chemical composition and physical state of lipid deposits in atherosclerosis. *Atherosclerosis* 56:93-110.

86. Small, D. M., M. G. Bond, D. Waugh, M. Prack, and J. K. Sawyer. 1984. Physicochemical and histological changes in the arterial wall of nonhuman primates during progression and regression of atherosclerosis. *J. Clin. Invest.* 73:1590-1605.

87. Rapp, J. H., W. E. Connor, D. S. Lin, T. Inahara, and J. M. Porter. 1983. Lipids of human atherosclerotic plaques and xanthomas: clues to the mechanism of plaque progression. *J. Lipid Res.* 24:1329-1335.

88. Shio, H., N. J. Haley, and S. Fowler. 1979. Characterization of lipid-laden aortic cells from cholesterol-fed rabbits. III. Intracellular localization of cholesterol and cholesteryl ester. *Lab. Invest.* 41:160-167.

89. Jerome, W. G. and J. C. Lewis. 1985. Early atherogenesis in White Carneau pigeons. II. Ultrastructural and cytochemical observations. *Am. J. Pathol.* 119:210-222.

90. Tabas, I. 1997. Free cholesterol-induced cytotoxicity. A possible contributing factor to macrophage foam cell necrosis in advanced atherosclerotic lesions. *Trends Cardiovasc. Med.* 7:256-263.

91. Tabas, I. 1997. Phospholipid metabolism in cholesterol-loaded macrophages. *Curr. Opin. Lipidol.* 8:263-267.

92. Shiratori, Y., A. K. Okwu, and I. Tabas. 1994. Free cholesterol loading of macrophages stimulates phosphatidylcholine biosynthesis and up-regulation of CTP:phosphocholine cytidylyltransferase. *J. Biol. Chem.* 269:11337-11348.

93. Shiratori, Y., M. Houweling, X. Zha, and I. Tabas. 1995. Stimulation of CTP:phosphocholine cytidylytransferase by free cholesterol loading of macrophages involves signalling through protein dephosphorylation. *J. Biol. Chem.* 270:29894-29903.

94. Tabas, I., S. Marathe, G. A. Keesler, N. Beatini, and Y. Shiratori. 1996. Evidence that the initial up-regulation of phosphatidylcholine biosynthesis in free cholesterol-loaded macrophages is an adaptive response that prevents cholesterol-induced cellular necrosis. Proposed role of an eventual failure of this response in foam cell necrosis in advanced atherosclerosis. *J. Biol. Chem.* 271:22773-22781.

DIRECT EVIDENCE FOR STEROL CARRIER PROTEIN-2 (SCP-2) PARTICIPATION IN ACTH STIMULATED STEROIDOGENESIS IN ISOLATED ADRENAL CELLS

R.F. Chanderbhan[1], A.T. Kharroubi[1], A.P. Pastuszyn[2], L.L. Gallo[1] and T.J. Scallen[2]

[1]Department of Biochemistry, The George Washington University, Washington, DC 20037 and [2]Department of Biochemistry, Health Sciences Center, University of New Mexico, Albuquerque, NM 87131

KEY WORDS: sterol carrier protein-2 (SCP-2), intracellular cholesterol transport, regulation of steroidogenesis

----

ABSTRACT

Intact, dispersed adrenal fasiculata cells were fused with liposomal entrapped anti-sterol carrier protein-2 IgG, washed and subsequently exposed to adrenocorticotropic hormone (ACTH). The steroidogenic response (measured as corticosterone production) of these cells was inhibited by 45-60%, compared to cells fused with liposomally entrapped non-immune IgG or buffer. Furthermore, the degree of inhibition was shown to be dependent on the amount of antibody utilized. Fusion of cells with liposomally entrapped antibody to fatty acid binding protein (FABP) had no effect on ACTH-induced steroidogenesis. The incorporation of liposomal SCP-2 into adrenal fasiculata cells pre-treated with affinity purified anti-

SCP-2 IgG resulted in a concentration dependent release of the inhibition of ACTH-induced steroidogenesis caused by the antibody. It was also demonstrated indirectly that the fusion of liposomal anti-SCP-2 IgG had no effect on ACTH binding to adrenal cells. Finally, cells treated with liposomal anti-SCP-2 IgG and subsequently exposed to ACTH in the presence of aminoglutethimide, accumulated unesterified cholesterol in their cytoplasmic lipid inclusion droplets. These results, taken together, establish that an important physiological function for SCP-2 in adrenal cells is the transfer of unesterified cholesterol from the cytoplasmic lipid inclusion droplets to mitochondria. This translocation is generally considered to be the rate-limiting step in steroid hormone biosynthesis.

CONTENTS

INTRODUCTION

The adrenal cortex of many species contains esterified cholesterol in high concentrations (1-3). The cholesterol esters are found mainly in cytoplasmic lipid inclusion droplets (1, 4, 5). During stress or elevations of circulating ACTH, these esters are hydrolyzed by a "hormone sensitive" neutral sterol ester hydrolase (6, 7) and provide a major source of the cholesterol substrate required for steroidogenesis (8). The translocation of this cholesterol from the lipid droplets to the cytochrome $P450_{scc}$ located on the matrix face of the inner mitochondrial membrane is generally considered the rate-limiting step in steroid hormone biosynthesis (9, 10).

Until recently little was known of the mechanism(s) by which this cholesterol is transferred from the lipid droplets to the mitochondrial site of cholesterol side chain cleavage ($P450_{scc}$). We have reported that pure rat liver sterol carrier protein-2 (SCP-2) enhanced the movement of cholesterol from adrenal lipid inclusion droplets to adrenal mitochondria (11). Further, SCP-2 or an adrenal cytosolic fraction stimulated mitochondrial utilization of membrane cholesterol for pregnenolone synthesis, but did not directly influence the interaction of cholesterol with cytochrome $P450_{scc}$; this stimulating effect could be abolished by pre-treatment of the SCP-2 or adrenal cytosol with anti-SCP-2 IgG (12). We have demonstrated that SCP-2 enhances the transfer of cholesterol from the outer mitochondrial membrane to the mitoplast (inner membrane of the mitochondria) (13). Western blotting techniques also established the presence in adrenals of a protein with similar electrophoretic and antigenic properties as authentic hepatic SCP-2 (14) While these earlier studies provided evidence for a possible mechanism of cholesterol translocation via SCP-2 sequestration, they utilized in vitro subcellular model systems.

In the present article we present studies providing direct evidence of SCP-2 involvement in adrenal steroidogenesis. These studies utilized intact isolated adrenal fasiculata cells fused with liposomally encapsulated anti-SCP-2 IgG and/or SCP-2.

EXPERIMENTAL PROCEDURES

Materials

Male Wistar rats (174-225 g) were obtained from Charles Rivers (Wilmington, MA). Rat liver SCP-2 was purified to homogeneity as previously reported (15). Preparation and characterization of anti-SCP-2 IgG has been described previously (12, 16). Corticosterone radioimmunoassay kits were supplied by Radioassay Systems Labs, Inc. (Carson, CA). Phospholipids and bovine serum albumin were purchased from Sigma Chemical Co. (St. Louis, MO). $ACTH_{1-39}$ was obtained from Armour Pharmaceutical Co. (Kanakee, IL). Sephadex G-150 was purchased from Pharmacia (Piscataway, NJ). Aminoglutethimide was kindly provided by Ciba-Geigy Corp. Econo columns were supplied by Bio-Rad Labs (Richmond, CA). Anti-fatty acid binding protein (FABP) IgG was generously provided by Dr. N.M. Bass. All other chemicals were of the highest purity and were purchased from Fisher Scientific Co. (Columbia, MD).

Preparation, liposomal fusion and incubation of dispersed rat adrenal fasiculata cells

Detailed methods for the preparation of rat dispersed adrenal fasiculata cells have been published elsewhere (8). Isolated adrenal cells (4.0 ml, 2-4 x 10^5 cells/ml) in Krebs-Ringer bicarbonate buffer (pH 7.4) were incubated with 1.0 ml of the appropriate liposomal preparation at $37^{\circ}C$ under 95% O_2-5% CO_2 for 60 min in a Dubnoff metabolic shaker (90 oscillations/min). Cells were reisolated by centrifugation at 150 x g, 4° C for 20 min and washed once with Krebs-Ringer bicarbonate buffer. If necessitated by experimental protocol, a second fusion with 1 ml of the appropriate liposomal preparation was carried out under the exact conditions described above. Reisolation and washing were also as described above. Aliquots (1.90 ml) of these freshly washed cells were incubated for 25 min (or as indicated) after the addition of 500 μunits of ACTH or 0.5 mM dibutryl cAMP. ACTH was added in 0.01% bovine serum in 0.9% NaCl. Again, incubations were conducted under the conditions described above. Cell viability was checked by Trypan blue exclusion and was always above 89% viable.

Corticosterone assay

At the end of the incubation, 0.2 ml aliquots of the cell suspension were extracted with 15 ml of methylene chloride. Aliquots of the methylene chloride extract were dried under N_2 and assayed for corticosterone by radioimmunoassay. No differences in corticosterone levels detected were seen, whether the cell suspensions were frozen (in liquid N_2) and thawed prior to extraction or extracted immediately (no freezing and thawing) after incubation.

Sterol carrier protein-2 (SCP-2) assay

SCP-2 was measured using an [^{125}I]SCP-2 radioimmunoassay as described previously (14).

Preparation of adrenal subcellular fractions and cytoplasmic lipid droplets

Adrenal subcellular fractionation and cytoplasmic lipid droplet isolation has been described in detail previously (11).

Preparation of liposomes

Liposomes containing anti-SCP-2 IgG, non-immune IgG (control), homogeneous SCP-2, or buffer (sodium phosphate, 5 mM, pH 7.6) were prepared as described by Hall et al. (17). The liposomally encapsulated samples were separated from non-encapsulated material by Sephadex G-150 gel filtration. Briefly, the columns (Econo columns, 0.7 x 20 cm) were packed and eluted with potassium phosphate buffer (10 mM, pH 7.4). The protein or buffer containing vesicles were eluted in the void volume and were further purified by centrifugation at 100,000 x g at 4° C for 60 min. The pellet (liposomes) was resuspended in phosphate buffered saline (pH 7.4) and kept at 4° C. The physical characteristics of these liposomes were essentially the same as those reported previously (17). Approximately 10-15% of the starting protein was recovered in the final purified pellet. Phospholipid recovery as measured by phosphorous content (18) was approximately 70%. Anti-SCP-2 IgG, non-immune IgG or SCP-2 did not bind or associate with preformed liposomes (containing only buffer) to any appreciable extent. Radioisotopic measurement of liposomal [^{125}I]anti-SCP-2 IgG incorporated into cells after fusions and several washings gave an estimated efficiency of cellular incorporation of antibody or protein (SCP-2) of ~20-25%. Protein was determined by the method of Lowry et al. (19).

RESULTS

Liposomal fusions with isolated adrenal cells

Data in Table 1 shows SCP-2 concentrations in rat adrenal subcellular fractions and a comparison of its distribution with the distribution of liposomally incorporated $[^{125}I]$anti-SCP-2 IgG.

Table 1 Subcellular distribution of SCP-2 and liposomal $[^{125}I]$-SCP-2 in rat adrenals.

Subcellular fraction	ng SCP-2/100 μg protein	Distribution of SCP-2	Distribution of ^{125}I-anti-SCP-2 IgG
		%	%
Nuclei	26.0 ± 4.1	14.5 ± 3.2	22.4
Mitochondria	53.0 ± 0.8	45.5 ± 2.9	43.5
Microsomes	56.7 ± 2.5	21.9 ± 0.8	20.7
Cytosol	17.2 ± 3.5	18.1 ± 3.8	13.4

Values are mean ± Standard error of triplicate samples. Adrenals for subcellular fractionation were collected into three separate pools, each consisting of the adrenals from six rats. SCP-2 was assayed by radioimmunoassay.

As can be seen, the percent distribution of the labeled antibody closely matches that of the SCP-2 antigen. This clearly indicates that the liposomes are delivering antibody into the cells, and that this antibody is distributing and localizing into areas of the cell consistent with where the SCP-2 antigen is concentrated, as reflected by these subcellular fractions.

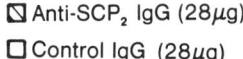

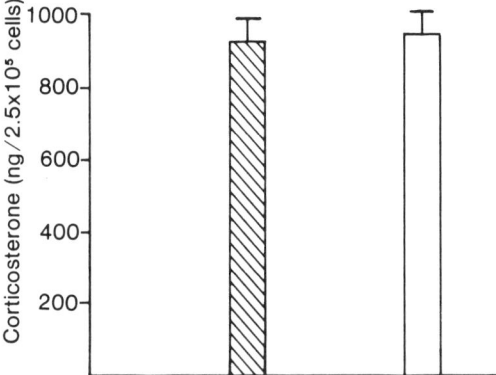

Fig. 1. **Effect of non-liposomal anti-SCP-2 IgG and control IgG on ACTH-stimulated steroidogenesis in isolated adrenal fasiculata cells.** Cells were isolated as described previously (8). Purified anti-SCP-2 IgG and control IgG were added in the quantities indicated. After incubation for 1 hour at 37° C under 95% O_2-5% CO_2, the cells were washed with Krebs-Ringer bicarbonate buffer, resuspended and incubated for 45 min in the presence of 500 μunits ACTH. Corticosterone was measured by radioimmunoassay. Results represent the mean ± S.D. of triplicate samples

When isolated adrenal cells were incubated in the presence of non-liposomal anti-SCP-2 IgG and subsequently stimulated with ACTH, no effect on steroidogenesis (as measured by corticosterone output) was detected. These data are shown in Fig. 1. That the antibody had no effect is most likely due to its inability to enter the cells without the aid of liposomes. In marked contrast, the influence of liposomal anti-SCP-2 IgG on ACTH induced steroidogenesis is shown in Table 2.

Table 2 Effect of liposomal anti-SCP-2 IgG on adrenal corticosterone production

Liposomal treatment of adrenal cells	Corticosterone production (ng/2.5 x 10^6 cells)		
	- ACTH	+ ACTH	% Control
Liposomes (buffer)	25.5 ± 0.91	920 ± 25.8	100
Liposomes (non-immune IgG, 27 μg)	16.6 ± 1.31	865 ± 28.1	94
Liposomes (anti-SCP-2 IgG, 27 μg)	14.2 ± 0.82	455 ± 12.5	49

Adrenal fasiculata cell isolation, liposome encapsulation of anti-SCP-2 IgG antibody and fusion with cells were done as described in "Experimental Procedures". When present ACTH = 500 μunits per incubation as described in "Experimental Procedures". Corticosterone was measured by radioimmunoassay. Values are means ± standard deviation of triplicate incubations.

In the absence of ACTH, adrenal cells pretreated with liposomal buffer, non-immune IgG or anti-SCP-2 IgG showed no significant differences in basal steroid output. In the presence of 500 μunits of ACTH the adrenal cells, which were pretreated with liposomal buffer or with liposomal non-immune IgG, both showed a typical steroidogenic response reflected by their 40- to 50-fold increase in steroid output. However, the ACTH-stimulated steroidogenic response of adrenal cells pretreated with liposomal anti-SCP-2 IgG was inhibited by 49%. These data indicate that anti-SCP-2 IgG is in some way inhibiting the response of these cells to ACTH stimulation. A reasonable explanation, based upon previous work (11, 12, 13, 16) and complemented by the data presented in Table 1, would be that during pre-treatment liposomes were delivering anti-SCP-2 antibody to the interior of the cell where it binds to SCP-2 and inhibits its ability to bind with and transport the unesterified cholesterol made available during subsequent ACTH stimulated hydrolysis of cytoplasmic lipid droplet cholesterol esters (4, 5, 8). If these assumptions are correct, it should be possible to restore the steroidogenic competence of these cells by reestablishing a viable intracellular pool of SCP-2.

To assess this possibility isolated adrenal fasiculata cells previously treated with liposomal anti-SCP-2 IgG were subjected to a second fusion with liposomes containing SCP-2, before being stimulated with ACTH. As shown in Table 3 we again see no effect of liposomes containing only buffer, a 42% inhibition of steroidogenesis by liposomal anti-SCP-2 IgG and a striking release of this inhibition by a second fusion with liposomal SCP-2. These cells were restored to approximately 90% of their steroidogenic competence compared to controls. These results indicate that, in the intact adrenal cell, SCP-2 plays a crucial role in the ACTH stimulated steroidogenic process.

Table 3 Effect of liposomal SCP-2 on ACTH stimulated steroidogenesis of adrenal cells pretreated with liposomal anti-SCP-2 IgG

Liposomal treatments of adrenal cells	Corticosterone production (ng/2.5 x 10^6 cells)	
	- ACTH	+ ACTH
Liposomes containing buffer (two fusions)	22.6 ± 0.84	840 ± 20.4
Liposomes containing anti-SCP-2 IgG (first fusion), liposomes containing buffer (second fusion)	12.4 ± 0.56	489 ± 17.5
Liposomes containing anti-SCP-2 IgG (first fusion), liposomes containing SCP-2 (second fusion)	28.2 ± 1.22	768 ± 24.3

Liposomal fusions were carried out as described in "Experimental Procedures"; ACTH = 500 μunits per incubation as described in "Experimental Procedures"; anti-SCP-2 IgG = 2.9 μg/ml liposomes; SCP-2 = 10.4 μg/ml liposome suspension. Values are means ± standard deviation of triplicate samples. Cell viability (Trypan blue exclusion) at end of all incubations = 85%. Incubation time after ACTH addition = 20 min.

Effect of varying liposomal anti-SCP-2 IgG on corticosterone production

To more fully examine the role of SCP-2 in the steroidogenic pathway, the effect of varying amounts of liposomal anti-SCP-2 IgG on corticosterone production by isolated adrenal fasiculata cells was investigated (Fig. 2). Liposomal anti-SCP-2 IgG (1, 2, 5, 10, 20, and 35 μg) inhibited ACTH stimulated corticogenesis in a concentration dependent manner, 25% at 1 μg and by 60% at the highest level of antibody (35 μg). In similar experiments using as much as 75 μg of liposomal anti-SCP-2 antibody, a 62% inhibition of steroidogenesis was achieved (data not shown). That higher doses of antibody could not produce an inhibition much beyond 60% may be due to the isolated cells' capacity to assimilate only a limited amount of liposomes and/or liposomally encapsulated antibody. An alternative explanation might be that a portion of the endogenous SCP-2 may be inaccessible to the antibody, e.g., inside either the mitochondria or the peroxisomes (20). Another alternative explanation might be that cytosolic factors in addition to SCP-2 contribute to steroidogenesis. These may include arachidonic acid metabolites (21), an ACTH stimulated, cycloheximide sensitive, labile peptide-steroidogenic activator polypeptide (SAP) (22, 23), cytosolic phospholipids (24, 25) which have been shown to be elevated by ACTH (24) and cytoskeletal structures (17). It has been shown that ACTH can induce SCP-2 in adrenal cells (26); SCP-2 may act in conjunction with other cytosolic factors which are sensitive to ACTH (27, 28).

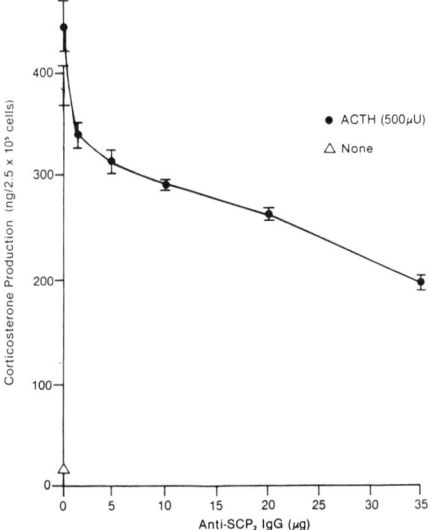

Fig. 2. Dose-dependent effect of anti-SCP-2 IgG on adrenal corticosterone production. Equivalent aliquots of adrenal fasiculata cells were incubated with varying concentrations of liposomal anti-SCP-2 IgG as described in "Experimental Procedures". Note that the quantities of antibody represented on the x-axis represent the amount added during the formation of liposomes. Liposomes routinely encapsulated 10-15% of the added protein. Corticosterone was assayed by radioimmunoassay. Incubations with ACTH, after cells were washed and resuspended in fresh buffer, were for 25 min. Values are means of triplicate samples ± S.D.

Effect of varying liposomal-SCP-2 on the reversal of anti-SCP-2 IgG inhibition of adrenal steroidogenesis

The studies shown in Table 3 indicate that it is possible to rescue anti-SCP-2 IgG inhibited steroidogenesis in adrenal calls with liposomal SCP-2. To assess the possibility of completely restoring steroidogenic competence to adrenal cells inhibited by liposomal anti-SCP-2 antibody, a study was conducted using a second fusion of varying concentrations of liposomal SCP-2. Isolated adrenal cells were subjected to an initial fusion (60 min) with liposomes containing buffer or liposomes containing 27 µg of anti-SCP-2 IgG. This concentration was chosen since previous studies indicated that this amount of antibody gave approximately 50% inhibition of steroidogenesis (Fig. 2). After reisolating and washing the cells, a second fusion with the indicated amounts of liposomal SCP-2 was performed. The cells were again reisolated and washed and, after resuspension, were exposed to 500 µunits of ACTH for 45 min. The results are shown in Fig. 3. Both control (no IgG) and anti-SCP-2 IgG treated cells showed increasing corticosterone output with increasing amounts of liposomal SCP-2, whether stimulated with ACTH (Panel A) or not (Panel B). However, the cells that were initially treated with anti-SCP-2 IgG all showed greater corticosterone output compared to those treated with liposomal buffer only (no anti-SCP-2 IgG). This difference in steroid hormone production in

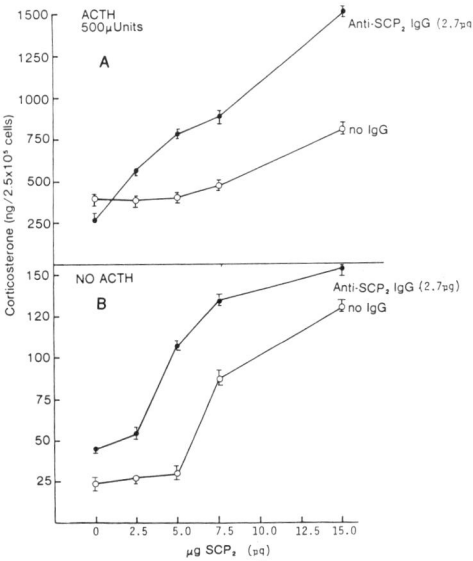

Fig. 3. **Dose-dependent liposomal SCP-2 reversal of anti-SCP-2 inhibition of steroidogenesis in adrenal fasiculata cells.** Equivalent aliquots of cells were treated with either liposomes containing antibody (anti-SCP-2 IgG) or liposomes containing buffer. After fusions were performed as described in "Experimental Procedures", cells were washed and subjected to a second fusion with liposomes containing the concentrations of SCP-2 shown. At the end of the second fusion the cells were washed, resuspended in fresh buffer, and ACTH 500 µunits (Panel A) or vehicle (Panel B) was added and the cells incubated for 45 min as described in "Experimental Procedures" Corticosterone was measured by radioimmunoassay. Values are means ± S.D. of duplicate samples.

the non-ACTH and ACTH-treated cells probably reflects the prior inhibition of SCP-2 (by the antibody) from functioning in its proposed role of transporting and translocating cholesterol from cytoplasmic lipid droplets to the mitochondria, across the mitochondrial membrane (11,13) to the cytochrome $P450_{scc}$ which is located on the inner face of the inner membrane (9,10). The addition of SCP-2 in liposomes replenishes the supply of intracellular SCP-2 protein which can now "shuttle" the accumulated unesterified cholesterol (higher than normal, basal levels even in the non-ACTH cells due to lengthy handling and manipulation in this study) resulting in increased steroid synthesis. The much greater difference in steroid output seen in Fig. 3 (Panel B) is probably due not only to the much greater unesterified cholesterol pool caused by ACTH-induced hydrolysis of sterol esters stored in cytoplasmic lipid droplets (7,29), but also to increases in SAP (22,23), arachidonic acid metabolites (21,30,31), phospholipids (24,25) and cytoskeletal structures (17) which may act synergistically with SCP-2 (28). The production of the StAR protein would also be increased in these cells (see Chapter by Stocco and Straus).

With respect to phospholipids it has been observed that acidic phospholipids, e.g., phosphatidylserine, phosphatidylinositol or cardiolipin, markedly increase the sterol transfer capacity of SCP-2 (32). All of these compounds possess a net negative charge at physiological pH. This finding might explain the findings of McNamara and Jefcoate (27) who found that cytosol contained a factor or factors which enhanced SCP-2 activity.

Effect of cyclic-AMP on anti-SCP-2 treated adrenal cells

In order to investigate the possibility that liposomal fusion and/or anti-SCP-2 may be disrupting the cell membrane and the ability of ACTH to generate second messenger thereby leading to reduced steroidogenesis, adrenal fasiculata cells treated with liposomal anti-SCP-2 IgG and liposomal buffer were incubated in the presence of dibutryl-cAMP (0.5 mM).

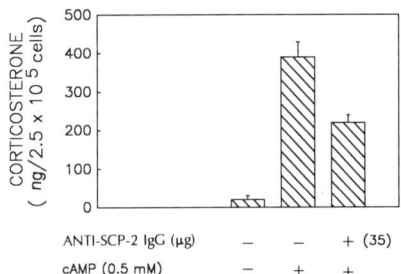

Fig. 4. **Effect of cyclic-AMP on steroidogenesis in adrenal fasiculata cells treated with anti-SCP-2 IgG.** Isolated adrenal fasiculata cells were subjected to liposomal anti-SCP-2 IgG or liposomal buffer as previously described. After washing and resuspending in fresh buffer, the addition of dibutryl cylic-AMP was made to the designated flasks. Where no cyclic-AMP was added, an equivalent aliquot of buffer was added. Incubations were conducted for 25 min under the same conditions used for ACTH incubations (see "Experimental Procedures). Corticosterone was assayed by RIA. Values are means ± S.D. of triplicate samples.

Fig. 4 shows that the steroidogenic response of cells (treated with anti-SCP-2) to cAMP is inhibited by 45%, compared to cells treated with liposomes containing only buffer. If a disruption of ACTH action on adrenal cells treated with liposomes (anti-SCP-2) or liposomes (buffer only) had occurred, one would not expect that the stimulatory effect of cAMP would not be inhibited.

Effect of anti-SCP-2 IgG on cholesterol distribution between cytoplasmic lipid droplets and adrenal mitochondria

Previous studies utilizing isolated adrenal organelles have shown that SCP-2 is required for the delivery of cholesterol from adrenal cytoplasmic lipid droplets to mitochondria (11). Studies were therefore conducted using adrenal cells and liposomal fusion to test whether or not SCP-2 plays a role in the translocation of cholesterol from adrenal cytoplasmic lipid droplet stores to mitochondria during ACTH stimulation. To accomplish this type of study aminoglutethimide was utilized; this compound causes cholesterol to accumulate in the mitochondria of adrenals, due to its inhibitory effect on cholesterol-P450$_{SCC}$ complex formation (33). All incubations contained 0.75 mM aminoglutethimide at all times.

Table 4 Effect of anti-SCP-2 on the levels of cholesterol in lipid droplets and mitochondria.

Incubation conditions	Chol (μg)	Lipid droplets (μg choll/ μg phos)	CE / phos (μg/μg)	Mitochondria (μg chol/μg prot.)
Non-immune IgG	128 ± 1.84	2.25 ± 0.03	42.9 ± 1.09	57.9 ± 0.85
Anti-SCP-2 IgG	143 ± 0.57	2.75 ± 0.01	44.3 ± 0.86	47.9 ± 0.42

Liposomal fusions were carried out as described in "Experimental Procedures". All incubations contained aminoglutethimide (0.75 mM). Non-immune IgG = 2.7 μg/ml liposome. ACTH in both incubations was 500 μunits. At the end of the ACTH incubation period (45 min) the cells were reisolated by centrifugation, homogenized and lipid droplets and mitochondria were isolated as previously described (11). Extraction was conducted by the method of Folch (38) and cholesterol and cholesterol esters were separated and quantified by gas liquid chromatography (8). Cholesterol and cholesterol esters are expressed per unit phospholipid phosphorous to correct for procedural losses, since lipid droplet phospholipids do not change during ACTH action (8). Values are means ± standard deviation of triplicate samples.

As shown in Table 4 cytoplasmic lipid droplet unesterified cholesterol from cells treated with anti-SCP-2 IgG **increased** by 22% over cells treated with non-immune IgG during ACTH stimulation. Concurrently, mitochondria in the anti-SCP-2 IgG treated cells showed an almost identical **decrease** (21%) in cholesterol content over the non-immune IgG treated cells. These results definitely establish that a role for SCP-2 in these steroidogenic cells is the translocation of unesterified cholesterol from cytoplasmic lipid droplets to mitochondria. These studies do not indicate whether the accumulated cholesterol is located in the outer or inner membrane of the mitochondria. The similar levels of cholesterol esters in the cytoplasmic lipid droplets of both groups of cells imply that SCP-2 probably has no effect on the ACTH-mediated activation of sterol ester hydrolase (34, 35).

Effect of liposomal anti-FABP on ACTH stimulated adrenal steroidogenesis

Although it has been shown that SCP-2 and fatty acid binding protein (FABP) (36) are separate and distinct, both structurally (36) and functionally (16), we decided to test whether or not FABP has any SCP-2-like activity in adrenal steroidogenesis utilizing this whole cell-liposomal antibody fusion model. The results shown in Fig. 5 demonstrate that anti-FABP treated cells are just as responsive to ACTH stimulation as control cells (liposomes containing buffer). A reasonable conclusion from this result is that FABP probably has no direct affect on the ACTH-stimulated steroidogenic pathway in rat adrenal cells.

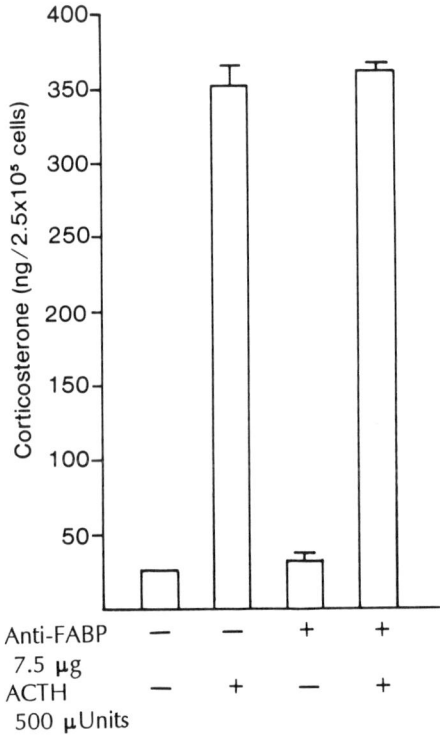

Fig. 5. Effect of liposomal anti-FABP on ACTH stimulated steroidogenesis in isolated adrenal fasiculata cells. Isolated adrenal cells were treated with liposomally encapsulated FABP or liposomal buffer as indicated. Fusions were done as described in "Experimental Procedures". After washing and resuspension in fresh buffer, ACTH (500 μunits) or vehicle (see "Experimental Procedures") was added as shown and the cells incubated for 25 min. Corticosterone was assayed by RIA. Values are means ± S.D. of triplicate samples.

A physiologic role for SCP-2 in rat adrenal cells

The present studies, combined with those reported earlier (11-14, 16) provide convincing proof for at least one major role for SCP-2 in adrenal steroidogenesis. Specifically, SCP-2 participates in Steps 6, 7 and 8 as indicated in Fig. 6, based on the present and previous studies (11,12,14,16). In addition Mendis-Handagama et al. have shown electron microscopic evidence for peroxisomal SCP-2 in cholesterol transport from cytoplasmic lipid droplets to mitochondria (39). Also, Stocco (40) has presented substantial evidence which supports an important role for a labile mitochondrial protein, steroidogenic acute regulatory (StAR) protein, as a mediator in ACTH-dependent steroidogenesis. The function of StAR in this process is not known at the present time, but it may well act in concert with SCP-2 in providing or regulating the delivery of cholesterol to cytochrome $P450_{scc}$ for pregnenolone synthesis.

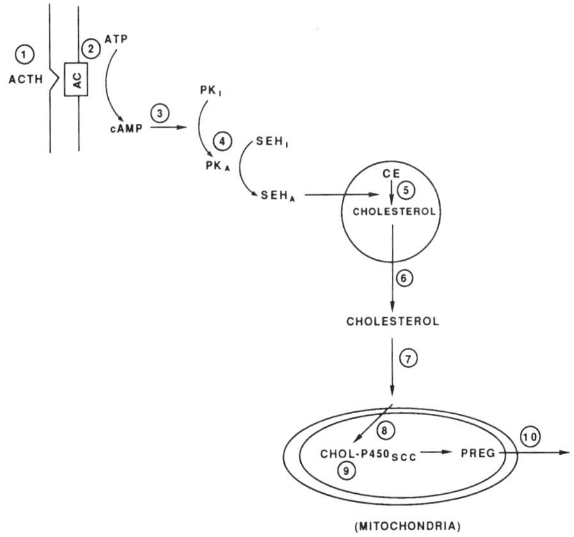

Fig. 6. **Schematic representation of steps involved in ACTH-stimulated adrenal steroidogenesis.** ACTH, adrenocorticotropic hormone; AC, adenylate cyclase: PK_I, protein kinase inactive; PK_A, protein kinase active: SEH_I, sterol ester hydrolase inactive; SEH_A, sterol ester hydrolase active; CE, cholesterol ester; cytochrome P450 dependent cholesterol side chain cleavage enzyme, CHOL-$P450_{scc}$; PREG, pregnenolone.

CONCLUSION

In summary evidence presented here establishes that a major physiologic role for SCP-2 is the translocation of unesterified cholesterol from cytoplasmic lipid droplets to mitochondria in intact adrenal cells. This process is known to be the rate-limiting step in steroidogenesis (34-36).

REFERENCES

1. Gidez LI, Feller E. Effect of the stress of unilateral adrenalectomy on the depletion of individual cholesterol esters in the adrenal. Lipid Res 1969;10:656-59

2. Sayers GM, Sayers A, Fry EY, White A, Long GNH. The effect of the adrenotropic hormone of the anterior pituitary on the cholesterol content of the adrenals, with a review of the literature on adrenal cholesterol. Yale J. Biol. Med. 1944;16:361-92

3. Glick D, Ochs MJ. Studies in histochemistry: quantitative histological distribution of cholesterol in adrenal glands of the cow, rat, and monkey and effects of stress conditions, adrenocorticotropin (ACTH), cortisone, and deoxycorticosterone. Endocrinology 1955; 56: 285-95

4. Moses HL, Davis WW, Rosenthal AS, Garren LD. Adrenal cholesterol: localization by electron-microscope autoradiography. Science 1969;163:1203-05

5. Boyd GS, Trzeciak, WH. Cholesterol metabolism in the adrenal cortex: studies on the mode of action of ACTH. Ann N. Acad Sci 1973;212:361-77

6. Beckett GJ, Boyd GS. Purification and control of bovine adrenal cortical ester hydrolase and evidence for the activation of the enzyme by a phosphorylation. Eur J Biochem 1977;72:223-33

7. Naghshineh S, Treadwell CR, Gallo LL, Vahouny GV. Protein kinase-mediated phosphorylation of a purified sterol ester hydrolase from bovine adrenal cortex. J Lipid Res 1978;19: 561-69

8. Vahouny GV, Chanderbhan R, Hinds R, Hodges VA, Treadwell CR. ACTH-induced hydrolysis of cholesteryl esters in rat adrenal cells. J Lipid Res 1978;19:570-77

9. Yago N, Ichii S. Submitochondrial distribution of components of the steroid 11 beta-hydroxylase and cholesterol sidechain-cleaving enzyme systems in hog adrenal cortex. J Biochem (Tokyo) 1969;65:215-24

10. Churchill PF, Kimura T. Topological studies of cytochromes $P-450_{scc}$ and P-450 11 beta in bovine adrenocortical inner mitochondrial membranes. Effects of controlled tryptic digestion. J Biol Chem 1979;254:10443-48

11. Chanderbhan R, Noland BJ, Scallen TJ, Vahouny GV. Sterol carrier protein$_2$. Delivery of cholesterol from adrenal lipid droplets to mitochondria for pregnenolone synthesis. J Biol Chem 1982;257:8928-34

12. Vahouny GV, Chanderbhan R, Noland BJ, Irwin D, Dennis P, Lambeth JD, Scallen TJ. Sterol carrier protein$_2$. Identification of adrenal sterol carrier protein$_2$ and site of action for mitochondrial cholesterol utilization. J Biol Chem 1983;258:11731-37

13. Vahouny GV, Dennis P, Chanderbhan R, Fiskum G, Noland BJ, Scallen TJ. Sterol carrier protein$_2$ (SCP$_2$)-mediated transfer of cholesterol to mitochondrial inner membranes. Biochem Biophys Res Commun 1984;122:509-15

14. Chanderbhan RF, Kharroubi AT, Noland BJ, Scallen TJ, Vahouny GV. Sterol carrier protein$_2$: further evidence for its role in adrenal steroidogenesis. Endocrine Res 1986;12:351-70

15. Noland BJ, Arebalo RE, Hansbury E, Scallen TJ. Purification and properties of sterol carrier protein$_2$. J Biol Chem 1980;255:4282-89

16. Scallen TJ, Noland BJ, Gavey KL, Bass NM, Ockner RK, Chanderbhan RF, Vahouny GV. Sterol carrier protein$_2$ and fatty acid-binding protein. Separate and distinct physiological functions. J Biol Chem 1985;260;4733-39

17. Hall PF, Charpponnier C, Nakamura M, Gabbiani G. The role of microfilaments in the response of adrenal tumor cells to adrenocorticotropic hormone. J Biol Chem 1979;193:265-75

18. Rouser G, Siakotos AN, Fleischer S. Quantitative analysis of phospholipids by thin-layer chromatography and phosphorous analysis of spots. Lipids 1966;1:85-86

19. Lowry OH, Rosebrough NJ, Farr AL, Randall RJ. Protein measurement with the folin phenol reagent.. J Biol Chem 1951;193: 265-75

20. Keller GA, Scallen TJ, . Clarke D, Maher PA, Krisans SK, Singer SJ. Subcellular localization of sterol carrier protein-2 in rat hepatocytes: its primary localization to peroxisomes. J Cell Biol 1989;108: 1353-61

21. Hirai A, Tahara K, Tamura Y, Saito H, Terano T, Yoshida S. Involvement of 5-lipoxygenase metabolites in ACTH-stimulated corticosteroidogeneis in rat adrenal glands. Prostaglandins 1985;30:749-67

22. Pedersen RC, Brownie AC. Cholesterol side-chain cleavage in the rat adrenal cortex: isolation of a cycloheximide-sensitive activator peptide. Proc Natl Acad Sci USA 1987; 80:1882-86

23. Pedersen RC, Brownie, AC. Steroidogenesis-activator polypeptide isolated from a rat Leydig cell tumor. Science 236;1987:188-90

24. Farese RV, Sabir AM. Polyphosphoinositides: Stimulator of mitochondrial cholesterol side chain cleavage and possible identification as an adrenocorticotropin-induced, cycloheximide-sensitive, cytosolic, steroidogenic factor. Endocrinology 1980;106:1869-78

25. Igaraski Y, Kimura J. Importance of the unsaturated fatty acyl group of phospholipids in their stimulatory role on rat adrenal mitochondrial steroidogenesis. Biochemistry 1986;25: 6461-66

26. Trzeciak WH, Simpson ER, Scallen TJ, Vahouny GV, Waterman MR. Studies on the synthesis of sterol carrier protein-2 in rat adrenocortical cells in monolayer culture. Regulation by ACTH and dibutryl cyclic 3', 5'-AMP. J Biol Chem 1987; 262:3713-17

27. McNamara BC, Jefcoate CR. The role of sterol carrier protein 2 in stimulation of steroidogenesis in rat adrenal mitochondria by adrenal cytosol. Arch Biochem Biophysics 1989;275:53-63

28. Xu TS, Bowman EP, Glass DB, Lambeth JD. Stimulation of adrenal mitochondrial cholesterol side-chain cleavage by GTP, steroidogenesis activator polypeptide (SAP), and sterol carrier protein 2. GTP and SAP act synergistically. J Biol Chem. 1991; 266: 6801-07

29. Vahouny GV, Chanderbhan R, Noland BJ, Scallen TJ. Cholesterol ester hydrolysis and sterol carrier proteins. Endocrine Res 1984-1985;10:473-505

30. Saruta T, Kaplan, NM Adrenocortical steroidogenesis: the effects of prostaglandins. J Clin Invest 1972; 51:2246-51

31. Warner W, Rubin RP. Evidence for a possible prostaglandin link in ACTH-induced steroidogenesis. Prostaglandins 1969;9:83-95

32. Butko P, Hapala I, Scallen TJ, Schroeder F. Acidic phospholipids strikingly potentiate sterol carrier protein 2 mediated intermembrane sterol transfer. Biochemistry 1990;29: 4070-77

33. Privalle CT , Crivello JF, Jefcoate CR. Regulation of intramitochondrial cholesterol transfer to side-chain cleavage cytochrome P-450 in rat adrenal gland. Proc Natl Acad Sci. USA 1983;80:702-06

34. Garren LD, Gill GN, Masui H, Walton GM. On the mechanism of action of ACTH. Recent Prog Hormone Res 1971;27:433-78

35. Trzeciak WH, Boyd GS. The effect of stress induced by ether anesthesia on cholesterol content and cholesteryl-esterase activity in rat-adrenal cortex. Eur J Biochem 1973;37:327-33

36.   Gordon JI, Alpers DH, Ockner RK, Strauss AW. The nucleotide sequence of the rat liver fatty acid binding protein mRNA. J Biol Chem 1983;258:3356-63

37.   Schroeder F, Butko P, Nemecz G, Scallen TJ. Interaction of fluorescent delta 5, 7, 9(11), 22-ergostatetraen-3β–ol with sterol carrier protein-2. J Biol Chem 1990;265:151-57

38.   Folch J, Lees M, Sloane-Stanley GH. A simple method for the isolation and purification of total lipids from animal tissues. J Biol Chem 1957;226:497-509

39.   Mendis-Handagama SM, Aten RF, Watkins PA, Scallen TJ, Berhman HR. Peroxisomal sterol carrier protein-2 in luteal cell steroidogenesis: a possible role in cholesterol transport from lipid droplets to mitochondria. Tissue and Cell 1995;27:483-90

40.   Stocco DM. A review of the characteristics of the protein required for the acute regulation of steroid hormone biosynthesis: the case for the steroidogenic acute regulatory (StAR) protein. 1998;217:123-9

INTRACELLULAR STEROL BINDING PROTEINS: CHOLESTEROL TRANSPORT AND MEMBRANE DOMAINS

Friedhelm Schroeder[1,*], Andrey Frolov[1], Jonathan K. Schoer[1], Adalberto M. Gallegos[1], Barbara P. Atshaves[1], Neal J. Stolowich[2], A. Ian Scott[2], & Ann B. Kier[3]

[1]Departments of Physiology and Pharmacology, [2]Chemistry, and [3]Pathobiology, Texas A&M Univ, College Station, TX 77843-4466; e-mail: fschroeder@cvm.tamu.edu

KEY WORDS: cholesterol, traffick, membrane, domains.

ABSTRACT

Regulation intracellular cholesterol transport is not well understood. Part of the difficulty is that not only is the intracellular distribution of cholesterol not uniform, but even within membranes cholesterol has an asymmetric transbilayer and lateral distribution. Although both vesicular and protein mediated pathways for cholesterol movement are recognized, the role of intracellular cholesterol binding proteins in cytosolic cholesterol trafficking remains largely unresolved. Recent work from this and other laboratories demonstrates that these cholesterol binding proteins are involved in cholesterol uptake and intracellular trafficking. Immunocytochemical evidence supports the presence of significant amounts of proteins such as the sterol carrier protein-2 outside of peroxisomes.

CONTENTS

INTRODUCTION

The mechanism(s) of intravascular cholesterol transport are the subject of intense interest.

The use of fluorescent sterols and other technologies have allowed rapid advances in three aspects of this problem (rev. in 1-4). First, cholesterol is distributed into kinetic and structural domains in both the cell surface as well as intracellular membranes. Second, while intracellular sterol trafficking involves vesicular transfer, this alone does not account for the extent and specificity of lipid trafficking in the cell. As vesicles are transported through the cytoplasm they differentially lose and add protein and lipid component independently (rev. in 5). Cholesterol moves to the plasma membrane by a faster route than does protein (6). Thus, selective processes must be available to modify the lipid composition of vesicles originally imposed by synthesis and vesicular formation (rev. in 4). Third, intracellular cholesterol trafficking may be mediated by intracellular cholesterol binding proteins [i.e. sterol carrier protein-2 (SCP-2), liver fatty acid binding protein (L-FABP), caveolin, or steroidogenic acute regulatory protein (StAR)]. Unfortunately, the often conflicting data on intracellular localization have led some biologists to essentially eliminate a role for proteins such as SCP-2 in intracellular cholesterol transport. The lack of quantitative evaluation of immunocytochemical data and/or the lack of utilization of subcellular fractionation results has at times led to the conclusion that SCP-2 is exclusively peroxisomal and does not function in intracellular cholesterol trafficking. However, quantitative analysis suggests that nearly half of cellular SCP-2 may be extra-peroxisomal. How intra- and extra-peroxisomal SCP-2 or other cholesterol binding proteins influence intracellular cholesterol trafficking remains a mystery. L-FABP and SCP-2 require a cholesterol binding site for activity. Whether this is also true for caveolin and StAR is not known. L-FABP and SCP-2 apparently bind to the membrane (donor or acceptor) to bind sterol therein and subsequently elicit sterol desorption/entry. In contrast, preliminary data suggest that the caveolin-heat shock protein-immunophilin-cholesterol complex may act as an aqueous cholesterol carrier.

INTRACELLULAR CHOLESTEROL BINDING PROTEINS

Liver fatty acid binding protein (L-FABP): L-FABP is present in high concentration (2-3%) in the cytosol of liver and of intestinal enterocytes. L-FABP binds a diverse group of hydrophobic ligands (7,8) including cholesterol (9). *In vitro* binding assays with sterol added in micellar/monomer form show that native rat liver L-FABP binds sterols with dissociation constants near 0.2-1.0 µM and 1:1 molar stoichiometry (10-12). In contrast, model membrane competition (13) or size exclusion column (13,14) assays were unsuccessful in demonstrating binding of sterol to L-FABP. Most likely, the very low 20 nM aqueous solubility of cholesterol as well as its high affinity for membranes (liposomes) and surfaces (the matrix material of size exclusion columns, Lipidex, etc) complicate the latter assays. Finally, the L-FABP cholesterol binding site is essential for intermembrane sterol transfer activity of L-FABP (15).

Sterol Carrier Protein-2-(SCP-2): With the exception of steroidogenic tissues, SCP-2 is generally present in most tissues at levels one tenth that of liver. SCP-2 also has broad ligand binding specificity (16-18) including cholesterol (14,19). Micellar/monomer sterol binding, but not liposomal or Lipidex competition binding (20), assays show that rat liver SCP-2 binds fluorescent sterols with a dissociation constant as high as 1.2-1.6 µM and apparent 1:1 molar stoichiometry (21,22). A submicellar NBD-cholesterol binding assay showed a saturation binding curve, with a dissociation constant as low as 6 nM (Fig. 1) (23).

The use of fluorescent sterols and other technologies have allowed rapid advances in three aspects of this problem (rev. in 1-4). First, cholesterol is distributed into kinetic and structural domains in both the cell surface as well as intracellular membranes. Second, while intracellular sterol trafficking involves vesicular transfer, this alone does not account for the extent and specificity of lipid trafficking in the cell. As vesicles are transported through the cytoplasm they differentially lose and add protein and lipid component independently (rev. in 5). Cholesterol moves to the plasma membrane by a faster route than does protein (6). Thus, selective processes must be available to modify the lipid composition of vesicles originally imposed by synthesis and vesicular formation (rev. in 4). Third, intracellular cholesterol trafficking may be mediated by intracellular cholesterol binding proteins [i.e. sterol carrier protein-2 (SCP-2), liver fatty acid binding protein (L-FABP), caveolin, or steroidogenic acute regulatory protein (StAR)]. Unfortunately, the often conflicting data on intracellular localization have led some biologists to essentially eliminate a role for proteins such as SCP-2 in intracellular cholesterol transport. The lack of quantitative evaluation of immunocytochemical data and/or the lack of utilization of subcellular fractionation results has at times led to the conclusion that SCP-2 is exclusively peroxisomal and does not function in intracellular cholesterol trafficking. However, quantitative analysis suggests that nearly half of cellular SCP-2 may be extra-peroxisomal. How intra- and extra-peroxisomal SCP-2 or other cholesterol binding proteins influence intracellular cholesterol trafficking remains a mystery. L-FABP and SCP-2 require a cholesterol binding site for activity. Whether this is also true for caveolin and StAR is not known. L-FABP and SCP-2 apparently bind to the membrane (donor or acceptor) to bind sterol therein and subsequently elicit sterol desorption/entry. In contrast, preliminary data suggest that the caveolin-heat shock protein-immunophilin-cholesterol complex may act as an aqueous cholesterol carrier.

INTRACELLULAR CHOLESTEROL BINDING PROTEINS

Liver fatty acid binding protein (L-FABP): L-FABP is present in high concentration (2-3%) in the cytosol of liver and of intestinal enterocytes. L-FABP binds a diverse group of hydrophobic ligands (7,8) including cholesterol (9). *In vitro* binding assays with sterol added in micellar/monomer form show that native rat liver L-FABP binds sterols with dissociation constants near 0.2-1.0 μM and 1:1 molar stoichiometry (10-12). In contrast, model membrane competition (13) or size exclusion column (13,14) assays were unsuccessful in demonstrating binding of sterol to L-FABP. Most likely, the very low 20 nM aqueous solubility of cholesterol as well as its high affinity for membranes (liposomes) and surfaces (the matrix material of size exclusion columns, Lipidex, etc) complicate the latter assays. Finally, the L-FABP cholesterol binding site is essential for intermembrane sterol transfer activity of L-FABP (15).

Sterol Carrier Protein-2-(SCP-2): With the exception of steroidogenic tissues, SCP-2 is generally present in most tissues at levels one tenth that of liver. SCP-2 also has broad ligand binding specificity (16-18) including cholesterol (14,19). Micellar/monomer sterol binding, but not liposomal or Lipidex competition binding (20), assays show that rat liver SCP-2 binds fluorescent sterols with a dissociation constant as high as 1.2-1.6 μM and apparent 1:1 molar stoichiometry (21,22). A submicellar NBD-cholesterol binding assay showed a saturation binding curve, with a dissociation constant as low as 6 nM (Fig. 1) (23).

216

SCP-2 binding of sterols was further confirmed by use of ^{13}C-cholesterol and NMR spectroscopy (23). The ^{13}C NMR spectrum of SCP-2 (Fig. 2, bottom) shows a series of prominent peaks due to the natural abundance ^{13}C signal of the protein. Addition of 4-^{13}C-cholesterol elicits a new peak (Fig.2, top curve, arrow) at 40.7 ppm. This chemical shift differs significantly from that of the ^{13}C signal of 4-^{13}C-cholesterol in either aqueous solution or organic solvents (23). Finally, the SCP-2 cholesterol binding site is essential for intermembrane sterol transfer activity of SCP-2 (24).

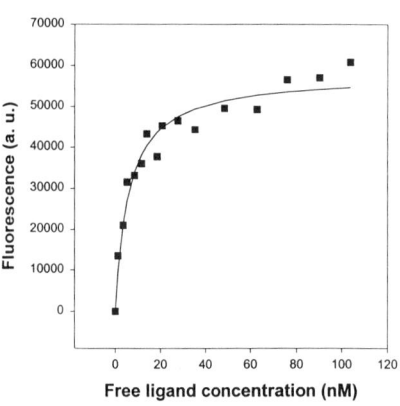

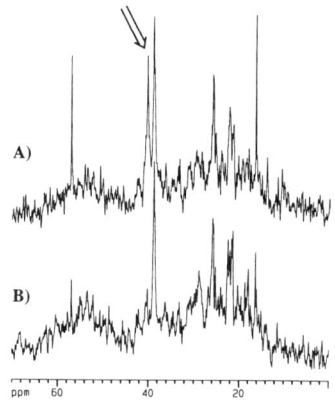

Fig. 1. Sterol Carrier Protein-2 Binding of NBD-cholesterol. NBD-cholesterol was added to 11 nM SCP-2 as described (23). The data fit a single binding site, r^2=0.96, with a K_d=6 nM.

Fig. 2. ^{13}C NMR detects interaction of 4-^{13}C-cholesterol sterol carrier protein-2 The upfield portion of the 125 MHZ ^{13}C NMR spectrum of SCP-2 with (Curve A) and without (Curve B) ^{13}C-cholesterol. The arrow shows a new peak at 40.7 ppm indicative of binding (23).

Other Sterol Binding Proteins: The rate limiting step in steroidogenesis is the transfer of cholesterol from the outer to the inner mitochondrial membrane, thought to be catalyzed by steroidogenesis acute regulatory protein (StAR) (rev. in 25,26). StAR binds NBD-cholesterol with high affinity with dissociation constant in the nanomolar range (27). Caveolin, an integral plasma membrane protein (cytofacial leaflet) involved in cholesterol transport between endoplasmic reticulum and plasma membrane caveolae (28), forms homo- and hetero-oligomers that bind cholesterol (29) and exists not only membrane bound but in the cytosol (30). In the cytosol caveolin appears as a complex of caveolin-heat shock protein 56, cyclophilin 40, cyclophilin A, and cholesterol (30).

CHOLESTEROL BINDING PROTEINS & CHOLESTEROL TRANSFER IN VITRO

In vitro studies with model membranes, liposomes. A new fluorescent sterol exchange assay not requiring separation of donor and acceptor liposomes yielded important new observations (rev. in 1,2,31,32): (i). Intermembrane sterol transfer is dependent on both the rate of desorption from the donor membrane and on acceptor membrane properties. (ii) Membranes have multiple kinetic cholesterol domains/pools. (iii) SCP-2, but not L-FABP, altered sterol domain size in model membranes. (iii) SCP-2, but not L-FABP, markedly stimulates the rate of sterol transfer when the model membranes contain negatively charged phospholipids. These findings are consistent with the basic SCP-2 (pI 8.5-9.0) binding to negatively charged model membrane surfaces and thereby interacting with membrane sterols to enhance their desorption and transfer to acceptor membranes.

In vitro studies with biomembranes: The intramembrane distribution as well as transfer of cholesterol between plasma membranes, microsomes, and mitochondria is not uniform (rev. in 1,2,26,31,33-35). (i). Intermembrane sterol transfer is dependent on both the properties of the donor and acceptor biomembranes. (ii) These membranes also have multiple cholesterol domains, the largest of which is non-exchangeable. (iii) L-FABP selectively stimulated sterol transfer between plasma membranes and microsomes. (iv) SCP-2 stimulates intermembrane sterol transfer rate, but not domain size, between all plasma membrane, microsome, and mitochondrial donor acceptor pairs, the effect being greatest, 27-fold, for transfer between plasma membranes and microsomes. (v) SCP-2 mediated sterol transfer was vectorial, e.g. 6-fold greater from plasma membrane to microsomes than from microsomes to plasma membranes.

In contrast to the above membranes, very little is known regarding the properties of the lysosomal membrane through which the majority of exogenous cholesterol enters the cell. Spontaneous lysosomal membrane sterol transfer (Fig. 3) is extremely slow, at least 4 to 5 fold slower than from plasma membranes, microsomes, or mitochondria (36). SCP-2 stimulated the initial rate of lysosomal sterol transfer nearly 50-fold (Fig.3). This fold-stimulation was nearly 10-fold greater than that for sterol desorption from the plasma membrane (37). Thus, SCP-2 may provide a potential mechanism(s) for release of LDL derived cholesterol from the lysosomal membrane to intracellular sites.

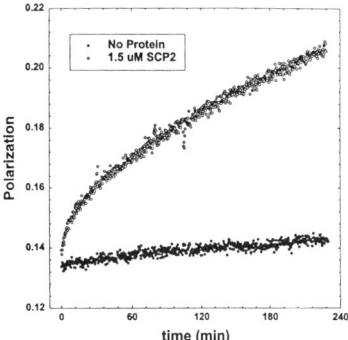

Fig. 3. Sterol transfer between lysosomal membranes in the presence and absence of exogenous cholesterol binding protein. Lower curve, spontaneous dehydroergosterol transfer; upper curve, dehydroergosterol transfer in the presence of 1.5 μM SCP-2. Dehydroergosterol transfer between donor and acceptor lysosomal membranes was determined as described (36).

CHOLESTEROL BINDING PROTEINS & CHOLESTEROL TRANSFER IN VIVO

Although data with cell lines showing mutations in proteins other than SCP-2 have largely not shown a role for SCP-2 (rev. in 2), strong positive correlative data suggest that SCP-2 expression is important for cholesterol trafficking in lung surfactant formation (38), diabetes (39), macrophage foam cell formation (40), intestinal cholesterol absorption (41,42), steroidogenesis (26), cholesterol oxidation (43,44), biliary cholesterol transport (43,45), and the action of hypocholesterolemic agents (46). In order to overcome the limitations of mutant cell lines with uncharacterized or gene defects other than SCP-2, investigators turned to the use of transfected cell lines overexpressing specific sterol binding proteins:

Overexpression of L-FABP in transfected L-cell fibroblasts. Expression of L-FABP stimulated cellular uptake of both ^{3}H-cholesterol and dehydroergosterol in addition to enhancing microsomal esterification of plasma membrane derived cholesterol (47,48). L-FABP overexpression resulted in equilibrium redistribution of cholesterol such that the cholesterol/phospholipid molar ratio in the plasma membranes was reduced by 50% while cholesterol ester mass was increased 56%. This markedly modulated plasma membrane fluidity and Na,K-ATPase activity (49,50).

Overexpression of 15 kDa pro-SCP-2 in CHO cells: cholesterol transfer to mitochondria and steroidogenesis. The SCP-2 gene encodes for the 15 kDa pro-SCP-2 which is completely post-translationally cleaved to form the mature 13 kDa SCP-2 and for a 58 kDa SCP-x, which is partially post-translationally cleaved to form 13 kDa SCP-2 Transfected monkey kidney COS cells coexpressing the 15 kDa pro-SCP-2 and P450scc enzyme displayed a 2-fold stimulation of mitochondrial cholesterol oxidation (51). Consistent with a role of SCP-2 in cholesterol oxidation, introduction of anti-SCP-2 antibodies into adrenal cells neutralized SCP-2 and inhibited corticosterone formation (52). This suggests that SCP-2 may (i) stimulate oxidation of cholesterol in non-steroidogenic tissues where StAR may not be present, and/or (ii) in steroidogenic tissues SCP-2 may enhance the relatively slower delivery/replenishment of outer mitochondrial membrane cholesterol which has been depleted by StAR in transient bursts of cholesterol transfer for inner mitochondrial membrane corticosteroid synthesis.

Overexpression of 15 kDa pro-SCP-2 in L-cell fibroblasts. Transfected L-cells overexpressing the 15 kDa pro-SCP-2 had 1.4-fold increased rate and maximal uptake of ^{3}H-cholesterol (53) and 1.4-fold increased ^{3}H-cholesterol-ester (54). In addition, 15 kDa SCP- overexpression (i) stimulated 1.6 fold the esterification of plasma membrane ^{3}H-cholesterol released by sphingomyelinase treatment, (ii) stimulated 1.6 fold the microsomal esterification of ^{3}H-oleic acid with plasma membrane derived cholesterol, (iii) increased 2-fold the equilibrium mass of intracellular cholesteryl esters and triglycerides (54). Thus, SCP-2 overexpression not only enhanced cholesterol uptake and transfer to endoplasmic reticulum, but increased its intracellular quantity as well.

Overexpression of 15 kDa pro-SCP-2 in hepatoma cells. In rat McA-RH7777 rat hepatoma cells transfected with cDNA encoding 15 kDa pro-SCP-2 the rate of newly synthesized cholesterol transfer from the endoplasmic reticulum to the plasma membrane as well as the rate of plasma membrane cholesterol internalization were each increased 4-fold, while the rate of plasma membrane conversion of desmosterol to cholesterol was also enhanced several fold (55). Although levels of microsomal acyl CoA cholesterol acyl transferase and neutral cholesterol ester hydrolase were not affected, cholesteryl ester synthesis and mass were reduced by 50%. HDL secretion was also decreased by 70% in the SCP-2 overexpressing cells. This suggests that SCP-2 overexpression enhanced cholesterol cycling in these cells with the net result being a 46% increase in plasma membrane cholesterol content (55). It was concluded that 15 kDa pro-SCP-2 overexpression in hepatoma cells enhanced cholesterol uptake and intracellular cycling (55).

Overexpression of 58 kDa SCP-x in L-cell fibroblasts. L-cells transfected with the cDNA encoding the 58 kDa SCPx had not only the expected increase in 58 kDa SCPx, but also 2-fold increased 13 kDa SCP-2 (56). The latter was the result of post-translational processing of the 58 kDa SCP-x. The transfected cells showed 1.9-fold increase in rate of exogenous ^{3}H-cholesterol uptake and 2.3- and 2.5- fold increase in cholesterol ester formation from exogenous ^{3}H-cholesterol and ^{3}H-oleic acid, respectively. Although cellular cholesterol mass was unaltered in transfected cells, there was a trend to decreased cholesterol ester mass, similar to that observed for hepatoma cells overexpressing 15 kDa pro-SCP-2 (55).

Decreased SCP-2 in human fibroblasts treated with antisense oligonucleotides. Normal human fibroblast transfer *de novo* synthesized cholesterol from the endoplasmic reticulum to the plasma membrane by at least two pathways: a rapid (10 min) cytoskeleton/Golgi independent pathway (SCP-2 mediated) and a slower, cytoskeleton-, Golgi- dependent pathway (vesicular) (57). In the normal human fibroblasts, the rapid SCP-2 mediated route predominates. Transfection of human fibroblasts with SCP-2 antisense oligonucleotides inhibits 5-fold the transfer of newly synthesized cholesterol from the endoplasmic reticulum to the plasma membrane via the rapid pathway (57).

Decreased SCP-2 in the rat treated with antisense oligonucleotides: biliary cholesterol secretion. Biliary free cholesterol comes from preformed as well as newly synthesized (20-50% of total) free cholesterol in the endoplasmic reticulum must be transferred to the bile canalicular (plasma) membrane for secretion (rev. in 58). Treatment of rats *in vivo* with SCP-2 antisense oligonucleotide of rats reduced liver SCP-2 by 60% and delayed biliary secretion of cholesterol (58). In contrast, Lith genes induce SCP-2 overexpression during gallstone formation in animals (45,59) and humans (43,60).

SCP-2 deletion in gene targeted mice. Deletion of the SCP-2 gene resulted in lack of both SCP-2 and SCP-x (61). Concomitantly, liver cholesteryl esters and triglycerides both decreased 2-fold. Nevertheless, it was suggested that SCP-2 did not play an obligatory role in intracellular cholesterol trafficking in knock-out mice (61). Consistent with this possibility the SCP-2 gene deleted mice showed 4-fold upregulation of L-FABP (another sterol binding/transfer protein (47,48), and of the cholesterol oxidative enzyme, cholesterol-7-alpha hydroxylase. These observations, as well as the fact that both SCP-2 and SCP-x were deleted, obscure clear cut interpretation of the role of SCP-2 in intracellular cholesterol trafficking in the gene ablated mice. Instead, the data suggest that compensatory mechanism(s) were activated to maintain normal phenotype in the SCP-2 gene-deleted mice.

Overexpression of StAR: mitochondrial steroidogenesis. StAR expression enhances steroidogenesis much more than SCP-2 or any of the other known protein (See chapter by Stocco et al; 62). However, it must be recognized that the cholesterol pool in the outer mitochondrial membrane would become rapidly depleted at the high rate of sterol utilization stimulated by StAR. Thus, other mechanism(s) for sterol transfer from intracellular donor sites (plasma membranes, lysosomes, lipid droplets, etc.) must exist to replenish the outer mitochondrial membrane cholesterol pool (see also chapter by Freeman et al in this volume; 63,64). Thus, by stimulating the latter process SCP-2 may play a supportive role in StAR mediated steroidogenesis.

Overexpression of caveolin: reverse cholesterol transport. MA104 cells overexpressing caveolin had 4-fold enhanced cholesterol transport from endoplasmic reticulum to the plasma membrane and 4-fold cholesterol enrichment in isolated caveolar fractions (28). Since caveolin can migrate from plasma caveolae to the endoplasmic reticulum, it is thought that caveolin may also be directly involved in rapid cholesterol transfer from endoplasmic reticulum to the plasma membrane caveolae with a half-time near 10 min (See chapter by Fielding and Fielding; 4,28). Overexpression of caveolin in L1210-JF cells led to formation of a cytosolic cholesterol transport complex comprised of caveolin-heat shock protein-immunophilins-cholesterol (30). Concomitantly, newly synthesized cholesterol was rapidly (10-20 min) transported from endoplasmic reticulum to plasma membrane caveolae independent of membrane vesicles.

Summary of the role of Sterol Binding Proteins in Cholesterol Trafficking/Metabolism in intact cells. The overwhelming impression from the above studies with transfected intact cells overexpressing L-FABP, SCP-2, caveolin, and StAR are consistent with a potential role of these proteins in the intracellular trafficking/metabolism of cholesterol. Despite these observations, some investigators maintain there is no role for L-FABP (13) or SCP-2 (rev. in 5,20,61,65) in cholesterol binding/metabolism/intracellular trafficking. The basis for the conflicting conclusions is primarily the failure of some *in vitro* binding assays (see above) and the use of mutant cell lines wherein the mutation is not in the SCP-2 gene. Substantial disagreements regarding the quantitative (as opposed to qualitative) intracellular localization of SCP-2 has additionally fueled the controversy. The following sections outline the essential issues.

INTRACELLULAR LOCATION OF STEROL BINDING PROTEINS

The physiological/functional significance and/or mechanism whereby the sterol binding proteins enhance intracellular sterol trafficking cannot be clearly resolved without adequate knowledge of the intracellular localization of these proteins. There is general agreement on the intracellular localization of several of the sterol binding proteins. For example, subcellular fractionation and immunocytochemical studies of bovine and rat indicate that the highest concentration of L-FABP is found in the cytosolic fraction, but significant amounts are associated with mitochondria, microsomes, and nuclei (rev. in 66). As indicated above, caveolin is localized to the cytoplasmic face of plasma membrane caveolae, the trans-Golgi, and cytosol. StAR rapidly translocates from endoplasmic reticulum to mitochondria (See chapter by Stocco et al; 62). In contrast, quantitative determination of intracellular localization/distribution of SCP-2 has been difficult. Qualitatively, SCP-2 is found in highest concentration in peroxisomes. Unfortunately, biologists are increasingly extending this qualitative observation to conclude that SCP-2 is quantitatively localized predominantly/exclusively in the peroxisome and therefore has no role in intracellular cholesterol trafficking. However, this **conclusion is not supported by (i) transfected cell data (see above), (ii) quantitative analysis of the immunocytochemistry (see below), or (iii) subcellular fractionation studies (see below)..**

SCP-2 intracellular localization: subcellular fractionation. While most subcellular fractionation studies agree that in the liver the 58 kDa SCP-x is almost exclusively peroxisomal, a substantial portion of liver 13 kDa SCP-2 may be cytosolic and/or bound to other intracellular membranes. **At least 4-fold more 13 kDa SCP-2 exists in the cytosolic fraction than can be accounted for by contamination with lysed peroxisomes** (67). Several other observations are not entirely consistent with 13 kDa SCP-2 being exclusively peroxisomal: First, if peroxisomal lysis during homogenization were the only consideration then a similar leakage of 58 kDa SCP-x and 13 kDa SCP-2 from the peroxisomal matrix might be expected. **However, the data show that the 58 kDa SCP-x did not leak out of putative lysed peroxisomes.** Subcellular fractionation of rat liver, rat testis, and CHO cells reveals that the **58 kDa SCP-x, but not the 13 kDa SCP-2, copurifies with catalase (a peroxisomal marker enzyme)** activity in subcellular fractionation studies (68,69). In fact, liver peroxisomes devoid of 13 kDa SCP-2, but containing the 58 kDa SCP-x have been isolated (70). Second, treatment of subcellular membrane fractions with proteases releases the 13 kDa SCP-2 (from the cytofacial face of these membranes), but not the 58 kDa SCP-x (68,69,71). Third, isoelectric focusing of 13 kDa SCP-2 released from the protease treated membranes revealed that the membrane bound 13 kDa SCP-2 was more basic than the 13 kDa SCP-2 present in the cytosolic fraction from rat liver (68,69). Fourth, **quantitative analysis of subcellular fractionation data shows that as much as half of 13 kDa SCP-2 is extra-peroxisomal.** For example, the concentration of SCP-2 in highly purified rat liver peroxisomes, microsomes, mitochondria, and cytosol was 10.4, 0.032, 0.17, and 0.59 µg/mg protein, respectively (67). In contrast, the **total amount of SCP-2 was 63, 2, 9, and 54 µg/g liver protein in peroxisomes, microsomes, mitochondria, and cytosolic fractions.**

In summary, **unlike the qualitative assessment of immunocytochemical and subcellular fractionation data, a quantitative assessment clearly shows that nearly half of cellular 13 kDa SCP-2 is extra-peroxisomal. While the organellar concentration of SCP-2 is highest in peroxisomes, the peroxisomes account for only 2.5% of liver protein (67) and 0.66% of typical cell volume (72). In contrast, although the concentration of SCP-2 in liver cytosol is 17-fold lower than in peroxisomes, cytosol accounts for 35% of cell protein and consequently total SCP-2 in cytosol is nearly equivalent to the total in peroxisomes (67).**

SCP-2 intracellular localization: immunocytochemical studies. While immunogold labeling of fixed tissues is very useful for detecting SCP-2 in organelles (peroxisomes) wherein it may be expressed at high levels, the technique is limited in providing quantitative evaluation of how much total SCP-2 is extra-peroxisomal and/or localized in the lesser stained organelles. For example, the 58 kDa SCP-x appears almost exclusively in the peroxisomal matrix (73). However, the 13 kDa SCP-2 is not exclusively found in liver peroxisomes. Immunogold labeling studies of liver and other tissues with polyclonal anti-13 kDa SCP-2 antibody (recognizes all SCP-2 gene products including 58 kDa SCP-x, 46 kDa thiolase, 15 kDa pro-SCP-2, and 13 kDa SCP-2) showed the highest accumulation of gold particles in the matrix of peroxisomes with **significant amounts of**

SCP-2 detected inside mitochondria and also definitely associated with the cytoplasmic face of endoplasmic reticulum and cytosol (65,67,70,74,75). Immunogold labeling did not detect SCP-2 in Golgi apparatus, lysosomes, or the nucleus. Immunofluorescence studies of L-cells transfected with cDNA encoding the 58 kDa SCP-x or the 15 kDa pro-SCP-2 both showed that anti-SCP-2 and anti-catalase (peroxisomal marker) only partly colocalized (56).

Quantitative analysis of immunocytochemical data shows that as much as half of 13 kDa SCP-2 is extra-peroxisomal. (i) The cell volume accounted for by these organelles in cells expressing high levels of SCP-2 (e.g. Leydig cells) is 12, 753, 237, and 564 μ^3 for peroxisomes, microsomes (endoplasmic reticulum), mitochondria, and cytosol (est. from total cell volume - membrane volume), respectively (72). Assuming that the protein/unit volume in these subcellular compartments is similar, multiplication of the SCP-2 concentration in each organelle by the approximate volume due to that fraction suggests that 125, 24, 40, and 333 μg SCP-2 are localized in peroxisomes, microsomes, mitochondria, and cytosol, respectively. (ii) Rat liver hepatoma McA-RH7777 transfected with cDNA encoding the 15 kDa SCP-2 they expressed primarily the 13 kDa SCP-2 at levels similar to those in liver (55). Immunogold labeling indicated that overexpression of SCP-2 in McA-RH7777 hepatoma cells resulted in an 8-fold and 1.3-fold higher level of 13 kDa SCP-2 in the peroxisomes and extraperoxisomal areas, respectively, as compared to mock transfected cells. The ratio of nonspecific to specific SCP-2 immunogold labeling was 8.0 and 2.1 in peroxisomal and extraperoxisomal areas of McA-RH7777 hepatoma cells. If it is assumed that the transfected hepatoma cell peroxisomes account for 2.5% of total protein as they do in liver (72), this would suggest that nearly 75% of total SCP-2 was extraperoxisomal. Consequently, even when the peroxisomal concentration of SCP-2 is 4-fold higher than in extraperoxisomal areas, total SCP-2 was quantitatively 3-fold higher in extraperoxisomal areas of the transfected hepatoma cells. (iii). Quantitative analysis of immunofluorescence of L-cells transfected with cDNA encoding the 58 kDa SCP-x or the 15 kDa pro-SCP-2 both showed that the correlation coefficient of anti-SCP-2 and anti-catalase (peroxisomal marker) was less than 0.5 (56). Complete colocalization would have yielded a coefficient of 1.0.

The immunocytochemical data are in disagreement regarding the intraperoxisomal (matrix vs matrix side of peroxisomal membrane vs cytoplasmic face of peroxisomal membrane) localization of 13 kDa SCP-2 within liver peroxisomes (67,70,73).

In summary, quantitative analysis of immunocytochemical data reveal that SCP-2 is not exclusively peroxisomal. In fact, half or more of cellular anti-SCP-2 immunoreactivity does not colocalize with peroxisomal markers and appears extra-peroxisomal.

POTENTIAL PROBLEMS IN INTRACELLULAR LOCALIZATION

A variety of complexities/potential problems urge caution in interpretation of immunolocalization of proteins such as SCP-2 to specific subcellular sites/organelles. First, since the SCP-2 gene codes for multiple proteins, a single polyclonal antiserum or monoclonal antiserum seldom detects the presence of only a single SCP-2 gene product. This is further complicated by the fact that most polyclonal anti-SCP-2 antisera are not affinity purified. Second, antisera against SCP-2 cross react with several additional proteins that are not SCP-2 gene products (73,76). Since all SCP-2 gene transcripts contain a C-terminus

SKF consensus sequence, a peroxisomal PTS-1 targeting sequence, and the 58 kDa SCP-x also contains a PTS-2 peroxisomal targeting sequence in the amino terminus (77), polyclonal antisera against these proteins may contain some antibodies recognizing these targeting epitopes may be common to many peroxisomal proteins. Finally, the 20 amino acid amino terminus of the 15 kDa pro-SCP-2 shares features common to mitochondrial targeting sequences (78). Which targeting sequence is dominant, the mitochondrial vs peroxisomal, is not known. By analogy, alanine:glyoxylate aminotransferase is another protein which has both mitochondrial and peroxisomal targeting sequences and the mitochondrial targeting sequence is dominant (79). Third, with rare exception (80) SCP-2 gene products are post-translationally modified (55,56) in intracellular location(s) as yet not completely clear (rev. in 81-83). In fact, some data show that the native 13 kDa SCP-2 isolated from bovine and rat liver do not contain the C-terminal residue(s) necessary for peroxisomal targeting (84,85). Fourth, it must be recognized that the transcription and/or translation of the 15 kDa pro-SCP-2 and 58 kDa SCP-x transcript/translation may be regulated independently in the cell (86-89). Thus, down -regulation of expression of one or both 15 kDa pro-SCP-2 and 58 kDa SCP-x may explain observations of the lack of one or both of these proteins in some mutant cell lines rather than lack of peroxisomes for the proteins to be imported into and post-translationally processed.

In summary, a major complication in SCP-2 subcellular localization is the specificity of antisera to SCP-2. In most previous studies polyclonal antisera to the 13 kDa SCP-2 have been used in both Western blots of isolated membrane fractions and in intracellular localization studies by immunocytochemistry. As described above, the use of such antisera has significant limitations that may lead to erroneous conclusions about the intracellular localization of SCP-2. Coupled with the fact that both cellular subfractionation studies and immunocytochemical studies are limited by the fragility of the peroxisomes to homogenization and harsh processing conditions, respectively, this would result in either case in potential release of intra-peroxisomal proteins which could subsequently associate with other subcellular fractions. These observations urge that caution must be excercised in the interpretation of immunological data on SCP-2 intracellular localization.

MECHANISM(S) OF SCP-2 MEDIATED STEROL TRAFFICKING

The following sections describe several potential mechanisms reconciling the majority of data on intracellular localization of cholesterol binding proteins with the intact and/or transfected cell data clearly showing a role for SCP-2 as well as other cholesterol binding proteins in intracellular sterol trafficking and metabolism.

Aqueous carriers of cholesterol? Based on exchange data obtained *in vitro* and on the effects of sterol utilizing enzymes *in vitro,* it was originally thought that the 13 kDa SCP-2 and other sterol binding proteins acted as aqueous carriers of cholesterol. However, a growing consensus indicates that, although a cholesterol binding site is required for activity SCP-2 or L-FABP, both proteins can also enhance intermembrane sterol transfer without acting as aqueous cholesterol carriers/transporters (24,90). Recent reports demonstrated that caveolin binds cholesterol *in vitro* and, more importantly, a caveolin-cholesterol-heat shock protein complex occurs in cytosol (44), suggesting that this complex may act as an aqueous cholesterol carrier between the trans-Golgi network and caveolae. While StAR has been shown to bind sterol (27), there is as yet no direct evidence showing that StAR is an aqueous soluble cholesterol carrier.

Direct interaction with membranes? Both L-FABP and SCP-2 can bind to plasma membranes to enhance intermembrane sterol transfer through a dialysis membrane whether or not the dialysis membrane is permeant to SCP-2 or L-FABP (24,90). A membrane interaction mechanism might involve formation of a transient SCP-2-cholesterol (or L-FABP-cholesterol) complex which rapidly dissociates (hopping mechanism) or remains membrane bound (scooting mechanism). Selective desorption and specific targeting of cholesterol within the cell may be accomplished by SCP-2 or L-FABP through their selective binding to donor and/or acceptor membranes. This possibility was demonstrated by *in vitro* exchange assays between dissimilar donor and acceptor biomembrane pairs (18,37). In such assays selectively enhanced intermembrane sterol transfer between certain donor/acceptor pairs much more than others. Such selectivity can also account for transfer of cholesterol up a concentration gradient.

Role in vesicular cholesterol trafficking. By binding to specific intracellular vesicles SCP-2 or other proteins may mediate intracellular trafficking of cholesterol rich vesicles. The possibility of cytosolic lipid transfer protein mediated vesicular trafficking has been demonstrated for phosphatidylinositol transfer between endoplasmic reticulum and the plasma membrane of yeast (rev. in 5). Likewise, caveolin is a cholesterol binding protein that binds to both cholesterol acceptor (specific plasma membrane domains, i.e. caveolae) and donor (trans-Golgi network) membranes (rev. in 4).

Role of peroxisomal SCP-2 in intracellular cholesterol trafficking and/or metabolism? As indicated in the above sections a very high concentration of SCP-2 is found in peroxisomes. Since both 13 kDa SCP-2 and 58 kDa SCP-x can transfer cholesterol between membranes *in vitro* (91), several mechanisms may be considered:

First, peroxisomal 13 kDa SCP-2 (peroxisomal matrix, the matrix leaflet, or the cytofacial leaflet) and 58 kDa SCP-x (matrix and/or matrix leaflet) are localized in the peroxisome. Since the transbilayer migration rate of cholesterol in biological membranes is rapid (rev. in 2), translocation of cholesterol across the peroxisomal membrane is likely to be rapid. Thus, SCP-2 bound on either face of the peroxisome could enhance efflux/transfer of intraperoxisomal cholesterol to other organelles. Alternately, the peroxisomal membrane and its associated SCP-2 could act as an intermediate membrane for transfer of cholesterol between other membranes where a direct pathway may not predominate. By analogy, one pathway for cholesterol trafficking is from the lysosome to the plasma membrane and thence to the endoplasmic reticulum (see chapter by Liscum). Consistent with these possibilities is the observation of an inverse relationship between peroxisomal SCP-2 levels and free cholesterol content (92). Furthermore, peroxisomes are in close proximity to sites of cholesterol storage such as lipid droplets (93), endocytic vesicles (4,94,95), endoplasmic reticulum (rev. in 81), and apical plasma membranes (96). This provides the peroxisome with ready access to multiple cellular cholesterol compartments. While immunocytochemical evidence suggests that peroxisome associated SCP-2 does not bind to lipid droplets, it has been pointed out that SCP-2 may enhance/target cholesterol transfer by interacting with cholesterol in either the donor or acceptor membrane or both (rev. in 2). Finally, in certain genetic disorders peroxisomes are absent from cells and the cellular level of SCP-2 is much reduced (rev. in 81,97). Such peroxisome deficient fibroblasts exhibit impaired LDL uptake/metabolism such that cellular cholesterol mass is reduced by half and cellular cholesterol biosynthesis is reduced as much as 75% (98).

Second, peroxisomes are essential for the formation of the bile acids, e.g. cholic acid. The enzyme, trihydroxycholestanoyl-CoA oxidase, is responsible for formation of cholic acid from trihydroxycholestanoic acid is peroxisomal (rev. in 81). *In vitro* studies show that the 13 kDa SCP-2 stimulates the 7α-hydroxylation of cholesterol, the key regulatory step in endoplasmic reticulum bile acid synthesis (rev. in 31,34). Thus, SCP-2 may similarly stimulate the entry of extraperoxisomal intermediates of bile acids and/or the conversion of such intermediates to cholic acid. Interestingly, the 58 kDa SCP-x is also involved in peroxisomal oxidation reactions (91,99,100). Since SCP-2 is activated by the presence of anionic lipids in membranes (rev. in 2,101), the insertion of bile acids into membranes may also serve to stimulate/activate SCP-2's ability to transfer sterols and/or bile acids into and out of peroxisomes.

Third, peroxisomes contain most of the enzymes necessary for cholesterol synthesis (rev. in 81,102,103). Purified peroxisomes contain all the necessary enzymes to convert mevalonate to farnesyl pyrophosphate. The 13 kDa SCP-2 binds several of these intermediates including dimethylallyl-PP, geranyl-PP, and farnesyl-PP (17). Furthermore, the 13 kDa SCP-2 stimulates the isopentyl transferase enzyme involved in chain elongation of such isoprenoids in peroxisomes (104). Because it binds farnesyl-PP the peroxisomal 13 kDa SCP-2 may also be involved in the transport/efflux of farnesyl-PP from the peroxisome for further elongation/cyclization to form lanosterol in the endoplasmic reticulum. The microsomal conversion of lanosterol to cholesterol is stimulated *in vitro* by 13 kDa SCP-2 (31). Alternately, peroxisomal 13 kDa SCP-2 may facilitate transfer/entry/uptake of lanosterol back into the peroxisome wherein 13 kDa SCP-2 may again contribute to the enzymatic conversion of lanosterol to cholesterol (rev. in 102). Finally, peroxisomes can accomplish the entire conversion of mevalonate to cholesterol *in vitro* in the presence of a cytosolic fraction (102). The rate of this conversion was 87 pmol/mg/h vs microsomal rate of 135 pmol/mg/h. This suggests that peroxisomal cholesterol synthesis represents a significant proportion of total cellular cholesterol synthesis. Indeed, cellular cholesterol biosynthesis is reduced as much as 75% in peroxisome deficient cells (98).

CONCLUSIONS AND FUTURE DIRECTIONS

Since the majority of exogenous cholesterol enters the cell via the LDL receptor endoytic-lysosomal pathway, mechanism(s) must exist for cholesterol to (i) efflux from the lysosome and (ii) become targeted to specific organelles/sites within the cell. The observation that the spontaneous desorption of sterols from lysosomes is very slow, coupled with a nearly 50-fold enhancement by physiological levels of SCP-2 suggests that cytosolic sterol transfer proteins can be very important for sterol exit from the lysosomal membrane (36). Likewise, cholesterol efflux from the lysosome must be specifically targeted since intracellular cholesterol distribution is asymmetric. Most membrane bound cholesterol is found in the plasma membrane and in intracellular lipid storage droplets (rev. in 1,2,31). Much lower quantities are in intracellular membranes such as microsomes and mitochondria. Both *in vitro* and transfected cell data suggest that several intracellular sterol binding proteins are involved in cholesterol trafficking/metabolism and targeting. Some of these proteins appear to bind to specific membrane cholesterol containing domains wherein they enhance cholesterol dynamics by acting as 'keys' to enable cholesterol to enter and/or efflux. Although an intact cholesterol binding site appears essential for this activity, the exact mechanism whereby these sterol binding proteins act in the intact cell are not known.

While there is general agreement that the peroxisomes are the intracellular organelle with the highest concentration of SCP-2, **quantitative analysis of total SCP-2 is consistent with a substantial fraction of total 13 kDa SCP-2 being extra-peroxisomal.** In contrast, it is generally agreed that the 58 kDa SCP-x appears not only enriched in peroxisomes but quantitatively present primarily in peroxisomes. Finally, the existing functional data obtained *in vitro*, in intact transfected cells, and in animal studies can be interpreted in terms of mechanism(s) that accommodate both peroxisomal as well as extra-peroxisomal SCP-2 as playing role(s) in intracellular cholesterol trafficking.

The finding that cells contain multiple cholesterol binding proteins presents a tremendous challenge and opportunity to examine the interactions between potentially overlapping functions of some of these proteins. At least three cholesterol binding proteins (L-FABP, SCP-2, and caveolin) have been shown to mediate rapid (10-20 min) cholesterol transport between endoplasmic reticulum and plasma membranes, independent of vesicles. Furthermore, multiple vesicular pathways also exist. Thus, studies resolving the individual contributions of these different cholesterol binding proteins as well as vesicular pathways are particularly important if we are to fully realize the potential of overexpressing or gene ablated animals wherein only a single one of the cholesterol binding proteins has been genetically manipulated. As shown with SCP-2 antisense cDNA and with SCP-2 gene ablated mice, if one cholesterol binding proteins is down regulated then other cholesterol binding proteins and/or vesicular pathways become upregulated to compensate and yield an apparent normal phenotype in cholesterol trafficking/metabolism.

Despite all the excellent work on intracellular cholesterol trafficking, to date it has been extremely difficult to visualize this process in intact cells by imaging microscopy. The present lack of adequate methods in this regard illustrates that perhaps the most significant technological advance that would greatly aid resolution of the intracellular cholesterol trafficking pathways would be the development of such techniques in intact cells. Until recently the only way to visualize cholesterol within the cell was through use of indirect histochemical stains, such as filipin (which does not bind to all cholesterol rich domains), or synthetic fluorescent sterols such as pyrene or NBD labeled cholesterol. (which differ structurally from cholesterol in geometry, polarity, rate of desorption from membranes, etc.) (rev. in 105,106). These difficulties have essentially been eliminated through three much needed technical advances. **First, the development of a fluorescent sterol, dehydroergosterol, that structurally resembles cholesterol much more than previous sterol probe molecules.** Dehydroergosterol is a unique fluorescent sterol in that it is naturally occurring sterol normally found as a substantial (as much as 20% of total sterol) membrane component in several animal cells (yeast and sponge). Furthermore, dehydroergosterol can be incorporated into cultured fibroblasts, esterified therein, codistributes with cholesterol in fibroblast membranes, replace >80% of endogenous sterol, and not significantly affect a variety of lipid and membrane functional (Na,K-ATPase, receptor function, etc.) parameters (rev. in 1,2,31,105-107). As cited in the above references, dehydroergosterol exchanges between membranes with essentially the same half-times as radiolabeled cholesterol *in vitro*. **Second, the use of single-photon fluorescence imaging of dehydroergosterol.** Dehydroergosterol's specific excitation and emission properties allow visualization of its intracellular distribution in fibroblasts by fluorescence deconvolution imaging (Fig. 4A), where the resulting three-dimensional image shows a number of bright structures resembling lipid droplets. These data represent the first visualization of a naturally occurring fluorescent sterol in a mammalian cell. Unfortunately, single photon excitation of dehydroergosterol at ultraviolet wavelengths results in severe photobleaching which essentially precludes kinetic measurements and determination of

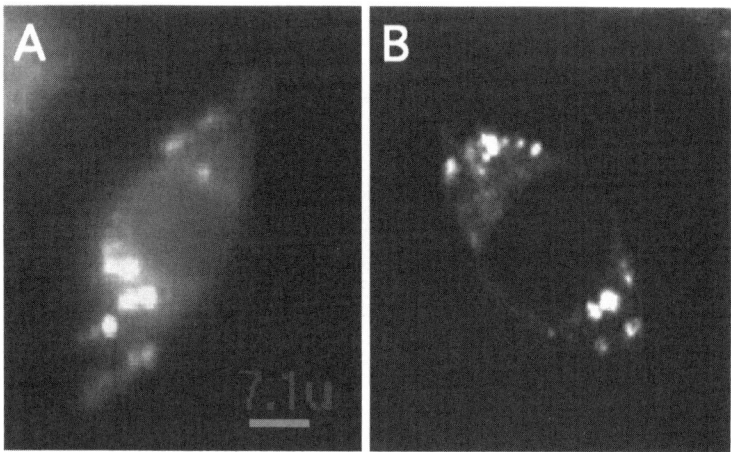

Fig. 4. Fluorescence Imaging of a Naturally Occurring Fluorescent Sterol (Dehydroergosterol) in Cultured L-cell Fibroblasts. L-cell fibroblasts were cultured on cover slips overnight in medium with 10% FBS and 20 µg/ml dehydroergosterol. Panel A. One-photon excitation (near the dehydroergosterol excitation maximum at 325 nm) imaging was done by the Texas A&M, CVM Image Analysis Laboratory by fluorescence deconvolution (CELLscan, Scanalytics Inc., Billerica, MA). Panel B. Three-photon excitation of dehydroergosterol in the cells was done in collaboration with Drs. Enrico Gratton and Peter Tso at the U. of Illinois Laboratory for Fluorescence Dynamics. Three-photon excitation was at 960 nm with a Coherent Mira 900 Ti/Sa laser pumped by a 10 Watt Coherent argon ion laser while emission near 375 nm was visualized through a dichroic filter.

intracellular diffusion rates. **Third, the use of multiphoton fluorescence imaging of dehydroergosterol eliminates many of the photobleaching as well as other photoartifacts associated with standard single photon imaging.** Three-photon excitation shows the intracellular distribution of dehydroergosterol in a single plane through intact L-cell fibroblasts (108). In this technology a very small volume (0.01 μ^3) of the cell is excited with infrared radiation, fluorescence from this volume is detected, and the volume is moved throughout the scanning plane to yield an image (Fig. 4B). Clearly, dehydroergosterol is distributed inside the cell, with high concentrations in intracellular structures which, based on colocalization with Nile Red, are believed to be lipid droplets. These technological advances (dehydroergosterol and multiphoton fluorescence imaging) now provide exciting tools for resolving the kinetics, diffusion rates, and targeting of intracellular sterol trafficking in living cells. They will also allow the first accurate quantitation of the sterol's three-dimensional distribution in the cell without the photobleaching artifacts s inherent in standard single photon imaging. Thus, the future for resolving the intracellular pathways of cholesterol movement in intact cells looks very promising.

ACKNOWLEDGEMENTS

This work was supported in part by the USPHS NIH grant GM 31651. The generous assistance of Drs. Peter Tso and Enrico Gratton and the Laboratory for Fluorescence Dynamics, Univ. of Illinois, Champaign-Urbana , in obtaining the three-photon excitation image of dehydroergosterol in L-cells was much appreciated.

REFERENCES

1.   Schroeder F, Jefferson JR, Kier AB, et al. Membrane cholesterol dynamics: cholesterol domains and kinetic pools. Proc Soc Exp Biol Med 1991; 196:235-52.
2.   Schroeder F, Frolov AA, Murphy EJ, et al. Recent advances in membrane cholesterol domain dynamics and intracellular cholesterol trafficking. Proc Soc Exp Biol Med 1996; 213:150-77.
3.   Liscum L, Underwood KW. Intracellular cholesterol transport and compartmentation. J Biol Chem 1995; 270:15443-6.
4.   Fielding CJ, Fielding PE. Intracellular cholesterol transport. J Lipid Res 1997; 38:1503-21.
5.   Allan D, Kallen KJ. Transport of lipids to the plasma membrane in animal cells. Prog Lipid Res 1993; 32:195-219.
6.   DeGrella RF, Simoni RD. Intracellular transport of cholesterol to the plasma membrane. J Biol Chem 1982; 257:14256-62.
7.   Hubbell T, Behnke WD, Woodford JK, et al. Recombinant liver fatty acid binding protein interactions with fatty Acyl-Coenzyme A. Biochemistry 1994; 33:3327-34.
8.   Frolov A, Cho TH, Murphy EJ, et al. Isoforms of rat liver fatty acid binding protein differ in structure and affinity for fatty acids and fatty acyl CoAs. Biochemistry 1997; 36:6545-55.
9.   Rustow B, Risse S, Kunze D. [Endogenous lipid pattern, organ distribution and effect of diet on a fatty acid binding protein fraction of rat liver cytosol]. [German]. Acta Biolog Med Germ 1982; 41:439-45.
10.   Schroeder F, Dempsey ME, Fischer RT. Sterol and squalene carrier protein interactions with fluorescent delta 5,7,9(11)-cholestatrien-3 beta-ol. J Biol Chem 1985; 260:2904-11.
11.   Schroeder F, Butko P, Nemecz G, et al. Sterol carrier protein: a ubiquitous protein

in search of a function. In: Verna R, Blumenthal R, Frati L., eds. Bioengineered Molecules: Basic and Clinical Aspects. New York, NY: Raven Press, 1989: 29-45.

12. Fischer RT, Cowlen MS, Dempsey ME, et al. Fluorescence of delta 5,7,9(11),22-ergostatetraen-3 β-ol in micelles, sterol carrier protein complexes, and plasma membranes. Biochemistry 1985; 24:3322-31.

13. Rolf B, Oudenampsen-Kruger E, Borchers T, et al. Analysis of the ligand binding properties of recombinant bovine liver-type fatty acid binding protein. Biochim Biophys Acta 1995; 1259:245-53.

14. Sams GH, Hargis BM, Hargis PS. Identification of two lipid binding proteins from liver of *Gallus domesticus*. Comp Biochem Phys 1991; 99B:213-9.

15. Woodford JK, Behnke WD, Schroeder F. Liver fatty acid binding protein enhances sterol transfer by membrane interaction. Mol Cell Biochem 1995; 152:51-62.

16. Schroeder F, Myers-Payne SC, Billheimer JT, et al. Probing the ligand binding sites of fatty acid and sterol carrier proteins: effects of ethanol. Biochemistry 1995; 34:11919-27.

17. Frolov A, Miller K, Billheimer JT, et al. Lipid specificity and location of the sterol carrier protein-2 fatty acid binding site: A fluorescence displacement and energy transfer study. Lipids 1997; 32:1201-9.

18. Frolov A, Cho TH, Billheimer JT, et al. Sterol carrier protein-2, a new fatty acyl coenzyme A-binding protein. J Biol Chem 1996; 271:31878-84.

19. Chanderbhan R, Noland BJ, Scallen TJ, et al. Sterol carrier protein2. Delivery of cholesterol from adrenal lipid droplets to mitochondria for pregnenolone synthesis. J Biol Chem 1982; 257:8928-34.

20. Gadella TW, Jr., Wirtz KW. The low-affinity lipid binding site of the non-specific lipid transfer protein. Implications for its mode of action. Biochim Biophys Acta 1991; 1070:237-45.

21. Schroeder F, Butko P, Nemecz G, et al. Interaction of fluorescent delta 5,7,9(11),22-ergostatetraen-3β-ol with sterol carrier protein-2. J Biol Chem 1990; 265:151-7.

22. Colles SM, Woodford JK, Moncecchi D, et al. Cholesterol interactions with recombinant human sterol carrier protein-2. Lipids 1995; 30:795-804.

23. Stolowich NJ, Frolov A, Scott AI, et al. ^{13}C-NMR Elucidation of the sterol carrier protein-2 cholesterol and fatty acid binding sites. 1998; submitted.

24. Woodford JK, Colles SM, Myers-Payne S, et al. Sterol carrier protein-2 stimulates intermembrane sterol transfer by direct membrane interaction. Chem Phys Lipids 1995; 76:73-84.

25. Stocco DM. Acute regulation of Leydig cell steroidogenesis. In: Anita H. Payne, Matthew P. Hardy, Lonnie D. Russell., eds. The Leydig Cell. Vienna, IL: Cache River Press, 1996: 241-58.

26. Pfeifer SM, Furth EE, Ohba T, et al. Sterol carrier protein 2: a role in steroid hormone synthesis? J Steroid Biochem Mol Biol 1993; 47:167-72.

27. Liu Z, Frolov A, Schroeder F, et al. Does cholesterol bind to the steroidogenic acute regulatory (STAR) protein? Biology of Reproduction 1996; 54:194

28. Smart EJ, Ying Y, Donzell WC, et al. A role for caveolin in transport of cholesterol from endoplasmic reticulum to plasma membrane. J Biol Chem 1996;

230

271:29427-35.

29. Murata M, Peranen J, Schreiner R, et al. VIP21/caveolin is a cholesterol-binding protein. Proc Natl Acad Sci 1995; 92:10339-43.

30. Uittenbogaard A, Ying YS, Smart EJ. Characterization of a Cytosolic Heat-shock Protein-Caveolin Chaperone Complex. Biological Chemistry 1998; 273:6525-32.

31. Moncecchi DM, Nemecz G, Schroeder F, et al. The participation of sterol carrier protein-2 (SCP-2) in cholesterol metabolism. In: Patterson GW, Nes WD., eds. Physiology and Biochemistry of Sterols. Champaign, IL: Am. Oil Chem. Soc. Press, 1991: 1-27.

32. Hapala I, Kavecansky J, Butko P, et al. Regulation of membrane cholesterol domains by sterol carrier protein-2. Biochemistry 1994; 33:7682-90.

33. Billheimer JT, Reinhart MP. Intracellular trafficking of sterols. Sub-Cell Biochem 1990; 16:301-31.

34. Vahouny GV, Chanderbhan R, Kharoubi A, et al. Sterol carrier and lipid transfer proteins. Adv Lipid Res 1987; 22:83-113.

35. Zilversmit DB. Lipid transfer proteins. J Lipid Res 1984; 25:1563-9.

36. Schoer J, Gallegos A, Petrescu A, et al. Lysosomal membrane sterol transfer: contribution of SCP-2 and pro-SCP-2 to intermembrane sterol transfer. 1998; submitted.

37. Frolov A, Woodford JK, Murphy EJ, et al. Spontaneous and protein-mediated sterol transfer between intracellular membranes. J Biol Chem 1996; 271:16075-83.

38. Batenburg JJ, Ossendorp BC, Snoek GT, et al. Phospholipid-transfer proteins and their mRNAs in developing rat lung and in alveolar type-II cells. Biochem J 1994; 298:223-9.

39. McLean MP, Billheimer JT, Warden KJ, et al. Differential expression of hepatic sterol carrier proteins in the streptozotocin-treated diabetic rat. Endocrinology 1995; 136:3360-8.

40. Hirai A, Kino T, Tokinaga K, et al. Regulation of sterol carrier protein 2 (SCP2) gene expression in rat peritoneal macrophages during foam cell formation. J Clin Invest 1994; 94:2215-23.

41. Lipka G, Schulthess G, Thurnhofer H, et al. Characterization of lipid exchange proteins isolated from small intestinal brush border membrane. J Biol Chem 1995; 270:5917-25.

42. Schulthess G, Lipka G, Compassi S, et al. Absorption of monoacylglycerols by small intestinal brush border membrane. Biochemistry 1994; 33:4500-8.

43. Kawata S, Imai Y, Inada M, et al. Modulation of cholesterol 7-α hydroxylase activity by nsLTP in human liver -- possibly alters regulation of cytosolic level in patients with gallstones. Clin Chim Acta 1991; 197:201-8.

44. Bun-ya M, Maebuchi M, Kamiryo T, et al. Thiolase Involved in Bile Acid Formation. J Biochem 1998; 123:347-52.

45. Fuchs M, Lammert F, Wang DQH, et al. Sterol carrier protein-2 but not phosphatidylcholine-gene expression is enhanced during cholesterol gallstone formation in mice. FASEB J 1997; 11:A-1267

46. Baum CL, Kansal S, Davidson NO. Regulation of sterol carrier protein-2 gene expression in rat liver and small intestine. J Lipid Res 1993; 34:729-39.

47. Jefferson JR, Powell DM, Rymaszewski Z, et al. Altered membrane structure in transfected mouse L-Cell fibroblasts expressing rat liver fatty acid-binding protein. J Biol Chem 1990; 265:11062-8.

48. Jefferson JR, Slotte JP, Nemecz G, et al. Intracellular sterol distribution in transfected mouse L-cell fibroblasts expressing rat liver fatty acid binding protein. J Biol Chem 1991; 266:5486-96.

49. Incerpi S, Jefferson JR, Wood WG, et al. Na pump and plasma membrane structure in L-cell fibroblasts expressing rat liver fatty acid binding protein. Arch Biochem Biophys 1992; 298:35-42.

50. Incerpi S, Vito PD, Luly P, et al. The Sodium Pump and Plasma Membrane Structure: Effect of Insulin and Aging. In: R. Verna, Y. Nishizuka., eds. Biotechnology of Cell Regulation, Serono Symposia Series Adv. in Exptl. Med. NY: Raven Press, 1991: 409-23.

51. Yamamoto R, Kallen CB, Babalola GO, et al. Cloning and expression of a cDNA encoding human sterol carrier protein 2. Proc Natl Acad Sci USA 1991; 88:463-7.

52. Chanderbhan R, Kharroubi A, Noland BJ, et al. Sterol carrier protein 2: Further evidence for its role in adrenal steroidogenesis. Endocrine Res 1986; 12:351-70.

53. Moncecchi DM, Murphy EJ, Prows DR, et al. Sterol carrier protein-2 expression in mouse L-cell fibroblasts alters cholesterol uptake. Biochim Biophys Acta 1996; 1302:110-6.

54. Murphy EJ, Schroeder F. Sterol carrier protein-2 mediated cholesterol esterification in transfected L-cell fibroblasts. Biochimica et Biophysica Acta; 1997; 1345:283-92.

55. Baum CL, Reschly EJ, Gayen AK, et al. Sterol carrier protein-2 overexpression enhances sterol cycling and inhibits cholesterol ester synthesis and high density lipoprotein cholesterol secretion. J Biol Chem 1997; 272:6490-8.

56. Atshaves BP, Petrescu A, Starodub O, et al. Expression and Intracellular Processing of the 58 kDa Sterol Carrier Protein 2/3-Oxoacyl-CoA Thiolase in Transfected Mouse L-cell Fibroblasts. J Biol Chem 1998; submitted.

57. Puglielli L, Rigotti A, Greco AV, et al. Sterol carrier protein-2 is involved in cholesterol transfer from the endoplasmic reticulum to the plasma membrane in human fibroblasts. J Biol Chem 1995; 270:18723-6.

58. Puglielli L, Rigotti A, Amigo L, et al. Modulation on intrahepatic cholesterol trafficking: Evidence by in vivo antisense treatment for the involvement of sterol carrier protein-2 in newly synthesized cholesterol transfer into bile. Biochem J 1996; 317:681-7.

59. Fuchs M, Lammert F, Wang DQH, et al. Lith genes induce overexpression of sterol carrier protein 2 during cholesterol gallstone formation. FASEB J 1997; 11:A1060

60. Ito T, Kawata S, Imai Y, et al. Hepatic Cholesterol Metabolism in Patients With Cholesterol Gallstones: Enhanced Intracellular Transport of Cholesterol. Gastroenterology 1996; 110:1619-27.

61. Seedorf U, Raabe M, Ellinghaus P, et al. Defective peroxisomal catabolism of branched fatty acyl coenzyme A in mice lacking the sterol carrier protein-2/sterol carrier protein-x gene function. Genes and Development 1998; 12:1189-201.

62. Lin D, Sugawara T, Strauss JFI, et al. Role of steroidogenic acute regulatory

protein in adrenal and gonadal steroidogenesis. Science 1995; 267:1828-31.

63. Nagy L, Freeman DA. Effect of cholesterol transport inhibitors on steroidogenesis and plasma membrane cholesterol transport in cultured MA-10 Leydig tumor cells. Endocrinology 1990; 126:2267-76.

64. Freeman DA. Plasma membrane cholesterol: removal and insertion into the membrane and utilization as substrate for steroidogenesis. Endocrinology 1989; 124:2527-30.

65. Reinhart MP, Avart SJ, Dobson TO, et al. The presence and subcellular distribution of sterol carrier protein-2 in embryonic chick tissue. Biochem J 1993; 295:787-92.

66. Bordewick U, Heese M, Borchers T, et al. Compartmentation of hepatic fatty-acid-binding protein in liver cells and its effect on microsomal phosphatidic acid biosynthesis. Biol Chem Hoppe-Seyler 1989; 370:229-38.

67. Tsuneoka M, Yamamoto A, Fujiki V, et al. Nonspecific lipid transfer protein (Sterol carrier protein$_2$) is located in rat liver peroxisomes. J Biochem 1988; 104:560-4.

68. van Heusden GPH, Bos K, Wirtz KWA. The occurrence of soluble and membrane-bound non-specific lipid transfer protein (sterol carrier protein 2) in rat tissues. Biochim Biophys Acta 1990; 1046:315-21.

69. van Heusden GPH, Bos K, Raetz CR, et al. Chinese hamster ovary cells deficient in peroxisomes lack the nonspecific lipid transfer protein (sterol carrier protein 2). J Biol Chem 1990; 265:4105-10.

70. Van der Krift TP, Leunissen J, Teerlink T, et al. Ultrastructural localization of a peroxisomal protein in rat liver using the specific antibody against the nonspecific lipid transfer protein (sterol carrier protein -2). Biochim Biophys Acta 1985; 812:387-92.

71. van Haren L, Teerds KJ, Ossendorp BC, et al. Sterol carrier protein 2 (non-specific lipid transfer protein) is localized in membranous fractions of Leydig cells and Sertoli cells but not in germ cells. Biochim Biophys Acta 1992; 1124:288-96.

72. Mendis-Handagama SM, Watkins PA, Gelber SJ, et al. Leydig cell peroxisomes and sterol carrier protein-2 in luteinizing hormone-deprived rats. Endocrinology 1992; 131:2839-45.

73. Ossendorp BC, Voorhout WF, Van Amerongen A, et al. Tissue-specific distribution of a peroxisomal 46-kDa protein related to the 58-kDa protein (sterol carrier protein X; sterol carrier protein 2/3-oxoacyl-CoA thiolase). Arch Biochem Biophys 1996; 334:251-60.

74. Keller GA, Scallen TJ, Clarke D, et al. Subcellular localization of sterol carrier protein-2 in rat hepatocytes: its primary localization to peroxisomes. J Cell Biol 1989; 108:1353-61.

75. Mendis-Handagama SM, Watkins PA, Gelber SJ, et al. Luteinizing hormone causes rapid and transient changes in rat Leydig cell peroxisome volume and intraperoxisomal sterol carrier protein-2 content. Endocrinology 1990; 127:2947-54.

76. Pu L, Foxworth WB, Kier AB, et al. Characterization of 30 kDa SCP-2 related protein from rat liver. Protein Exp Purif 1998; in press:

77. Terlecky SR, Wreiner EAC, Nuttley WM, et al. Signals, receptors, and cytosolic factors involved in peroxisomal protein import. In: Reddy JK, Suga T, Mannaerts GP, Lazarow PB, Subramani S., eds. Peroxisomes: Biology and Role in Toxicology and Disease. New York: Ann. N.Y. Acad. Sci. 1996: 11-20.

78. Pfeifer SM, Furth EE, Ohba T, et al. Sterol carrier protein 2: A role in steroid hormone synthesis? J Steroid Biochem Molec Biol 1993; 47:167-72.

79. Oda T, Funai T, Ichiyama A. Generation from a single gene of two mRNAs that encoded the mitochondrial and peroxisomal serine: pyruvate aminotransferase of rat liver. J Biol Chem 1990; 265:7513-9.

80. Johnson WJ and Reinhart MP. Lack of requirement for sterol carrier proein-2 in the intraclelular trafficking of lysosomal cholesterol . J Lip Res 1994; 35:563-573.

81. van den Bosch H, Schutgens RBH, Wanders RJA, et al. Biochemistry of peroxisomes. Ann Rev Biochem 1992; 61:157-97.

82. Wirtz KWA. Phospholipid transfer proteins revisited. Biochem J 1997; 324:353-60.

83. Heikoop JC, Ossendorp BC, Wanders RJ, et al. Subcellular localization and processing of non-specific lipid transfer protein are not aberrant in Rhizomelic Chondrodysplasia Punctata fibroblasts. FEBS Ltr 1992; 299:201-4.

84. Westerman J, Wirtz KW. The primary structure of the nonspecific lipid transfer protein (sterol carrier protein 2) from bovine liver. Biochem Biophys Res Comm 1985; 127:333-8.

85. Pastuszyn A, Noland BJ, Bazan F, et al. Primary sequence and structural analysis of sterol protein-2 from rat liver: homology with immunoglobulins. J Biol Chem 1987; 262:13219-27.

86. Ohba T, Holt JA, Billheimer JT, et al. Human Sterol Carrier Protein x/Sterol Carrier Protein 2 Gene Has Two Promoters. Biochemistry 1995; 34:10660-8.

87. Seedorf U, Raabe U, Assmann G. Cloning, expression and sequence of SCP-X encoding cDNAs and a related pseudogene. Gene 1993; 123:165-72.

88. Seedorf U, Assmann G. Cloning, expression and nucleotide sequence of rat liver sterol carrier protein-2 cDNAs. J Biol Chem 1991; 266:630-6.

89. Ohba T, Rennert H, Pfeifer SM, et al. The structure of the human sterol carrier protein X/sterol carrier protein 2 gene (SCP2). Genomics 1994; 24:370-4.

90. Brdiczka D. Contact sites between mitochondrial envelope membranes. Structure and function in energy-and protein-transfer. Biochim Biophys Acta 1991; 1071:291-312.

91. Seedorf U, Brysch P, Engel T, et al. Sterol carrier protein X is peroxisomal 3-oxoacyl coenzyme A thiolase with intrinsic sterol carrier and lipid transfer activity. J Biol Chem 1994; 269:21277-83.

92. Lyons HT, Kharroubi A, Wolins N, et al. Elevated cholesterol and decreased sterol carrier protein-2 in peroxisomes from AS-30D hepatoma compared to normal rat liver. Arch Biochem Biophys 1991; 285:238-45.

93. Mendis-Handagama SM, Aten RF, Watkins PA, et al. Peroxisomes and sterol carrier protein-2 in luteal cell steroidogenesis: a possible role in cholesterol transport from lipid droplets to mitochondria. Tissue Cell 1995; 27:483-90.

94. Hamilton RL, Moorehouse A, Havel RJ. Isolation and properties of nascent

lipoproteins from highly purified rat hepatocytic Golgi fractions. J Lipid Res 1991; 32:529-43.

95.   Lange Y. Tracking cell cholesterol with cholesterol oxidase. [Review]. J Lipid Res 1992; 33:315-21.

96.   Wouters FS, Markman M, de Graaf P, et al. The immunohistochemical localization of the non-specific lipid transfer protein (sterol carrier protein-2) in rat small intestine enterocytes. Biochim Biophys Acta 1995; 1259:192-6.

97.   Liscum L, Dahl NK. Intracellular cholesterol transport. J Lipid Res 1992; 33:1239-54.

98.   Mandel H, Getsis M, Rosenblat M, et al. Impaired cholesterol synthesis rate in fibroblasts and reduced cellular uptake of LDL's derived from peroxisome deficient patients cause cellular cholesterol deficiency in peroxisome deficient fibroblasts. In: Reddy JK, Suga T, Mannaerts GP, Lazarow PB, Subramani S., eds. Peroxisomes: Brotozy and Role in Toxicology and Disease. New York: Ann. New York Acad. Sci. USA, 1996: 752-5.

99.   Wanders RJA, Denis S, Wouters F, et al. Sterol carrier Protein X (SCPx) is a peroxisomal branched-chain β-ketothiolase specifically reacting with 3-Oxo-pristanoyl-CoA: A new, unique role for SCPx in branched-chain fatty acid metabolism in peroxisomes. Biochem Biophys Res Com 1997; 236:565-9.

100.   Antonenkov VD, Van Veldhoven PP, Waelkens E, et al. Substrate specificities of 3-oxoacyl-CoA thiolase A and sterol carrier protein 2/3-oxoacyl-CoA thiolase purified from normal rat liver peroxisomes. J Biol Chem 1997; 272:26023-31.

101.   Leonard AN, Cohen DE. Submicellar Bile Salts Stimulate Phosphatidylcholine Transfer Activity Of Sterol Carrier Protein 2. J Lipid Res, In Press 1998;

102.   Krisans SK. Cell compartmentalization of cholesterol biosynthesis. In: Reddy JK, Suga T, Mannaerts GP, Lazarow PB, Subramani S., eds. Peroxisomes: Biology and Role in Toxicology and Disease. New York: Annals of the New York Acad. Sci. 1996: 142-64.

103.   Appelkvist EL, Reinhart M, Fischer R, et al. Presence of individual enzymes of cholesterol biosynthesis in rat liver peroxisomes. Arch Biochem Biophys 1990; 282:318-25.

104.   Ericsson J, Scallen T, Chojnacki T, et al. Involvement of sterol carrier protein-2 in dolichol biosynthesis. J Biol Chem 1991; 266:10602-7.

105.   Schroeder F. Fluorescent sterols: probe molecules of membrane structure and function. [Review]. Prog Lipid Res 1984; 23:97-113.

106.   Schroeder F, Nemecz G. Transmembrane Cholesterol Distribution. In: Esfahani M, Swaney J., eds. Advances in Cholesterol Research. Caldwell, NJ: Telford Press, 1990: 47-87.

107.   Gimpl G, Burger K, Fahrenholz F. Cholesterol as modulator of receptor function. Biochemistry 1997; 36:10959-74.

108.   Frolov A, Schroeder F. Multiphoton imaging of a naturally occurring fluorescent sterol in cultured fibroblasts. 1998; submitted.

FUNCTIONAL ANALYSIS OF STEROL CARRIER PROTEIN-2 (SCP2) IN THE SCP2 KNOCKOUT MOUSE

Udo Seedorf

Institute for Arteriosclerosis Research and Institute for Clinical Chemistry and Laboratory Medicine (Zentrallaboratorium), Westfalian Wilhelms-University, D-48129 Münster, Germany; e-mail: seedorf@ear002.uni-muenster.de

KEY WORDS: cholesterol trafficking, peroxisomal lipid metabolism, β-oxidation, bile acid synthesis.

ABSTRACT

Sterol carrier protein-2 (SCP2) and sterol carrier protein-x (SCPx, peroxisomal 3-ketoacyl-CoA thiolase with intrinsic non-specific lipid transfer activity) are expressed from a fused gene (designated *Scp2*) which is similarly organized in all vertebrates and could be traced back to *Drosophila melanogaster*. Numerous *in vitro* studies were performed over the past two decades. They have led to the concept that SCP2 functions intracellularly in cholesterol transport to diverse subcellular locations. On the other hand, apparent lack of transport specificity and the well established localization of SCP2 within peroxisomes has made it difficult to understand how the protein might carry out this function in the intact cell. This chapter summarizes phenotypic abnormalities, that were detected in a murine strain harboring a targeted *Scp2* null mutation. Whereas the SCP2/SCPx-deficient strain revealed no major abnormality in cholesterol trafficking, profound effects were found with regard to the metabolism of several methyl-branched acyl-CoAs that are metabolized in peroxisomes. Along with the metabolic abnormalities, the gene disruption led to diet-dependent alterations in gene expression, peroxisome proliferation, hypolipidemia, impaired body weight control and neuropathy.

CONTENTS

TARGETED DISRUPTION OF THE SCP2/SCPx GENE - EVIDENCE AGAINST
AN ESSENTIAL ROLE OF SCP2 IN STEROIDOGENESIS AND CHOLESTEROL
SYNTHESIS IN MICE

ABNORMAL PEROXISOMAL METABOLISM OF METHYL-BRANCHED
FATTY ACYL-COAs IN *SCP2* KNOCKOUT MICE

EFFECTS OF THE GENE DISRUPTION ON PEROXISOMES AND GENE
EXPRESSION

ABNORMAL BILE ACID SYNTHESIS IN SCP2-KO MICE

CONCLUSIONS

THE STEROL CARRIER HYPOTHESIS

Sterol carrier protein 2 (SCP2) was isolated originally as a "cytosolic" factor required for efficient *in vitro* conversion of 7-dehydrocholesterol to cholesterol, catalyzed by membrane bound sterol-Δ^7-reductase in the ER[1]. SCP2 is identical with the non-specific lipid transfer protein (ns-LTP) which was purified based on its ability to catalyze the exchange of a variety of phospholipids between membranes *in vitro*[2]. Subsequently, it was shown that the cholesterol transfer activity of SCP2 could be used *in vitro* to stimulate acyl-CoA cholesterol acyltransferase (ACAT)-mediated esterification of free cholesterol by several-fold[3,4]. Presence of the protein stimulates *in vitro* dolichol biosynthesis[5,6] and the hydroxylation of the cholesterol nucleus by the membrane bound enzymes that catalyze the initial steps of bile acid synthesis in the ER[7,8]. More recent studies employing antisense oligonucleotides and overexpression of SCP2 cDNA in cell lines suggested participation of SCP2 in cytosolic cholesterol transport to the plasma membrane in these cells[9,10]. Other studies provided several lines of indirect evidence that supported a role of SCP2 in adrenal and ovarian steroidogenesis (reviewed in[11]): SCP2 is abundant in steroidogenic glands and trophic hormones stimulate steroidogenesis along with SCP2 gene expression[12,13]. SCP2 enhanced the movement of cholesterol between vesicles and isolated mitochondria *in vitro*, which corresponded to increased pregnenolone synthesis in the *in vitro* system[14,15]. Moreover, over-expression of SCP2 in COS cells engineered to produce progestins increased steroid formation[16]. These results led to the hypothesis that SCP2 functions as potentially important mediator of cholesterol transport through the cytosol to various locations in the cell, such as the plasma membrane, mitochondria and the ER.

As might be expected for a cytosolic cholesterol carrier, considerable amounts of SCP2 were detected in the cytosolic fraction in early studies that employed subcellular fractionation techniques to investigate the intracellular localization of SCP2[17-19]. However, cloning and sequencing of SCP2 cDNAs showed that the protein comprises a highly conserved C-terminal S/AKL peroxisomal targeting signal[16,20-24] and immunocytochemical studies demonstrated the pre-dominant localization of SCP2 within peroxisomes[25-28]. This localization was further confirmed by the observation that absence of peroxisomes in Zellweger disease or in peroxisome-deficient Chinese hamster ovary cells was invariably associated with severe SCP2-deficiency[29-31]. There are at least two other observations that appear to question the role of SCP2 in target-specific cholesterol transport: (1) *in vivo*, lipid compositions of membranes and transport vesicles are highly specific, asymmetric and controlled tightly. Conversely, SCP2 transfers a variety of lipids besides cholesterol, such as oxysterols, sitosterols, ergosterols, as well as all major phospholipids and glycolipids, with little or no discrimination between these lipid classes[5,32-34]. (2) It is known that high rates of SCP2-mediated sterol transfer depend on electrostatic interactions of SCP2 with the surface of the lipid phase. Inhibition of these interactions by increasing the ionic strength to physiologically relevant levels also leads to considerable inhibition of SCP2-mediated lipid transfer[35]. Thus, low expected activity under physiological conditions, apparent lack of specificity for cholesterol, and the predominant localization of SCP2 in peroxisomes made it difficult to understand how the protein might carry out its proposed role in cytosolic free cholesterol trafficking in the intact cell.

GENE STRUCTURE AND EVOLUTION OF THE SCP2/SCPx GENE

The SCP2-encoding gene (*Scp2*) comprises 16 exons, which span ~100 kb on human chromosome 1p32[36-38]. Transcription initiation is controlled by two distant promoters that were mapped immediately upstream of the first exon (P1) and exon 12 (P2) [39]. P2 is used to generate SCP2 encoding transcripts, which combine the coding information provided by exons 12-16. In addition, alternate transcription initiation at P1 leads to production of a second transcript that includes the coding information provided by exons 1-16. The respective gene product consists of 547 amino acids and was named sterol carrier protein-x (SCPx) [20]. SCPx represents a fusion protein between a thiolase domain, extending from amino acids 1-404, and SCP2 which is located at the carboxyl terminus[22]. The fused SCPx-encoding gene is present in all vertebrates and could be traced back to *Drosophila melanogaster*[24] (Genbank Accession No. X97685). In contrast, two separated genes for SCP2 and the thiolase were identified in *Caenorhabditis elegans* and several yeast species[40,41]. Interestingly, an ancient precursor of SCP2 could be identified even in the primitive methanogenic archaeon, *Methanococcus jannaschii*[42].

It is known from previous *in vitro* studies that SCPx has similar lipid transfer activity as SCP2 and that the substrate specificity of the SCPx thiolase shows a preference for straight medium-chain acyl-CoA substrates and tetramethyl-branched 3-ketopristanoyl-CoA[43,44]. Thus, the properties of the SCPx-associated thiolase differ from the initially identified peroxisomal thiolase that is assumed traditionally to play a prominent role in peroxisomal β-oxidation of most natural substrates, including bile acids and very-long chain fatty acids (VLCFA)[45,46]. SCPx is structurally similar to another peroxisomal protein in which a C-terminal SCP2-like domain is fused to a domain leading to acyl-CoA 2-enoyl hydratase / 3-OH-acyl-CoA dehydrogenase (peroxisomal bifunctional protein, PBE) activity, presumably involved in peroxisomal β-oxidation. The fused peptide is part of the 80-kDa precursor of 17β-hydroxysteroid dehydrogenase type IV (17β-HSD4), a multi-functional, multi-domain protein consisting of an amino terminal SCAD domain that harbors 17β-HSD and 3-OH-acyl-CoA dehydrogenase activity, followed by the acyl-CoA 2-enoyl-hydratase domain and the carboxyl terminal SCP2-like domain[47]. Although all the domains are functionally active in the fused protein[48], processing occurs after import into peroxisomes at the junction between the SCAD and acyl-CoA 2-enoyl-hydratase domains[49]. The SCP2-like domain seems to be required for import of the 80-kDa precursor into peroxisomes and confers a similar intrinsic lipid transfer activity to the PBE-SCP2 fusion protein as was demonstrated for SCPx[47]. However unlike SCPx, which is partly cleaved post-translationally at the thiolase-SCP2 junction[43], the SCP2-like domain is not separated from the acyl-CoA 2-enoyl-hydratase domain in 17β-HSD4. Also mechanisms leading to separate expression of the SCP2-like peptide (*i.e.* alternative splicing pathways or alternative transcription initiation) could not be demonstrated. Despite the differences that exist between the two genes, both are reminiscent of bacterial operons, in which distinct functions are combined in a common transcriptional unit. The presence of similarly organized SCP2-domains in two distinct enzymes with an expected function in peroxisomal β-oxidation suggested that they may act in the same metabolic pathway. One evident consequence of their structures is that the SCP2 domains are strictly co-expressed with the enzymatic activities. Coordinate expression of an SCP2 homolog with the peroxisomal β-oxidation system was also observed in *Candida tropicalis*, in whom two separated genes encode the SCP2 homolog (called PXP18) and the thiolase. In this yeast, PXP18 is co-induced with the peroxisomal β-oxidation system after addition of oleic acid to the culture medium[41]. It appears noteworthy that purified SCP2 bound most fatty acids and fatty acyl-CoAs with similar or even higher affinity than sterols in a recently published study[50].

TARGETED DISRUPTION OF THE SCP2/SCPx GENE - EVIDENCE AGAINST AN ESSENTIAL ROLE OF SCP2 IN STEROIDOGENESIS AND CHOLESTEROL SYNTHESIS IN MICE

The experimental approaches, which are summarized in the previous sections, have led to rather broad information regarding structure, evolution and properties of SCP2 and SCPx. On the other hand, convincing conclusions regarding their functions could not be obtained, neither for SCP2 or the thiolase nor why evolution has favored gene fusion between the two, obviously distinct functions. Extensive efforts in our and other laboratories to identify human inherited diseases that would result from *Scp2* mutations were not successful. Therefore, we were interested to investigate the biological function of *Scp2* by employing gene targeting in mice. Meanwhile, the phenotypic abnormalities of the *Scp2(-/-)* KO mice have revealed presumably important functions of the two gene products. Although most of the results have been published recently[51], the purpose of this section is to give a brief summary and add supplementary information, as well as some of our more recent data which have not yet been published and, despite their somewhat preliminary nature, may be of interest to the topic.

In our model, SCP2- and SCPx-deficiency was obtained by introducing a gene disruption at the exon 14 region of the gene. The correctly targeted allele led to exon 14 skipping and thus abnormal splicing of *Scp2* transcripts. Although the gene disruption did not completely eliminate synthesis of *Scp2* transcripts, the abnormal RNAs were barely detectable and encode truncated SCP2 and SCPx peptides which should be functionally inactive and were not detected in *Scp2(-/-)* mice. Homozygous transgenes also lacked the previously identified peroxisomal 44-kDa thiolase-like peptide, that was considered to result from proteolytic processing of SCPx[43]. Thus, the targeted allele behaved like a null allele, leading to complete absence of SCP2, SCPx and the *Scp2*-encoded thiolase in homozygous transgenes.

Neither hetero- (+/-) nor homozygous (-/-) transgenes revealed a clear-cut pathology under standard laboratory conditions (apart from the subjective impression of slightly reduced general fitness, such as poor competition for food when (-/-) mice were kept together with wild-type C57Bl/6 controls or that they were more easy to catch than controls). We did not observe differences in the incidence of (+/+), (+/-) and (-/-) mice from the Mendelian distribution. *Scp2(-/-)* males and females reached fertility at the normal age of app. 6 weeks and interbreeding between (-/-) males and (-/-) females gave rise to viable progeny. The litters were of comparable sizes as that found in (+/+) or (+/-) interbreeding. Adrenal morphology and storage of neutral lipids was normal and adrenal hyperplasia could not be detected. Testosterone and glucocorticoid concentrations were within the normal range in 6 to 8 weeks old males (Tab. 1).

Table 1:   Laboratory Values in C57Bl/6 and *Scp2*(-/-) mice.

	Unit	Scp2(+/+)	Scp2(-/-)
testosterone (serum)	nmol/L	3.0 ±1.8	2.0 ±0.9
progesterone (serum, ♀)	nmol/L	11.2 ±6.9	10.4 ±3.0
corticosteroids (serum)	ng/dL	188 ±32	204 ±35
insulin (serum)	ng/dL	161 ±34	207 ±34
cholesterol (serum)	mg/dL	71 ±11	66 ±17
triglycerides (serum)	mg/dL	89 ±3	105 ±4*
free fatty acids (serum)	mM	1.12 ±0.09	0.72 ±0.04*
glucose (serum)	mg/dL	116 ±13	81 ±3
phospholipids (liver)	mg/g	20.5 ±1.2	19.3 ±2.6
cholesterol (liver)	mg/g	3.2 ±0.3	2.9 ±0.4
cholesterol ester (liver)	mg/g	0.50 ±0.12	0.25 ±0.06*
triglycerides (liver)	mg/g	66.2 ±8.5	32.8 ±6.9*

Values represent means ±SEM; *p ≤ 0.05 (comparison to controls with the paired t-test; n ≥ 5). Progesterone was measured in non-pregnant eight- to twelve weeks-old females. All other values are from eight- to twelve weeks-old males.

No differences between the two strains were also found for progesterone in non-pregnant females under baseline conditions. Whereas plasma insulin and cholesterol concentrations were normal, triglycerides were slightly higher and free fatty acid and glucose concentrations were moderately lower in (-/-) mice. As expected from normal cholesterol levels, 7-dehydrocholesterol was non-detectable in homozygous transgenes. In addition, sex-matched *Scp2(-/-)* mice and C57BL/6 mice had similar body weights at birth and subsequent weight gain did also not differ between both strains, thus excluding a mild variant of a Smith-Lemli-Opitz syndrome-like abnormality.

Normal adrenal morphology, essentially normal plasma lipids, absence of developmental abnormalities or salt wasting and the fact that (-/-) mice had no abnormalities affecting fertility differed very clearly from all known human inherited diseases affecting steroidogenesis, cholesterol synthesis or cholesterol trafficking. Therefore, absence of corresponding abnormalities in the murine model of complete SCP2- and SCPx-deficiency seemed to exclude an obligatory role of SCP2 in these processes. It is known from studies with apolipoprotein AI (apoAI) knockout mice that adrenal cholesterol uptake depends importantly on selective cholesterol ester uptake from HDL[52]. Therefore, we generated an *Scp2*-apoAI double knockout line in order to stress the adrenal cholesterol mobilizing pathway. Although the detailed characterization of this line has not yet been completed, the doubly knocked out strain had no evident abnormalities. Again, we found Mendelian distribution of the two defective alleles and homozygosity for both targeted alleles was associated with normal fertility in males and females (Seedorf, in preparation).

We also evaluated potential effects of the gene disruption on intracellular lipid distribution. Cholesterol ester storage pools were indeed markedly depleted in livers from knockout mice (Tab. 1). However, the effect was not specific for cholesterol because the same result was obtained also for free fatty acid and triglyceride concentrations. Thus, depletion of cholesterol esters seemed to relate to decreased availability of free fatty acids for intracellular lipid esterification rather than a specific abnormality in cytosolic free cholesterol trafficking. Intestinal lipid absorption was normal, as judged by monitoring intestinal uptake of radiolabeled cholesterol or palmitic acid. Food intake was even slightly higher in *Scp2(-/-)* mice (256 ±12.9 mg / [day x g of body weight]) than in controls (196 ±10.7 mg / [day x g of body weight]) and we did not detect abnormal liver function, as indicated by normal GOT, GPT, γGT and bilirubin serum levels. A likely explanation for the effect of the gene disruption on neutral lipid storage in the liver consisted of enhanced *in vivo* activation of the peroxisome proliferator activated receptor-α (PPARα), as will be discussed in another section of this chapter.

ABNORMAL PEROXISOMAL METABOLISM OF METHYL-BRANCHED FATTY ACYL-COAs IN *SCP2* KNOCKOUT MICE

Given the fact that SCP2 and SCPx are mainly peroxisomal activities, we next studied potential effects of the gene disruption on fatty acid metabolism. Gas chromatographic fatty acid analyses with saponified lipid extracts from plasma and liver revealed no significant differences regarding the relative levels of the straight long-chain saturated, monounsaturated, polyunsaturated or very long chain fatty acids. However, 3,7,11,15-tetramethylhexadecanoic acid (phytanic acid) was significantly elevated in (-/-) mice compared with controls (Tab. 2).

Phytanic acid is a terpene fatty acid which is produced in heterotrophic organisms from plant-derived phytol (an isoprenoic alcohol esterified to ring IV of chlorophyll, compare Fig. 2 which illustrates phytol metabolism schematically). Since neither phytanic acid nor phytol are synthesized *de novo* in mammals, phytanic acid serum concentrations depend on dietary intake of preformed phytanic acid or its precursor phytol, storage of phytanic acid in cellular neutral lipids and the catabolic rate of phytanic acid[53]. Excessive storage of phytanic acid is known to occur also in Refsum disease, an autosomal recessive disorder associated with severe neurologic impairment in affected patients. The disease was recently shown to be caused by mutations in the gene encoding phytanoyl-CoA α-hydroxylase (*PHYH*)[54,55]

Table 2: Phytanic and Pristanic Acid in *Scp2(+/+)*, (+/-) and (-/-) Mice

	diet	Scp2(+/+)	Scp2(+/-)	Scp2(-/-)
phytanic acid	chow	1.4 ±0.4	4.8 ±2.6*	16 ±3*
	high-fat	16 ±6	33 ±6*	152 ±11*
	5 mg/g phytol	129 ±26	313 ±44*	1163 ±367*
pristanic acid	chow	<0.5	<0.5	<0.5
	high-fat	<0.5	<0.5	<0.5
	5 mg/g phytol	15 ±8	17 ±6	47 ±26

Results are expressed in µmoles/L serum as mean ±SD (n ≥ 5); *: paired t-test, p < 0.05. <0.5 µMol/L: not detectable.

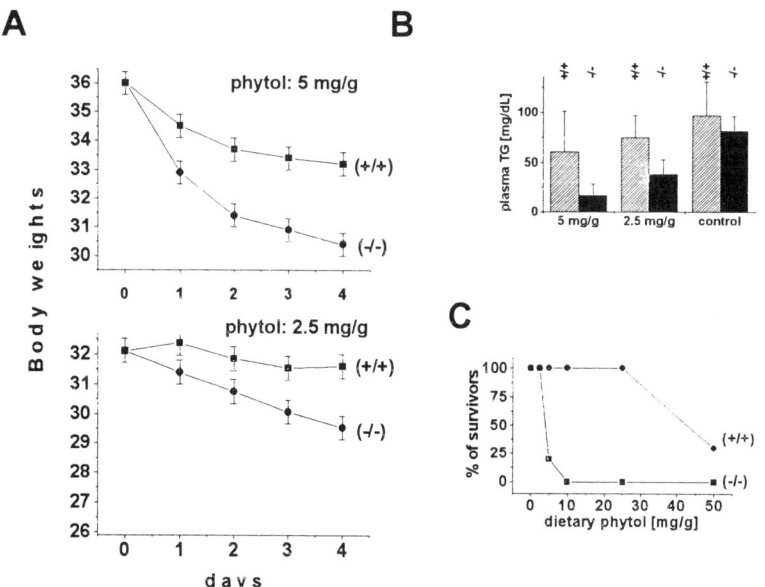

Figure 1: Effects of phytol on body weights (A), plasma lipids (B) and survival (C). Mice were fed standard diets supplemented with the indicated amounts of phytol for two (B) or three weeks (C). Dietary intake was similar in both strains during the first two weeks.

Because the phenotype may be partly masked by the low amounts of free phytol and phytanic acid that we detected in the normal laboratory diet, we performed feeding experiments with semi-synthetic diets supplemented with phytol. When *Scp2(-/-)* mice were exposed to a diet containing 5 mg/g of phytol for 7 days, the levels in serum of phytanic acid rose from 16 to 1163 μmoles/L, whereas in *Scp2(+/+)* mice, it rose from 1.4 μmoles/L to 129 μmoles/L (Tab. 2).

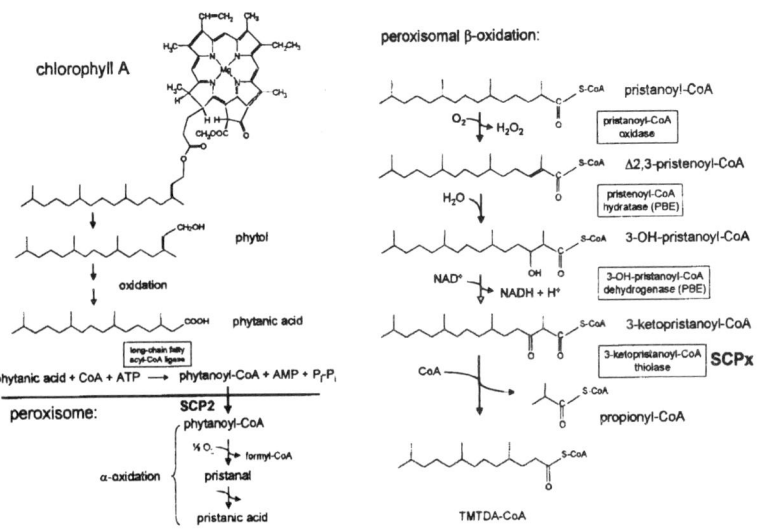

Figure 2: Schematic illustration of phytol metabolism.
The proposed roles of SCP2 and SCPx are indicated. Phytanic acid α- and pristanic acid β-oxidation has been assigned to peroxisomes. TMTDA-CoA, 4,8,12-trimethyl-tridecanoyl-CoA.

Diets containing up to 2.5 mg/g of phytol were tolerated rather well in *Scp2(-/-)* mice. Although body weights declined slowly by up to 25% within 6 weeks, we noticed no signs of toxicity. Apart from their skinny appearance, *Scp2(-/-)* mice looked healthy, they remained active, had no signs of neurological abnormalities and dietary intake was high until the end of the experiment at six weeks. Mobilization of body mass was maximal within the first few days and occurred in the absence of significant differences in food intake (Fig. 1A). In parallel, *Scp2(-/-)* mice developed more pronounced hypo-lipidemia than controls (Fig. 1B). Higher phytol enrichment of the diet (5 mg/g) led to much more severe abnormalities. Already beginning with the first day, (-/-) mice lost body weight extensively (Fig. 1A), that went on to decline rapidly until they reached close to 60% of their starting weights at the end of the second week, when we noticed unhealthy appearance, inactivity, reduced muscle tone, ataxia and peripheral neuropathy (uncoordinated movements, unsteady gait and trembling). Already after one week, *Scp2(-/-)* mice showed decreases of lipid and glucose levels in serum and depletion of fat tissue. The values obtained in serum for GPT, GOT and alkaline phosphatase were several-fold elevated and liver histology indicated pronounced liver disease.

The *Scp2(-/-)* mice died in the third week, presumably because the extensive neurological disturbances progressively disabled their food intake (Fig. 1C). In contrast, controls tolerated both diets well until the end of the experiment (six weeks). When 5 mg/g of phytol were added to the diet, we observed only a moderate decrease in body weight (-15%), mild hypolipidemia and hypoglycemia, and moderately elevated serum GPT levels without morphological signs of liver disease or reduced food intake. In contrast, supplementation of the diet with more than 50 mg of phytol per gram of food led to a similar range of abnormalities and premature death in controls, as could be observed with 5 mg/g in the transgenic strain (Fig. 1C).

As illustrated in Fig. 2, normal catabolism of phytol starts with conversion to phytanic acid, followed by activation to phytanoyl-CoA in the cytoplasm. Phytanoyl-CoA is then imported into peroxisomes followed by α-oxidation, which involves hydroxylation at the α-carbon position by phytanoyl-CoA hydroxylase (PHYH). Subsequently, 2-OH-phytanoyl-CoA is converted to pristanic acid and formyl-CoA[56]. Whereas formyl-CoA is further catabolized in mitochondria, pristanic acid is activated to pristanoyl-CoA, which is then subject to six cycles of peroxisomal β-oxidation. The intermediates of the first cycle are Δ2,3-pristenoyl-CoA (produced by pristanoyl-CoA oxidase), 3-OH-pristanoyl-CoA and 3-ketopristanoyl-CoA (produced by a peroxisomal bifunctional enzyme). Finally, 3-ketopristanoyl-CoA is substrate for thiolytic cleavage, catalyzed by a 3-ketopristanoyl-CoA thiolase, which yields the (n-3) lower homologue of pristanoyl-CoA (4,8,12-trimethyltridecanoyl-CoA) and propionyl-CoA (reviewed in[55]). Although studies on mice are not available, it now appears that this pathway operates similarly in rats and humans[57,58].

To pinpoint the precise block in phytol catabolism, we employed time-of-flight secondary-ion-mass-spectrometry (TOF-SIMS) which enabled detection of a wide range of intermediates. Evaluation of the signal intensities of the relevant ions indicated pronounced accumulation of phytanic acid in *Scp2(-/-)* mice (six- to eight-fold), which exceeded that of pristanic acid and the downstream catabolic intermediates considerably (Fig. 3). This may be explained by inhibition of initial phytanic acid activation, phytanoyl-CoA import into peroxisomes or phytanoyl-CoA α-hydroxylation. Abnormal activation was unlikely because recent studies demonstrated convincingly that phytanoyl-CoA ligation is mediated by a common long-chain fatty acyl-CoA ligase[59], whereas metabolism of long straight-chain fatty acids was not inhibited by the gene disruption. Since *PHYH* was expressed normally in *Scp2(-/-)* mice, we also excluded secondary down-regulation of *PHYH* expression. Based on these considerations, we hypothesized that the lipid carrier function of SCP2 may be involved in peroxisomal phytanoyl-CoA uptake, *i.e.* by acting as phytanoyl-CoA binding protein. The analysis, which was performed with a very specific FRET competition assay on the purified recombinant rat protein, revealed indeed much higher affinity for binding of phytanoyl-CoA than of pristanoyl-CoA, phytanic acid, pristanic acid or cholesterol. In addition, the K_d value was within a physiologically meaningful range (250 nM), thus supporting the postulated role of SCP2 in peroxisomal phytanoyl-CoA uptake indirectly[51].

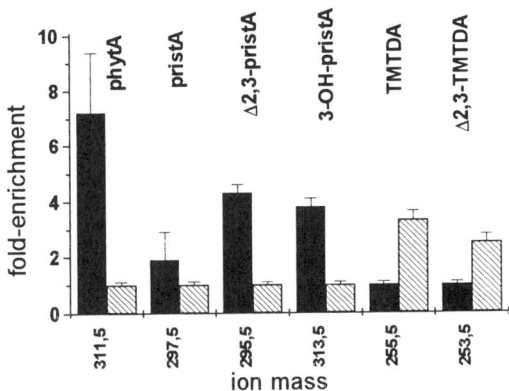

Figure 3: Defective phytanic acid breakdown in *Scp2* knockout mice. Shown are relative TOF-SIMS signal intensities obtained for the relevant ions (mean ±SD, n = 5) in phytol challenged male C57Bl/6 controls (hatched bars) and *Scp2(-/-)* KO mice (black bars). PhytA, phytanic acid; pristA, pristanic acid; Δ2,3-pristA, Δ2,3-pristenic acid; 3-OH-pristA, 3-hydroxypristanic acid; TMTDA, 4,8,12-trimethyltridecanoic acid; Δ2,3-TMTDA, 4,8,12-trimethyl-Δ2,3-tridecenoic acid.

Impaired phytanoyl-CoA import into peroxisomes would lead to the expectation that the production of downstream intermediates that are generated in peroxisomes from phytanoyl-CoA, should be repressed in *Scp2(-/-)* mice. In contrast, evaluation of TOF-SIMS signals, which corresponded to these intermediates, indicated even higher concentrations than in controls after challenging mice with high dosages of dietary phytol (Fig. 3). Although the two-fold increase in pristanic acid did not reach statistical significance, accumulation of Δ2,3-pristenic acid (four- to five-fold) and 3-OH-pristanic acid (three- to four-fold) was highly significant (Fig. 3). In contrast, 3-ketopristanic acid was not stable enough to withstand alkaline extraction and subsequent TOF-SIMS or GC-MS analyses. We could, however, detect a very significant 70% repression of the signals produced by the downstream products of the 3-ketopristanoyl-CoA thiolase reaction in *Scp2(-/-)* mice (4,8,12-trimethyltridecanoic acid and 4,8,12-trimethyl-Δ2,3-tridecenoic acid, Fig. 3). Together with enrichment of upstream intermediates in the pathway, the latter result supported very clearly inhibition at the level of 3-ketopristanoyl-CoA cleavage. These *in vivo* data corresponded to recent studies, which demonstrated high specific activity of recombinant rat SCPx to catalyze the thiolytic cleavage of 3-ketopristanoyl-CoA *in vitro*[44].

Thus, the data appeared to indicate a dual effect of the gene disruption: reduced peroxisomal phytanoyl-CoA import may lead to excessive phytanic acid accumulation whereas defective thiolytic cleavage of 3-ketopristanoyl-CoA may cause inhibition of the last step in the β-oxidation pathway. Whereas the first effect seems to relate to the phytanoyl-CoA carrier function of SCP2, the second may reflect the enzymatic activity associated with SCPx. This hypothesis appears compelling, because it would clarify why evolution has established a molecular basis for co-expression of the two *Scp2* encoded functions by fusing two originally separated SCP2- and thiolase genes into one common transcriptional unit. The fused gene is present in all vertebrates and could be traced back to *Drosophila melanogaster* (Genbank Accession No. X97685). In contrast, two separated genes were identified in *Caenorhabditis elegans* and several yeast species[24,40,40,41]. Interestingly, an ancient precursor of SCP2 could be identified even in the primitive methanogenic archaeon, *Methanococcus jannaschii*[42], in whom methyl-branched fatty acids play a prominent role.

Based on rates of phytanic acid removal from *Scp2* knockout mice, the residual activity for phytanic acid breakdown was calculated to app. 10 %. This was in line with the other findings, like ten-fold higher steady state concentrations of phytanic acid and roughly ten-fold increased phytol toxicity. However, whether the residual activity is due to compensatory up-regulation of peroxisomal straight-chain β-oxidation, the presence in our model of the SCP2-like activity associated with the 80-kDa precursor of 17β-HSD4, or an alternative pathway for phytanic acid oxidation, cannot be decided.

EFFECTS OF THE GENE DISRUPTION ON PEROXISOMES AND GENE EXPRESSION

When tissues from *Scp2(-/-)* mice were examined by light microscopy at various times after birth and compared with those of heterozygous and wild-type mice, we noticed more intense diaminobenzidine staining (DAB, specifically staining peroxisomes) in frozen liver sections from *Scp2(-/-)* mice than from controls. Enzyme activity levels of the peroxisomal marker catalase were 1.8-fold elevated in *Scp2(-/-)* liver. Likewise, peroxisomal palmitoyl-CoA oxidase (ACO), mitochondrial butyryl-CoA dehydrogenase and total 3-ketooctanoyl-CoA thiolase activities were all two- to three-fold higher in *Scp2(-/-)* mice than in controls.

In addition to peroxisome proliferation and induction of β-oxidation, the *Scp2* gene disruption had marked effects on hepatic gene expression. *Scp2(-/-)* mice showed increased expression of liver fatty acid binding protein (L-FABP, four-fold), peroxisomal 3-ketoacyl-CoA thiolase (pTHIOL, three- to four-fold), mitochondrial 3-ketoacyl-CoA thiolase (mTHIOL, two- to three-fold), peroxisomal acyl-CoA oxidase (ACO, two-fold) and cholesterol-7α-hydroxylase (CYP7α, four-fold). In contrast, no effect was observed on the level of glyceraldehyde-3-phosphate dehydrogenase (GAPDH), β-actin and sterol-27-hydroxylase (CYP27) expression, whereas phosphoenolpyruvate carboxykinase (PEPCK) expression was down-regulated in the *Scp2(-/-)* group, which corresponded to mild hypoglycemia in that group. These data pointed to induction of the mitochondrial and peroxisomal pathways of straight-chain fatty acid β-oxidation which could be a reason for depletion of the hepatic stores of cholesterol esters and triglycerides in *Scp2(-/-)* KO mice.

Peroxisome proliferation combined with up-regulation of β-oxidation-related gene expression could be the result of enhanced *in vivo* activation of peroxisome proliferator activated receptor-α (PPARα), a ligand-dependent nuclear receptor that was shown to mediate the effects of a class of hypolipidemic drugs called fibrates[60]. We evaluated this possibility by employing dietary phytol supplementation to modulate the concentrations of phytanic acid in C57Bl/6 and *Scp2(-/-)* KO mice and monitored PPARα-dependent gene expression in their livers. It was found that serum concentrations of phytanic acid correlated well with expression of peroxisomal ACO, PBE, 17β-HSD4, pTHIOL and L-FABP. Moreover, the phytol-induced pleiotropic effects on peroxisome proliferation and lipid metabolism could be mimicked more effectively with the PPARα agonist bezafibrate than with the RXRα agonist 9-*cis*-retinoic acid. These *in vivo* findings corresponded to binding of phytanic acid to a fused glutathione-S-transferase-murine-PPARα ligand binding domain with almost the same affinity as the strong artificial PPARα agonist WY 14,643 (Seedorf et al., submitted). Thus, enhanced PPARα activation seems to play a prominent role in the *Scp2(-/-)* KO mouse.

ABNORMAL BILE ACID SYNTHESIS IN SCP2-KO MICE

Phytanic acid belongs to the terpenes which are build by condensation of methyl-branched units called isoprenes. The cholesterol side-chain has a very similar structure and it is therefore conceivable that its peroxisomal β-oxidation, which is required for bile acid production, may depend on the presence of SCP2 and/or SCPx. Upon bile acid synthesis, 3α,7α,12α-trihydroxycholestanoyl-CoA, which is formed in the cytoplasm, must be imported into peroxisomes followed by removal of propionyl-CoA from the side-chain in one cycle of peroxisomal β-oxidation (reviewed in[61]). To clarify whether SCP2 and SCPx play a role in this pathway, we measured biliary bile acid concentrations and tried to identify intermediates of the bile acid synthetic pathway in *Scp2(-/-)* KO mice.

We found that homozygous knockout mice produced considerable amounts of all primary bile acids, including cholic acid, chenodeoxycholic acid and the three isomeric muricholic acids. Although the levels in bile appeared somewhat lower than in controls, the differences were barely significant. Nevertheless, we could identify two additional bile species which were present in bile from homozygous transgenes but were non-detectable in controls. Moreover, cholesterol-7α-hydroxylase gene expression was at least three-fold up-regulated in *Scp2(-/-)* mice which could account for partial compensation of the defect. When we exposed *Scp2(-/-)* mice to a diet containing the bile acid sequestrating drug cholestyramine, known to stress the bile acid synthetic pathway, both unusual bile compounds (labeled X1 and X2 in Fig. 4) increased pronouncedly whereas cholic acid remained at a low level. In contrast, cholic acid concentrations in bile increased close to two-fold in normal C57Bl/6 control mice and neither X1 nor X2 were detectable after ten days of cholestyramine feeding.

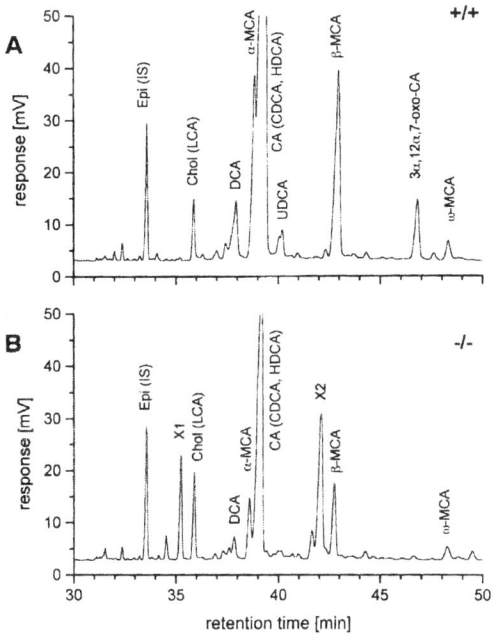

Figure 4: GC of carboxymethyl ester-trimethylsilane ether derivatives of bile from C57Bl/6 (A) and *Scp2*(-/-) KO mice (B). Male mice were fed a diet supplemented with 2% of cholestyramine for ten days.. Epicoprostanol (epi, IS) was used as internal standard. Abbreviations mean: Chol (cholesterol), DCA (deoxycholic acid), α-, β- and ω-MCA (α-, β-, and ω-muricholic acid), CA (cholic acid), CDCA (chenodeoxycholic acid), HDCA (hyocholic acid), UDCA (ursodeoxycholic acid). Levels of 3α,12α,7-oxo-CA varied in both strains. X1: 23-*nor*-cholic acid; X2: 3α,7α,12α-27-*nor*-trihydroxycholestane-24-one.

As can be seen in Fig. 4, the difference regarding bile composition that existed between both genotypes was quite remarkable. The mass spectra that were obtained for both compounds pointed to normal 3α,7α,12α-hydroxylated ring structures but abnormal side-chains. Comparison with a number of chemically synthesized standards led to identification of X1 as 23-*nor*-cholic acid (23-NCA), whereas the structure of X2 was resolved as 3α,7α,12α-trihydroxy-27-*nor*-cholestane-24-one. The presence of a 24-keto function in a molecule that was detected in extremely high amounts in bile from *Scp2*(-/-) KO mice, suggested very strongly a mechanism involving inefficient thiolytic cleavage of 3α,7α,12α-trihydroxy-24-ketocholestanoyl-CoA which is the substrate for the thiolase reaction in peroxisomal β-oxidation of the cholesterol side-chain. It may be assumed that the respective β-ketoacid (3α,7α,12α-trihydroxy-24-ketocholestanoic acid) is subject to spontaneous decarboxylation which would explain the high accumulation of $3\alpha,7\alpha,12\alpha$-trihydroxy-27-*nor*-cholestane-24-one in *Scp2(-/-)* KO mice. In contrast, accumulation of 23-NCA is less clear. Excessive accumulation of this unusual bile acid is also known to occur in the human disease of *cerebrotendinous xanthomatosis* (CTX) in which defective 27-hydroxylation leads to a block in normal side-chain shortening. Thus, the compound may result from activation of alternative side-chain oxidation pathways which lead to 23-hydroxylation and subsequent production of 23-NCA.

Thus, the gene disruption seemed to affect bile acid production distinct from phytanic acid catabolism: (1) the residual activity for cholic acid synthesis was rather high whereas it was relatively low for phytanic acid breakdown. (2) whereas the latter pathway was inactivated significantly already at an early step (peroxisomal import of the phytanoyl-CoA ester), cholic acid synthesis was affected specifically at the final step in the β-oxidation pathway (thiolytic cleavage). This would suggest that SCP2 may not be necessary for efficient peroxisomal import of $3\alpha,7\alpha,12\alpha$-trihydroxycholestanoyl-CoA (THCA-CoA, the immediate precursor of peroxisomal β-oxidation in bile acid synthesis). We currently believe that THCA-CoA does not require a carrier-mediated mechanism for its import into peroxisomes because the compound is much more soluble in aqueous solutions than phytanoyl-CoA. Alternatively, the defect may be compensated by presence of the SCP2-like function that is associated with 17β-HSD4 or another, still unknown member of the SCP2 gene family.

CONCLUSIONS

1. So far, the phenotypic characterization of the *Scp2(-/-)* KO mouse model has not provided convincing evidence for an obligatory role of this gene in intracellular cholesterol trafficking.

2. Our data appear to indicate that both gene products, SCP2 and SCPx, contribute to peroxisomal oxidation of certain naturally occurring methyl-branched acyl-CoAs in mice. The proposed roles of SCP2 and SCPx are consistent with their genetic organization, the well established peroxisomal localization of SCP2 and SCPx[25], the ability of SCP2 to bind phytanoyl-CoA *in vitro*, high 3-ketopristanoyl-CoA thiolase activity of the SCPx protein[44], and the expression pattern which correlates with lipid uptake of cells and thus phytanic acid exposition[16,20,21,62].

3. At present, detailed data are only available on phytol metabolism and bile acid synthesis in the *Scp2(-/-)* KO mouse model. However, a preliminary evaluation of our data regarding influences on other peroxisomal pathways that may be SCP2/SCPx-dependent, suggests that the gene disruption led to corresponding effects also on the metabolism of several retinoic acids and vitamin A. The latter effects may eventually provide further insights into the mechanisms which are responsible for the profound influences on gene expression that we observed in *Scp2(-/-)* KO mice.

249

ACKNOWLEDGMENTS

The author's work was supported by grants from the *Deutsche Forschungsgemeinschaft* (grant Se 459/2), the *Interdisziplinäres Klinisches Forschungszentrum* (Project A4) of the Medical Faculty, University of Münster, the *Boehringer Ingelheim Stiftung*, *Bristol Myers Squibb* and the *Bayer AG*.

REFERENCES

1. Noland BJ, Arebalo RE, Hansbury E, Scallen TJ. Purification and properties of sterol carrier protein-2. *J. Biol. Chem.* 1980;**255**:4282-9.
2. Bloj B, Zilversmit DB. Rat liver proteins capable of transferring phosphatidylethanolamine. Purification and transfer activity for other phospholipids and cholesterol. *J. Biol. Chem.* 1977;**252**:1613-9.
3. Trzaskos JM, Gaylor JL. Cytosolic modulators of activities of microsomal-enzymes of cholesterol-biosynthesis - purification and characterization of a non-specific lipid-transfer protein. *Biochim. Biophys. Acta* 1983;**751**:52-65.
4. Gavey KL, Noland BJ, Scallen TJ. The participation of sterol carrier protein-2 in the conversion of cholesterol to cholesterol ester by rat liver microsomes. *J. Biol. Chem.* 1981;**256**:2993-9.
5. van Amerongen A, Demel RA, Westerman J, Wirtz KWA. Transfer of cholesterol and oxysterol derivatives by the nonspecific lipid transfer protein (sterol carrier protein-2) - a study on its mode of action. *Biochim. Biophys. Acta* 1989;**1004**:36-43.
6. Ericsson J, Scallen TJ, Chojnacki T, Dallner G. Involvement of sterol carrier protein-2 in dolichol biosynthesis. *J. Biol. Chem.* 1991;**266**:10602-7.
7. Seltman H, Diven W, Rizk M, Noland BJ, Chanderbhan R, Scallen TJ, Vahouny G, Sanghvi A. Regulation of bile-acid synthesis - role of sterol carrier protein-2 in the biosynthesis of 7-alpha-hydroxycholesterol. *Biochem. J.* 1985;**230**:19-24.
8. Lidstrom-Olsson B, Wikvall K. The role of sterol carrier protein-2 and other hepatic lipid- binding proteins in bile-acid biosynthesis. *Biochem. J.* 1986;**238**:879-84.
9. Puglielli L, Rigotti A, Greco AV, Santos MJ, Nervi F. Sterol carrier protein-2 is involved in cholesterol transfer from the endoplasmic-reticulum to the plasma-membrane in human fibroblasts. *J. Biol. Chem.* 1995;**270**:18723-6.
10. Baum CL, Reschly EJ, Gayen AK, Groh ME, Schadick K. Sterol carrier protein-2 overexpression enhances sterol cycling and inhibits cholesterol ester synthesis and high density lipoprotein cholesterol secretion. *J. Biol. Chem.* 1997;**272**:6490-8.
11. Pfeifer SM, Furth EE, Ohba T, Chang YJ, Rennert H, Sakuragi N, Billheimer JT, Strauss JF. Sterol carrier protein-2 - a role in steroid-hormone synthesis. *J. Ster. Bioch. Mol. Biol.* 1993;**47**:167-72.
12. Trzeciak WH, Simpson ER, Scallen TJ, Vahouny GV, Waterman MR. Studies on the synthesis of sterol carrier protein-2 in rat adrenocortical-cells in monolayer-culture - regulation by ACTH and dibutyryl cyclic 3',5'-AMP. *J. Biol. Chem.* 1987;**262**:3713-7.
13. Rennert H, Amsterdam A, Billheimer JT, Strauss JF. Regulated expression of sterol carrier protein-2 in the ovary - a key role for cyclic-AMP. *Biochemistry* 1991;**30**:11280-5.
14. Chanderbhan R, Noland BJ, Scallen TJ, Vahouny GV. Sterol carrier protein-2 - delivery of cholesterol from adrenal lipid droplets to mitochondria for pregnenolone synthesis. *J. Biol. Chem.* 1982;**257**:8928-34.
15. Xu TS, Bowman EP, Glass DB, Lambeth JD. Stimulation of adrenal mitochondrial cholesterol side-chain cleavage by GTP, steroidogenesis activator polypeptide (SAP), and sterol carrier protein-2. *J. Biol. Chem.* 1991;**266**:6801-7.
16. Yamamoto R, Kallen CB, Babalola GO, Rennert H, Billheimer JT, Strauss JF. Cloning and expression of a cDNA-encoding human sterol carrier protein-2. *Proc. Natl. Acad. Sci. U.S.A.* 1991;**88**:463-7.
17. van Amerongen A, van Noort M, van Beckhoven JRCM, Rommerts FFG, Orly J, Wirtz KWA. The subcellular-distribution of the nonspecific lipid transfer protein (sterol carrier protein-2) in rat-liver and adrenal-gland. *Biochim. Biophys. Acta* 1989;**1001**:243-8.

18. van Heusden GPH, Bos K, Wirtz KWA. The occurrence of soluble and membrane-bound nonspecific lipid transfer protein (sterol carrier protein-2) in rat-tissues. *Biochim. Biophys. Acta* 1990;**1046**:315-21.
19. van Heusden GPH, Ossendorp BC, Wirtz KWA. Subcellular-distribution of nonspecific lipid transfer protein from rat-tissues. *Meth. Enzymol.* 1992;**209**:535-43.
20. Seedorf U, Assmann G. Cloning, expression, and nucleotide-sequence of rat-liver sterol carrier protein-2 cDNAs. *J. Biol. Chem.* 1991;**266**:630-6.
21. Ossendorp BC, van Heusden GPH, deBeer ALJ, Bos K, Schouten GL, Wirtz KWA. Identification of the cDNA clone which encodes the 58-kda protein containing the amino-acid-sequence of rat-liver nonspecific lipid- transfer protein (sterol-carrier protein 2) - homology with rat peroxisomal and mitochondrial 3-oxoacyl-CoA thiolases. *Eur. J. Biochem.* 1991;**201**:233-9.
22. Ossendorp BC, van Heusden GPH, Wirtz KWA. The amino-acid-sequence of rat-liver nonspecific lipid transfer protein (sterol carrier protein-2) is present in a high-molecular- weight protein - evidence from cDNA analysis. *Biochem. Biophys. Res. Commun.* 1990;**168**:631-6.
23. Mori T, Tsukamoto T, Mori H, Tashiro Y, Fujiki Y. Molecular-cloning and deduced amino-acid-sequence of nonspecific lipid transfer protein (sterol carrier protein-2) of rat-liver - a higher molecular mass (60-kDa) protein contains the primary sequence of nonspecific lipid transfer protein as its C-terminal part. *Proc. Natl. Acad. Sci. U.S.A.* 1991;**88**:4338-42.
24. Pfeifer SM, Sakuragi N, Ryan A, Johnson AL, Deeley RG, Billheimer JT, Baker ME, Strauss JF. Chicken sterol carrier protein-2 / sterol carrier protein-x - cDNA cloning reveals evolutionary conservation of structure and regulated expression. *Arch. Biochem. Biophys.* 1993;**304**:287-93.
25. Keller GA, Scallen TJ, Clarke D, Maher PA, Krisans SK, Singer SJ. Subcellular-localization of sterol carrier protein-2 in rat hepatocytes - its primary localization to peroxisomes. *J. Cell Biol.* 1989;**108**:1353-61.
26. Ossendorp BC, Wirtz KWA. The nonspecific lipid-transfer protein (sterol carrier protein-2) and its relationship to peroxisomes. *Biochimie* 1993;**75**:191-200.
27. Wouters FS, Markman M, deGraaf P, Hauser H, Tabak HF, Wirtz KWA, Moorman AFM. The immunohistochemical localization of the nonspecific lipid transfer protein (sterol carrier protein-2) in rat small-intestine enterocytes. *Biochim. Biophys. Acta-Lipids Lipid Met.* 1995;**1259**:192-6.
28. Deguchi J, Yamamoto A, Fujiki Y, Uyama M, Tsukahara I, Tashiro Y. Localization of nonspecific lipid transfer protein (nsLTP = sterol carrier protein-2) and acyl-CoA oxidase in peroxisomes of pigment epithelial-cells of rat retina. *J. Histochem. Cytochem.* 1992;**40**:403-10.
29. van Amerongen A, Helms JB, van der Krift TP, Schutgens RBH, Wirtz KWA. Purification of nonspecific lipid transfer protein (sterol carrier protein-2) from human-liver and its deficiency in livers from patients with cerebro-hepato-renal (Zellweger) syndrome. *Biochim. Biophys. Acta* 1987;**919**:149-55.
30. van Heusden GPH, Bos K, Raetz CRH, Wirtz KWA. Chinese hamster ovary cells deficient in peroxisomes lack the nonspecific lipid transfer protein (sterol carrier protein-2). *J. Biol. Chem.* 1990;**265**:4105-10.
31. Suzuki Y, Yamaguchi S, Orii T, Tsuneoka M, Tashiro Y. Nonspecific lipid transfer protein (sterol carrier protein-2) defective in patients with deficient peroxisomes. *Cell Struc. Func.* 1990;**15**:301-8.
32. Gadella TWJ, Wirtz KWA. Phospholipid-binding and transfer by the nonspecific lipid- transfer protein (sterol carrier protein-2) - a kinetic-model. *Eur. J. Biochem.* 1994;**220**:1019-28.
33. Gadella TWJ, Wirtz KWA. The low-affinity lipid-binding site of the nonspecific lipid transfer protein - implications for its mode of action. *Biochim. Biophys. Acta* 1991;**1070**:237-45.
34. Billheimer JT, Gaylor JL. Effect of lipid-composition on the transfer of sterols mediated by nonspecific lipid transfer protein (sterol carrier protein-2). *Biochim. Biophys. Acta* 1990;**1046**:136-43.
35. Butko P, Hapala I, Scallen TJ, Schroeder F. Acidic phospholipids strikingly potentiate sterol carrier protein-2 mediated intermembrane sterol transfer. *Biochemistry* 1990;**29**:4070-7.
36. Yamamoto R, Naylor SL, George H, Billheimer JT, Strauss JFI. Assignment of the gene encoding sterol carrier protein-2 to human chromosome-1pter-p21. *Cytogenet. Cell Genet.* 1991;**58**:1866-7.
37. Raabe M, Seedorf U, Hameister H, Ellinghaus P, Assmann G. Structure and chromosomal

assignment of the murine sterol carrier protein 2 gene (*Scp2*) and two related pseudogenes by in situ hybridization. *Cytogenet. Cell Genet.* 1996;**73**:279-81.

38. Ohba T, Rennert H, Pfeifer SM, He ZG, Yamamoto R, Holt JA, Billheimer JT, Strauss JF. The structure of the human sterol carrier protein-x / sterol carrier protein-2 gene (scp2). *Genomics* 1994;**24**:370-4.

39. Ohba T, Holt JA, Billheimer JT, Strauss JF. Human sterol carrier protein-x / sterol carrier protein-2 gene has 2 promoters. *Biochemistry* 1995;**34**:10660-8.

40. Bunya M, Maebuchi M, Hashimoto T, Yokota S, Kamiryo T. A second isoform of 3-ketoacyl-CoA thiolase found in *Caenorhabditis elegans*, which is similar to sterol carrier protein-x but lacks the sequence of sterol carrier protein-2. *Eur. J. Biochem.* 1997;**245**:252-9.

41. Tan H, Okazaki K, Kubota I, Kamiryo T, Utiyama H. A novel peroxisomal nonspecific lipid-transfer protein from *Candida tropicalis* - gene structure, purification and possible role in beta-oxidation. *Eur. J. Biochem.* 1990;**190**:107-12.

42. Bult CJ, White O, Olsen GJ, Zhou L, Fleischmann RD, Sutton GG, Blake JA, FitzGerald LM, Clayton RA, Gocayne JD, and others. Complete genome sequence of the methanogenic archaeon, *Methanococcus jannaschii*. *Science* 1996;**273**:1058-73.

43. Seedorf U, Brysch P, Engel T, Schrage K, Assmann G. Sterol carrier protein-x is peroxisomal 3-oxoacyl coenzyme-A thiolase with intrinsic sterol carrier and lipid transfer activity. *J. Biol. Chem.* 1994;**269**:21277-83.

44. Wanders RA, Denis S, Wouters F, Wirtz KA, Seedorf U. Sterol carrier protein-x (SCPx) is a peroxisomal branched-chain beta- ketothiolase specifically reacting with 3-oxo-pristanoyl-CoA: A new, unique role for SCPx in branched-chain fatty acid metabolism in peroxisomes. *Biochem. Biophys. Res. Commun.* 1997;**236**:565-9.

45. Schram AW, Goldfischer S, van Roermund CWT, Brouwer KEM, Collins J, Hashimoto T, Heymans HS, van den Bosch H, Schutgens RB, Tager JM, and others. Human peroxisomal 3-oxoacyl-coenzyme A thiolase deficiency. *Proc. Natl. Acad. Sci. U.S.A.* 1987;**84**:2494-6.

46. Hijikata M, Ishii N, Kagamiyama H, Osumi T, Hashimoto T. Structural analysis of cDNA for rat peroxisomal 3-ketoacyl-CoA thiolase. *J. Biol. Chem.* 1987;**262**:8151-8.

47. Leenders F, Tesdorpf JG, Markus M, Engel T, Seedorf U, Adamski J. Porcine 80-kda protein reveals intrinsic 17-beta-hydroxysteroid dehydrogenase, fatty acyl-CoA-hydratase/dehydrogenase, and sterol transfer activities. *J. Biol. Chem.* 1996;**271**:5438-42.

48. Seedorf U, Engel T, Assmann G, Leenders F, Adamski J. Intrinsic sterol- and phosphatidyl-choline transfer activities of 17 beta-hydroxysteroid dehydrogenase type IV. *J. Ster. Bioch. Mol. Biol.* 1995;**55**:549-53.

49. Markus M, Husen B, Leenders F, Seedorf U, Jungblut PW, Hall PH, Adamski J. Peroxisomes contain an enzyme with 17 beta-estradiol dehydrogenase, fatty acid hydratase/dehydrogenase, and sterol carrier activity. *Ann N. Y. Acad. Sci.* 1996;**804**:691-3.

50. Stolowich NJ, Frolov A, Atshaves B, Murphy EJ, Jolly CA, Billheimer JT, Scott AI, Schroeder F. The sterol carrier protein-2 fatty acid binding site: an NMR, circular dichroic, and fluores-cence spectroscopic determination. *Biochemistry* 1997;**36**:1719-29.

51. Seedorf U, Raabe M, Ellinghaus P, Kannenberg F, Fobker M, Engel T, Denis S, Wouters F, Wirtz KWA, Wanders RJA, and others. Defective peroxisomal catabolism of branched fatty acyl coenzyme A in mice lacking the sterol carrier protein-2 / sterol carrier protein-x gene function. *Gene. Dev.* 1998;**12**:in press

52. Plump AS, Erickson SK, Weng W, Partin JS, Breslow JL, Williams DL. Apolipoprotein A-I is required for cholesteryl ester accumulation in steroidogenic cells and for normal adrenal steroid production. *J. Clin. Invest.* 1996;**97**:2660-71.

53. Steinberg D. Refsum disease. In: Scriver CR, Beaudet AL, Sly WS, Valle D, editors. *The metabolic and molecular bases of inherited disease*. 7th ed. New York: McGraw-Hill; 1995. p 2351-70.

54. Jansen GA, Wanders RJ, Watkins PA, Mihalik SJ. Phytanoyl-coenzyme A hydroxylase deficiency -- the enzyme defect in Refsum's disease. *N. Engl. J. Med.* 1997;**337**:133-4.

55. Mihalik SJ, Morrell JC, Kim D, Sacksteder KA, Watkins PA, Gould SJ. Identification of PAHX, a Refsum disease gene. *Nature Genet.* 1997;**17**:185-9.

56. Croes K, van Veldhoven PP, Mannaerts GP, Casteels M. Production of formyl-CoA during peroxisomal alpha-oxidation of 3-methyl-branched fatty acids. *FEBS Lett.* 1997;**407**:197-200.

57. Watkins PA, Howard AE, Mihalik SJ. Phytanic acid must be activated to phytanoyl-CoA prior to its alpha-oxidation in rat liver peroxisomes. *Biochim. Biophys. Acta* 1994;**1214**:288-94.

252

58. Singh H, Poulos A. Substrate-specificity of rat-liver mitochondrial carnitine palmitoyl trans-ferase-I - evidence against alpha-oxidation of phytanic acid in rat-liver mitochondria. *FEBS Lett.* 1995;**359**:179-83.
59. Watkins PA, Howard AE, Gould SJ, Avigan J, Mihalik SJ. Phytanic acid activation in rat liver peroxisomes is catalyzed by long-chain acyl-CoA synthetase. *J. Lipid Res.* 1996;**37**:2288-95.
60. Lee SS, Pineau T, Drago J, Lee EJ, Owens JW, Kroetz DL, Fernandez Salguero PM, West-phal H, Gonzalez FJ. Targeted disruption of the alpha isoform of the peroxisome proliferator-activated receptor gene in mice results in abolishment of the pleiotropic effects of peroxisome proliferators. *Mol. Cell Biol.* 1995;**15**:3012-22.
61. Björkhem I, Boberg KM. Inborn errors in bile acid biosynthesis and storage of sterols other than cholesterol. In: Scriver CR, Beaudet AL, Sly WS, Valle D, editors. *The metabolic and molecular bases of inherited disease.* 7th ed. New York: McGraw-Hill; 1995. p 2073-102.
62. Hirai A, Kino T, Tokinaga K, Tahara K, Tamura Y, Yoshida S. Regulation of sterol carrier protein-2 (SCP2) gene-expression in rat peritoneal-macrophages during foam cell-formation - a key role for free-cholesterol content. *J. Clin. Invest.* 1994;**94**:2215-23.

SCAVENGER RECEPTORS, CAVEOLAE, CAVEOLIN, AND CHOLESTEROL TRAFFICKING

Eric J. Smart* and Deneys R. van der Westhuyzen

Departments of Physiology and Internal Medicine, University of Kentucky Medical Center, Medical Center MS 567, Lexington, KY 40536-0084; e-mail: ejsmart@pop.uky.edu

KEY WORDS: SR-BI, caveolae, caveolin, selective uptake, HDL, chaperone

---

ABSTRACT

The ability of cells to internalize HDL-derived cholesterol ester without internalizing the HDL particle is called "selective lipid uptake". The mechanism of selective uptake can be divided into three components: 1) association of HDL with the plasma membrane, 2) transport of cholesterol esters from HDL to the plasma membrane, and 3) trafficking of the cholesterol esters from the plasma membrane to the endoplasmic reticulum. A recent convergence of studies has yielded a greater understanding of the molecular mechanisms involved in cholesterol ester uptake. We discuss an integrated model of selective cholesterol ester uptake in the context of; 1) scavenger receptor BI as the HDL receptor, 2) caveolae as the initial sites of cholesterol ester uptake, and 3) the caveolin-chaperone complex as a vesicle-independent transport mechanism.

CONTENTS

THE HDL RECEPTOR, SR-BI

CAVEOLIN

CAVEOLIN AND CHOLESTEROL TRAFFICKING

SUMMARY AND CONCLUSIONS

REFERENCES

INTRODUCTION

The physiological need to maintain proper levels of cellular cholesterol cannot be overstated. Too much cholesterol or too little cholesterol can result in cell death. As one would expect for such an important molecule, a multitude of inter-dependent regulatory mechanisms exist to control its synthesis, metabolism, and cellular location. The recent characterization of scavenger receptor BI (SR-BI) as a receptor that mediates the uptake/efflux of HDL cholesterol ester/cholesterol and the localization of SR-BI to cholesterol-enriched plasma membrane microdomains, called caveolae, prompts a re-examination of the interface between plasma and cellular cholesterol homeostasis and intracellular cholesterol transport. This chapter will focus on the mechanisms for; 1) transporting cholesterol and cholesterol esters between high-density lipoproteins (HDL) and the plasma membrane and 2) transporting cholesterol and cholesterol esters between the plasma membrane and the endoplasmic reticulum (ER). We start by describing the process of "selective lipid uptake" and the concept of bi-directional flux of cholesterol and cholesterol esters between HDL particles and the ER. We then develop a possible mechanistic model for selective uptake in relation to; 1) caveolae, 2) SR-BI, and 3) caveolin, a putative cholesterol transport protein.

SELECTIVE UPTAKE HYPOTHESIS

Selective uptake refers to the internalization of HDL-derived cholesterol esters without the internalization of the HDL particle (1). This is in contrast to the classical receptor-mediated endocytosis mechanism responsible for the uptake of LDL-derived cholesterol (2). During endocytosis of LDL-cholesterol, the LDL particle and the LDL receptor are internalized and degraded in lysosomes. In the case of selective uptake, the major HDL apolipoproteins, apoA-I and A-II are not degraded and the receptor remains on the cell surface (1). The HDL-derived cholesterol ester initially appears in the plasma membrane as a "reversible pool" that can either move to the endoplasmic reticulum or efflux back to the HDL particle (3). Once the cholesterol ester reaches the ER it enters the cholesterol esterification/de-esterification cycle and can translocate back to the plasma membrane (1).

The mechanism of selective lipid uptake can be divided into two distinct steps; 1) transfer into the plasma membrane, and 2) translocation from the plasma membrane to the ER. The mechanism of transferring cholesterol esters from HDL to the plasma membrane is not clear, but two hypotheses have been put forth. First, uptake is receptor-mediated. In this model a specific receptor binds to HDL and somehow facilitates the flow of sterol from the particle to the membrane (4). Second, uptake occurs in a specific plasma membrane microdomain. In this model, a microdomain that is depleted of cholesterol serves as an acceptor of HDL sterol (1, 5). Furthermore, the same microdomain when enriched in cholesterol could promote efflux to HDL. Two mechanisms have also been proposed for the transfer of cholesterol esters from the plasma membrane to the ER: 1) vesicle-mediated, 2) carrier protein-mediated.

A convergence of recent work allows the generation of an integrative mechanistic model (Fig. 1) to explain the transfer of cholesterol esters from HDL to the plasma membrane and then to the ER. In this model the initial sites of cholesterol ester uptake are plasma membrane microdomains called caveolae. Caveolae are intimately involved in intracellular cholesterol trafficking (6-12) and are the primary location for scavenger receptor BI (SR-BI) (13). SR-BI has been shown to bind to HDL and to mediate the selective uptake of cholesterol esters (14). Finally, recent studies from our laboratories have demonstrated the existence of a chaperone protein-sterol complex that translocates cholesterol directly from the ER to caveolae (11, 12).

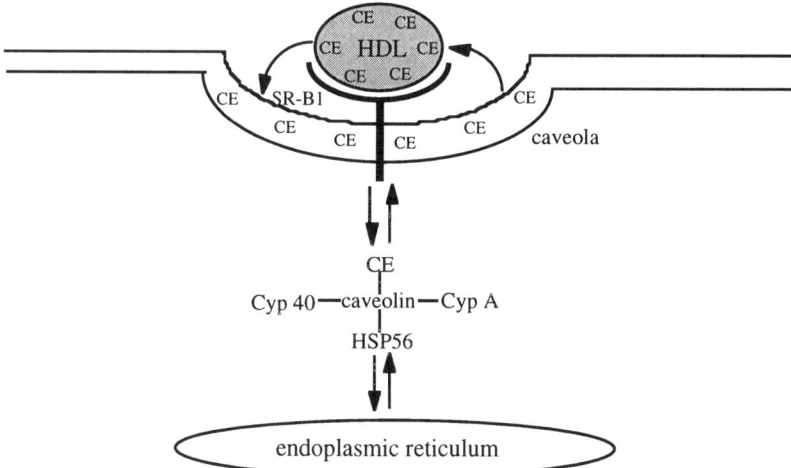

Fig. 1. **Hypothetical model for the uptake and intracellular trafficking of HDL-derived cholesterol esters.** In this model a sterol-rich HDL particle binds to scavenger receptor BI which is located in a caveola. Binding of the lipoprotein to its receptor facilitates the transfer of cholesterol esters from the HDL to the caveola. The cholesterol esters are then transported from the caveola to the endoplasmic reticulum in a protein-sterol chaperone complex. Presumably free cholesterol can traffic in the reverse direction.

CAVEOLA STRUCTURE

Caveolae were first described by electron microscopy in 1953 by George Palade (15) as non-coated, 50-300 nm, flask-shaped invaginations contiguous with the plasma membrane. For the next 35 years caveolae were extensively described by cell biologists using electron microscope methods but few functional or biochemical studies were performed. In the late 1980's a team of researchers lead by Richard G.W. Anderson "re-discovered" caveolae and ushered in the modern age of caveolae research. They developed functional biochemical assays (16-22) as well as methods to physically isolate and analyze the microdomain (23, 24). Caveolae or very similar microdomains have been found in every cell studied including; fibroblasts (25), endothelial cells (26), epithelial cells (20), muscle cells (27), astrocytes (28), adipocytes (29) and lymphocytes (11, 30). Importantly, caveolae have been recently identified in macrophages (31, 32) and hepatocytes (33), both of which play critical roles in the selective uptake of cholesterol esters.

In order to evaluate the potential role of caveolae in the selective uptake of cholesterol and in the intracellular trafficking of cholesterol, a basic understanding of caveolae structure and function is required. First, the classical definition of caveolae as small invaginations is inadequate and has been expanded to include both invaginated and flat microdomains. Although caveolae literally means "small caves", the bulk of data clearly demonstrates that caveolae can assume both invaginated and flattened morphologies. Rapid freeze, deep-etch electron microscopic methods have been used to demonstrate that flat caveolae exist within the plane of the membrane (Fig. 2) (25).

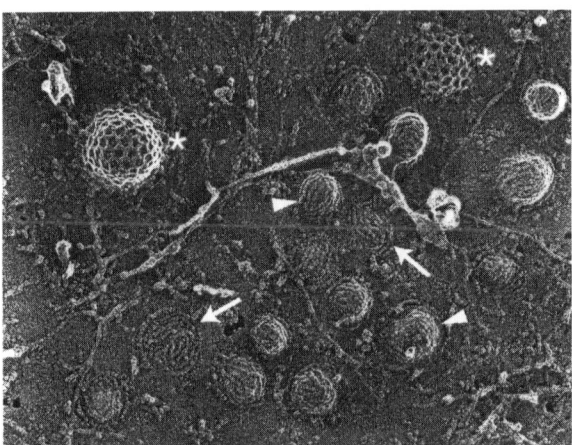

Fig. 2. **Rapid-freeze, deep-etch microscopy of cultured human fibroblasts.** The caveolae are decorated with a distinctive spiral coat. Both invaginated (arrowheads) and flat caveolae (arrows) can be visualized. Coated pits are marked with asterisks. Electron micrograph kindly provided by Dr. John Heuser.

Furthermore, microdomains indistinguishable from caveolae can be isolated from cells that do not contain invaginated caveolae (11, 34). In some cell types, such as smooth muscle cells, caveolae appear to be static structures (27, 35). However, in other cells such as MA104 cells, a monkey kidney epithelial cell line, caveolae appear to cycle between flat and invaginated states in a protein kinase C dependent manner (36, 37). In addition, caveolae in endothelial cells may also form plasma membrane-independent vesicles (26, 38).

Caveola Lipids

Caveolae are plasma membrane microdomains, consequently their lipid composition should relate directly to their structure and function. Surprisingly, the precise lipid composition of a caveola is not known. Table 1 lists the types and relative amounts of various lipid constituents of caveolae.

Table 1: Caveola Lipids

Lipid	Percent in Caveolae	Source
Cholesterol	26 %	MDCK (39)
	7-10 %	Fibroblast (10)
Sphingomyelin	96 %	MDCK (39)
	50-70 %	Fibroblast (40)
Ptd Ethanolamine	6 %	MDCK (39)
Ptd Serine	10 %	MDCK (39)
Ptd Choline	5 %	MDCK (39)
Ptd Inositol	5 %	MDCK (39)
Ptd Inositol P_2	50 %	A431 (41)
Ceramide	50 %	Fibroblast (40)
DAG	50 %	Fibroblast (40)

Although the exact lipid composition is unknown, work from many laboratories has shown that caveolae are highly enriched in cholesterol, with respect to the total plasma membrane. In human fibroblasts, caveolae constitute about 1% of the plasma membrane surface area (25) yet contain about 10% of the plasma membrane cholesterol (10). Cholesterol is the major lipid contributing to caveola morphology (20). When human fibroblasts are grown in the presence of lovastatin, total cellular cholesterol is reduced by about 60 % (20). Surprisingly, the number of morphologically invaginated caveolae is also decreased markedly (10 fold fewer than controls) (20). When the cholesterol-depleted fibroblasts are made cholesterol replete, caveolae invaginate to control levels (20). Acute treatment of cells with cholesterol-binding compounds such as filipin, also cause caveolae to flatten. Thus morphological changes are not merely an artifact of prolonged lovastatin treatment. These studies demonstrate that cholesterol-depleted caveolae are flat within the plane of the plasma membrane and maximally exposed to extracellular components, such as lipoproteins. However, when caveolae are cholesterol replete these microdomains are invaginated with a narrower opening exposed to extracellular components. Consequently, the amount of cholesterol in a caveola controls its morphology, exposure to extracellular components, and presumably its function.

Caveola Proteins

Many proteins have been localized to caveolae by subcellular fractionation and/or electron microscopy (Table 2-not comprehensive). Caveolae from different cell types contain different protein profiles and therefore not all proteins listed in Table 2 are to be found in any single caveola. The vast majority of proteins are actively excluded from caveolae.

258

Table 2: Caveola Proteins

Category	Protein	Cell Type
Predominantly in Caveolae	Caveolin(s)	Most cells
	Ha-Ras	Rat-1 (42)
	eNOS	Endothelial (43)
	EGF Receptor	Fibroblast (24)
	SR-BI	Transfected CHO (13)
	TFPI	Endothelial (44)
	CD-36	Platelets (45)
Enriched in Caveolae	Grb 2	Rat-1 (42)
	Gα	Smooth muscle (23)
	Src	Transfected (46)
Transiently in Caveolae	Raf-1	Rat-1 (42)
	α-Calmodulin	Endothelial (47)
	14-3-3	Rat-1 (42)
	IgE Receptor	Mast (48)
Excluded from Caveolae	Clathrin	Fibroblast (24)
	LDL Receptor	Fibroblast (24)
	Paxillin	Fibroblast (24)
	Transferrin Receptor	Fibroblast (24)
	Annexin I	Fibroblast (24)
	Integrin-β_3	Fibroblast (24)

endothelial nitric oxide synthase, eNOS; epidermal growth factor receptor, EGFR; scavenger receptor, SR; tissue factor pathway inhibitor, TFPI

A 22 kDa protein called caveolin is the best marker for caveolae. Generally, plasma membrane microdomains containing caveolin are defined as caveolae. However, as discussed below, caveolin can traffic between caveolae and the ER/Golgi. As a result, not all caveolae will contain caveolin all the time and other caveolae markers become useful. The quite large number of caveola-associated proteins can be divided into three broad categories; 1) proteins predominantly associated with caveolae; 2) proteins enriched in caveolae compared to the plasma membrane but not present predominately in the microdomain; and 3) proteins transiently recruited to caveolae. Three proteins predominantly associated with caveolae have been implicated in cholesterol metabolism/uptake: caveolin, cholesterol transport; SR-B1, receptor for HDL; CD36, receptor for modified LDL.

CAVEOLA FUNCTION

Three major functions have been proposed for caveolae: 1) signal transduction, 2) endocytosis, and 3) regulation of cholesterol trafficking. These divisions are artificial and ultimately inadequate because caveolae appear to integrate and regulate a multitude of events. When looked at superficially it is difficult to reconcile a role for caveolae in cholesterol trafficking with a role in signal transduction. However, at closer inspection many of the signaling proteins associated with caveolae depend on the lipid microenvironment for function (20, 42, 43, 49) and conversely, many of these same signaling proteins are known to influence cholesterol metabolism (36, 42, 43). In addition, caveolae in some cells may be specialized for cholesterol metabolism whereas caveolae in other cell types may be specialized for signal transduction. The role of caveolae in signal transduction has been reviewed elsewhere (50, 51).

Transcytosis

Caveolae were originally thought to be involved in transcytosis, the (15, 38) vesicular movement of material across a cell. Because caveolae are easily seen with an electron microscope as small invaginations, and because many transport vesicles are between 50-300 nm in size it was postulated that caveolae detach from the membrane and move to the other side of the cell. Cell biologists, primarily relying on microscopic evidence, hotly debated this concept for decades without resolution. Recently, Schnitzer et al (26) have suggested that caveolae in endothelial cells have the necessary machinery (NSF, SNAPs, SNAREs, etc) to form vesicles, however these observations have yet to be confirmed. Endothelial cells contain a large number of small vesicles, but their relationship, if any, to caveolae is unresolved. It is generally accepted, however, that for most cells caveolae remain attached to or closely associated with the plasma membrane. This distinction is of critical importance because in the process of selective lipid uptake only cholesterol esters are internalized, whereas the lipoprotein remnant remains extracellular.

Endocytosis

Caveolae have also been reported to be involved in endocytosis (31, 38, 52), a process in which plasma membrane is endocytosed from coated pits to form endosomes. Caveolae have been shown to specifically exclude the low-density lipoprotein receptor which is localized to coated pits (20). Maxfield and Mayor (53) suggested that the kinetics of folate uptake via the folate receptor is consistent with endocytosis. However, Anderson and colleagues (16, 17, 54) have used radiolabeled folate to demonstrate convincingly that caveola-localized folate receptors are not endocytosed. In addition, Ritter et al (55) have provided molecular genetic evidence that caveolae do not undergo endocytosis. There have been several reports showing that caveola-localized proteins can be found in endosomes (56, 57). These studies almost exclusively used electron microscopy and immunogold labeling techniques to show that a small percentage of various caveola proteins are in endosomes. Most likely this reflects uptake into endosomes as a result of normal membrane turnover. Furthermore, most proteins found in caveolae are not exclusively localized to the microdomain but rather are only enriched. The bulk of evidence therefore strongly supports a non-endocytic role for caveolae.

Potocytosis

One of the most recently proposed functions for caveolae is potocytosis (22). Potocytosis is a mechanism for the uptake of small molecules or solutes independent of an endocytoic process. A molecule binds to a receptor in a flat or open caveola. The caveola then invaginates and may transiently form a sealed compartment independent of the extracellular space but still contiguous with the plasma membrane. The formation of an invaginated/sealed microenvironment then facilitates the uptake of the molecules across the plasma membrane. The invaginated caveola then flattens/opens and the cycle is repeated. Although potocytosis has only been clearly demonstrated for the uptake of folic acid (22) the concept is entirely compatible with the selective uptake of cholesterol esters from HDL particles. HDL receptors are clustered in caveolae (see below) (13). Caveolae that are depleted of cholesterol are flat within the plasma membrane and maximally accessible to extracellular material (20). Invaginated/closed caveolae are highly enriched in cholesterol and are inaccessible to extracellular material.

THE HDL RECEPTOR, SR-BI

A number of plasma membrane receptors are known to bind and to mediate the cellular uptake of different classes of lipoproteins. These include members of the LDL receptor family of proteins: the LDL receptor, the VLDL receptor and the LDL receptor-like protein (LRP) (58). Other receptors, such as the class A scavenger receptor (SRA) and the class B scavenger receptor, CD36, recognize modified lipoproteins like oxidized LDL (59). In all these cases, receptors bind and mediate the endocytic uptake of whole lipoprotein particles. In contrast, the class B scavenger receptor, SR-BI, which functions as an HDL receptor, does not mediate endocytic uptake of particles but rather facilitates the selective delivery of the cholesterol ester component of HDL into cells. This functional property of SR-BI, first demonstrated by Krieger and coworkers (14, 60), suggested that SR-BI is responsible for the previously described selective delivery of HDL cholesterol ester to the liver and to steroidogenic tissues. Subsequent studies have confirmed that SR-BI plays a major role in HDL metabolism and selective lipid delivery. SR-BI-deficient mice showed a 125 % increase in HDL levels (61), whereas transgenic mice overexpressing SR-BI showed HDL levels that were decreased by more than 95 % (62). Two additional noteworthy features of SR-BI point to its importance in cellular and HDL cholesterol metabolism. First, unlike most other lipoprotein receptors, SR-BI on the plasma membrane is localized to caveolae microdomains (13). SR-BI is thus localized to the plasma membrane microdomain that appears to play a key role both in intracellular cholesterol trafficking and in cellular cholesterol efflux (6). Second, SR-BI, in addition to facilitating cholesterol uptake from HDL, in fact itself mediates efflux of free cholesterol from cells to HDL (63, 64).

Cloning and Structure of SR-BI

Interestingly, SR-BI was first cloned (as CLA-1) from human cells by virtue of its structural similarity to CD36, a plasma membrane receptor for oxidized LDL (65), and then by expression cloning as a protein able to mediate lipid uptake from acetylated LDL (66). Subsequently, it was recognized as an HDL receptor (14). Based on their structural similarity and their ability to recognize oxidized and modified lipoproteins, SR-BI and CD36 have been classified as class B scavenger receptors (66). SR-BI is a 509-amino acid integral transmembrane protein with two putative membrane-spanning domains. The transmembrane topology of SR-BI, as for CD36, has not been firmly established. One possible orientation is that SR-BI spans the membrane twice, giving rise to a large extracellular loop domain and two smaller cytoplasmic tail domains at the N- and C-termini (67). Alternatively, only the C-terminal hydrophobic sequence may anchor the protein, resulting in a single cytoplasmic tail at the C-terminus (13, 68). The large extracellular domain contains a number of cysteine residues and putative N-linked glycosylation sites that are conserved between SR-BI and CD36 (66). Glycosylation of the polypeptide core of SR-BI (~57 kDa) leads to the ~82 kDa mature receptor (13). A feature of the protein that is consistent with its localization in caveolae is that it undergoes fatty acylation in the form of palmitoylation as well as myristoylation (13). Three cysteine residues are present in the two putative cytoplasmic tail domains of both mouse and human SR-BI and represent potential palmitoylation sites, while the N-terminal amino acid glycine is the presumed site of myristoylation. Such fatty acyl chains may serve, as for certain other proteins, to localize SR-BI to caveolae (69).

The single SR-BI gene on human chromosome 12Q24.2-qter spans about 75 kilobase pairs and codes for 13 exons (70). The mouse SR-BI gene lies on chromosome 5, interestingly within a locus known to influence HDL levels (71). Initial studies on the 5'-upstream promoter region indicated the presence of a number of putative cis-acting elements. One of these is a binding site for SF-1, an orphan member of the nuclear hormone receptor gene family. SF-1 plays a key role in the regulation of steroidogenesis and its binding site was shown to be necessary for SR-BI promoter activity (70). The functional activity of other regulatory elements in this promoter are not known.

In addition to the major form of SR-BI described, a second and less prevalent isoform exists as a result of alternative mRNA exon splicing (72, 73). The minor form, termed SR-BII, has been found in mice and alternatively-spliced mRNA corresponding to this isoform has also been identified in human cells (72, 73). SR-BII is identical to SR-BI except in having an almost completely altered C-terminal cytoplasmic tail domain. SR-BII is localized to caveolae and mediates selective lipid uptake, although less efficiently than SR-BI (72, 73). An intriguing possibility is that alternative splicing represents a mechanism for regulating SR-BI activity. Alternatively, SR-BII may have another as yet unknown function. An additional alternatively-spliced mRNA was reported in humans (65), but is expressed at very low levels and gives rise to an inactive SR-BI isoform (unpublished, Connell and van der Westhuyzen).

SR-BI Mediates Selective Lipid Uptake from HDL

Unlike endocytic receptors such as the LDL receptor which mediate the uptake of whole lipoprotein particles, SR-BI acts by binding HDL and mediating the selective uptake of lipid from the core of the HDL particle into the cell. Lipid uptake is not accompanied by the uptake of the apolipoproteins. This process was first elucidated using transfected Chinese hamster ovary (CHO) cells expressing SR-BI (14), and represents a mechanism that can account for the known selective uptake of HDL cholesterol ester by various cells, particularly steroidogenic and liver cells (reviewed in (74)). Selective lipid uptake is analyzed using HDL in which either the protein or cholesterol ester components are radiolabeled and selective uptake is quantified as the difference between apolipoprotein and lipid uptake (14). When HDL was labeled with a fluorescent lipid (DiI), the internalized lipid showed a diffuse pattern of cellular uptake, with a concentration of fluorescent label at the periphery of cells (14), which is distinctly different from the punctuate pattern obtained when whole lipoprotein particles are taken up by endocytosis and delivered to lysosomes, as in the case of LDL uptake by the LDL receptor.

SR-BI binds HDL with high affinity. Intriguingly, the receptor exhibits a broad ligand specificity in also binding with high affinity LDL, VLDL, and oxidized LDL, as well as anionic phospholipids in liposomes (14, 66, 75). The significance of SR-BI binding LDL is not clear, but is not expected to significantly effect HDL binding to SR-BI since LDL competes relatively poorly with HDL for SR-BI binding (14, 75). Some of the broad ligand binding specificity may be accounted for by the fact that multiple apolipoproteins bind the receptor with high affinity (76). The apolipoproteins apoA-I, apoA-II and apoC-III were all shown to bind SR-BI when reconstituted into phospholipid/unesterified cholesterol complexes and were each able to compete for receptor binding of HDL (76). Although multiple apolipoproteins have been shown to mediate binding to SR-BI, it is noteworthy that selective lipid uptake into adrenal cells, studied in apoA-I-deficient mice, is specifically dependent on the presence of apoA-I (77). The explanation for this is not clear. One possibility is that apoA-I plays a more specific role as the SR-BI ligand than has been indicated in the binding and uptake studies carried out *in vitro*. Other possibilities are that apoA-I is required for the localization of HDL to the adrenocortical microvillar channels that are the probable site of lipid uptake or that apoA-I is responsible for structural features on HDL that permit selective lipid uptake to occur (77).

Expression and Regulation

SR-B1 in both humans and rodents is most highly expressed in steroidogenic tissues and the liver (14, 70, 78). These tissues are known to be the principal sites of selective lipid uptake from HDL (74). Generally, there is a correlation between SR-BI mRNA levels and protein expression, indicating that SR-B1 expression is controlled mainly at the level of transcription, although post-transcriptional regulation in some tissues has been indicated. For example, adipose tissue in mice was reported to lack SR-BI protein, despite having significant levels of mRNA (14, 66). SR-BI expression in steroidogenic cells is coordinately regulated with the rates of selective cholesterol ester uptake and steroidogenesis *in vivo,* providing strong evidence that it is this receptor that mediates the selective delivery of HDL cholesterol to cells, thereby providing cholesterol necessary for steroid synthesis. Hormonal stimulation of steroidogenesis and selective HDL cholesterol ester uptake into steroidogenic cells in the rat was accompanied by an increase in SR-BI expression. For example, SR-BI expression in adrenals and the ovary increased in response to a high-dose estrogen treatment (78), human chorionic gonadotrophin stimulated SR-BI in testicular Leydig cells (78), and adrenocorticotrophic hormone (ACTH) increased SR-BI in adrenocortical cells (79). Adrenal SR-BI mRNA was also induced in apoA-I and hepatic lipase knockout mice, further supporting the hypothesis that SR-BI expression in these tissues is regulated by the adrenal cholesterol (80, 81).

In contrast to steroidogenic cells, a high dose of estrogen decreases SR-BI in the liver (78), providing an explanation for decreased selective lipid uptake in response to such treatment (82). SR-BI expression in the liver is also decreased by high cholesterol diets. This decreased SR-BI activity in the liver correlates with the decreased rate of selective lipid uptake in liver parenchyma cells. An important observation was that SR-BI is also expressed in Kupffer cells of the liver and, intriguingly, that parenchyma and Kupffer cells respond in an opposite manner in their regulation of SR-BI expression and selective HDL lipid uptake (83). In contrast to parenchyma cells, Kupffer cells show a marked increase in uptake of about five fold in response to high-dose estrogen treatment or a high cholesterol diet (83).

Interestingly, SR-BI has been shown to be also expressed in extrahepatic macrophages. Macrophages play a central role in cholesterol metabolism and flux within the arterial wall. SR-BI mRNA is expressed in human monocytes and in the human monocyte cell lines U937 and THP-1(65, 84). However, expression in THP-1 cells is downregulated following PMA treatment and differentiation to macrophages (84). The murine monocytic lines, J774 and RAW, as well as peritoneal macrophages, also exhibit significant levels of SR-BI protein (64). The finding, by in situ hybridization, that SR-BI mRNA in mice is expressed within the thickened intima of an atheromatous aorta, most likely in macrophages, suggests that this receptor may play a role in regulating cholesterol flux between macrophages and lipoproteins, either by mediating sterol uptake or efflux (64). Regulation of SR-BI, as shown in Kupffer cells, might therefore have significant consequences on cholesterol flux between HDL and these cells.

Caveolae Are the Potential Sites of Selective Lipid Uptake

Although the mechanism by which SR-BI selectively transfers HDL lipid into cells is not known, evidence presented strongly suggests that uptake of cholesterol ester from HDL occurs first into the caveolae membranes. As discussed above, SR-BI on the plasma membrane is concentrated largely in caveolae and transfer is therefore likely to occur through this microdomain. Another possibility is that uptake is mediated by a fraction of SR-BI that is localized outside caveolae, either generally distributed or localized to another specific region on the plasma membrane. The small percentage of SR-BI outside caveolae and the relative rate or efficiency of the uptake process (assessed by the rate of uptake relative to the amount of HDL bound to the cell at any given time) makes this possibility unlikely.

Previous studies on selective lipid uptake into adrenal and HepG2 cells showed that HDL cholesterol esters were initially taken up into the plasma membrane (85, 86). Selective uptake occurred through a transient and rapidly reversible pool of cholesterol ester in the plasma membrane. This finding is supported in the case of SR-BI-mediated uptake in CHO cells by the diffuse cell surface distribution of fluorescent lipid taken up from HDL (14). In recent experiments we have assessed the kinetics of lipid uptake mediated specifically by SR-BI into both caveolae and caveolae-poor membrane fractions The results showed that cholesterol ester uptake by cells occurred first into caveolae and that caveolae uptake was followed by uptake into the intracellular membrane fraction (unpublished data). The cholesterol ester pool in caveolae was reversible and could account for the transient cholesterol ester pool previously described in the plasma membrane.

CAVEOLIN

Evidence presented strongly suggests that caveolae represent key membrane sites for cholesterol trafficking between HDL particles, the plasma membrane and the ER. One critical aspect of selective cholesterol uptake not yet discussed is the trafficking of the sterol from the plasma membrane to the ER. If caveolae are the sites of uptake then a mechanism must exist to coordinate cholesterol ester uptake with cholesterol trafficking to the ER. Recent studies on the caveola protein, caveolin, provide a theoretical framework in which to address this elusive mechanism. This section will focus on caveolin function, particularly in relation to its role in cholesterol trafficking.

Caveolin was originally cloned by Kurzchalia et al (87) as VIP21. Glenney et al (88) cloned the same protein as a 22 kDa phosphoprotein from v-src transformed chicken embryo fibroblasts This protein was shown to localize to the cytoplasmic side of caveolae and was later named caveolin (25). Although caveolin was originally described as a type II transmembrane protein (89) it is now generally accepted that caveolin does not cross the plasma membrane (89, 90). Both the N-terminal and C-terminal of caveolin are cytosolic and a relatively large hydrophobic domain inserts into the inner leaflet as a hairpin loop (89, 91). Caveolin is acylated with three palmitoyl groups, the function of which is unknown (92). A variety of kinases can phosphorylate caveolin also with unknown consequences (93-96). To date three different isoforms of caveolin have been cloned. Caveolin-1 and caveolin-2 have the broadest tissue distribution whereas caveolin-3 may be restricted to muscle cells (97-99). In most cells caveolin is found predominantly associated with caveolae, however smaller amounts are found in the ER and Golgi (10-12). Recently, a cytosolic pool of caveolin has been identified (12). Most importantly, the intracellular distribution of caveolin is a dynamic one. Caveolin can recycle between caveolae and the ER/Golgi, apparently in a vesicle independent manner (11, 12).

Caveolin Distribution

Caveolin is found in almost all cell types examined. In the past, the only marker for caveolae was an invaginated morphology. Now, the presence of caveolin is used as the "definitive" marker for the presence of caveolae, although conversely the absence of caveolin does not necessarily mean that caveolae or caveola-like domains do not exist. Reliance on negative data has led to the suggestion that macrophage and hepatocytes do not contain caveolae (6). Because the liver is a major organ involved in the uptake of HDL cholesterol esters, the hypothesis that caveolae are the sites of selective cholesterol uptake would be invalid if hepatocytes did not contain this microdomain. Recently two reports have described the presence of caveolin in macrophages (31, 32) and one report has provided evidence of caveolin in liver (33). Consequently, caveolin, caveolae, and SR-BI each appear to be present in hepatocytes, cells that together with steroidogenic cells, are the cell types most active in selective lipid uptake.

CAVEOLIN AND CHOLESTEROL TRAFFICKING

The exact function(s) of caveolin is not known. Three major functions have been considered. First, caveolin has been proposed to be the structural unit of the cytoplasmic coat which is seen on caveolae in electron micrographs (25). Second, caveolin has been proposed to be an anchor or organizing center for signal transduction molecules (100). Third, caveolin has been proposed to be a mediator of intracellular cholesterol trafficking (10-12, 101). To date it is unclear if different isoforms of caveolin perform these different functions or if a single species of caveolin can perform these seemingly diverse functions. The ability of caveolin to transport and organize cholesterol may contribute to all three potential functions. We will focus on the role of caveolin in cholesterol transport.

Caveolin Binds Cholesterol

An ability of caveolin to directly bind cholesterol would provide a solid basis for a proposal that caveolin serves to transport cholesterol in the cytoplasm. Several papers (10-12, 101) have suggested that caveolin is a sterol binding protein, but conclusive data was lacking. Definitive data was provided by Murata et al (102) who demonstrated that caveolin binds at least 1 mole of cholesterol per mole of protein. Two approaches were used to demonstrate this interaction. First, cholesterol co-purified with caveolin in a sucrose gradient containing 0.2% SDS which should have removed associated lipids. Co-purification of caveolin and cholesterol has recently been confirmed by Uittenbogaard et al (12). The second approach involved the association of purified caveolin with cholesterol-containing mixtures. One surprising result from this study was that E. coli expressed caveolin could not reconstitute into proteoliposomes without the addition of exogenous cholesterol. The authors speculated that since E. coli does not synthesize cholesterol, cholesterol is required to change the conformation of caveolin into a reconstitution-competent state. Not only does caveolin bind cholesterol but Trigatti et al (103) has shown that caveolin also binds free fatty acids. The significance of caveolin/fatty acid interactions are unknown.

Cholesterol Oxidase

The first suggestion that caveolin and trafficking of cholesterol are linked came from studies using the enzyme cholesterol oxidase (10). Cholesterol oxidase converts the alcohol at carbon 3 to a ketone. On live, unfixed cells the enzyme can only act on cholesterol exposed to the extracellular media (104). A fortuitous characteristic of cholesterol oxidase is that it works most efficiently on areas of the membrane that have high concentrations of cholesterol (105). Consequently, with the appropriate experimental conditions, the enzyme is specific for caveola cholesterol (10). A drawback with this tool is that it does not work on all cell types. Cholesterol oxidase appears to have access to the cholesterol in fibroblast caveolae but most likely the glycocalyx on many other cells prevents modification of the cholesterol (data not shown). Nevertheless, cholesterol oxidase has been an invaluable tool in studying caveolin-mediated cholesterol trafficking.

Incubation of human fibroblasts with cholesterol oxidase causes the complete oxidation of caveola cholesterol within one hour without affecting the bulk plasma membrane (10). Removal of the enzyme allowed a restoration of "normal" caveola cholesterol levels within one hour. The striking finding was that cholesterol oxidase treatment caused the majority of caveolin to leave the caveolae and translocate to the lumen of the ER before cycling back to the caveolae following the removal of the enzyme (Fig. 3). Several important points need to be emphasized. First, although cholesterol oxidase caused caveolin to leave the caveolae no significant decrease in the number of invaginated caveolae was detected. This observation suggests that caveolin is not trafficking through the cytosol in a caveola vesicle. Second, radiolabeled pulse-chase experiments in conjugation with cycloheximide were used to demonstrate that the caveolin at the cell surface was translocating to the ER and that the ER localized material was not the result of new protein synthesis. Third, caveolin which inserts into caveolae by a hairpin loop apparently leaves the membrane, transverses the cytosol and then crosses the ER membrane to reside in the lumen (10, 11, 101). Electron microscopy, protease protection, and differential solubility were used to confirm that caveolin was in fact in the lumen of the ER (11). Recently, the homeobox protein, engrailed, has been shown to display similar topological feats (106). Fourth, removal of the cholesterol oxidase permitted the very same caveolin molecules to cycle back to caveolae. Fifth, concomitant with the return of caveolin to the caveolae, the level of caveola cholesterol was restored to control levels. Sixth, because cholesterol oxidase does not cross or penetrate the membrane, all of the cholesterol in the caveolae must be in the outer leaflet and exposed to the extracellular surface. Consequently, the

266

cholesterol that replaced the damaged cholesterol must be able to rapidly move to the outer leaflet.

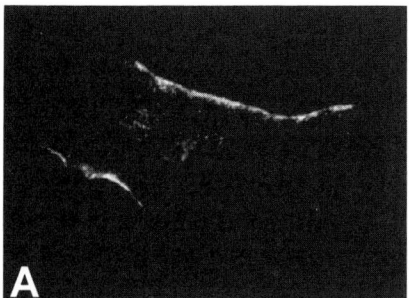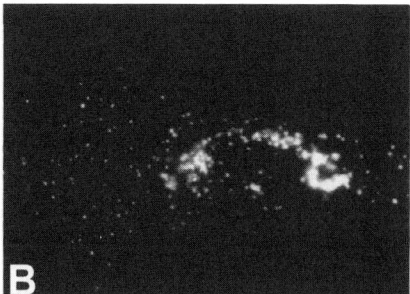

Fig. 3. **Cholesterol oxidase induces the redistribution of caveolin from caveolae to the ER/Golgi.**
Cultured human fibroblasts were treated in buffer only (A) or 0.5 U/ml of cholesterol oxidase (B) for 1 hr, 37°C. The cells were then fixed in paraformaldehyde and processed for indirect immunofluorescence using caveolin IgG.

Caveolin Minus Cells

The key study that clearly demonstrated a role for caveolin in intracellular cholesterol trafficking made use of a lymphocyte cell line, L1210-JF, that does not express caveolin (11, 12). The critical experiments directly followed the protocol of Kaplan and Simoni (107). In brief, cells incubated with ^{3}H-acetate at 14°C will synthesize radiolabeled cholesterol in the ER but the cholesterol will not traffick from the ER (107). Upon warming to 37°C, labeled cholesterol rapidly (10-20 min) translocates to the plasma membrane without moving through the Golgi (107). When these same experiments were repeated with human fibroblasts and coupled with subcellular fractionation to isolate caveolae, it was shown that the newly synthesized cholesterol moves directly to caveolae (11, 12). Surprisingly, the labeled cholesterol did not remain in caveolae but flowed into the bulk plasma membrane. Thus, newly synthesized cholesterol moved directly from the ER to caveolae then to the bulk plasma membrane. When identical experiments were conducted with L1210-JF lymphocytes, cells that lacked caveolin, the newly synthesized cholesterol did not move to caveolae and only very slowly (>60 min) moved to the bulk membrane (11, 12) (Fig. 4). The small amount of movement to the bulk plasma membrane was inhibited by Brefeldin A which suggested that this was cholesterol moving through the classic membrane secretion pathway (11, 12). L1210-JF cells expressing caveolin had dramatically different cholesterol trafficking kinetics (Fig. 4). The radiolabeled cholesterol was rapidly (~10 min) and specifically transported to caveolae. The transport was not inhibited by nocodozole or Brefeldin A suggesting a direct transport mechanism (11, 12). These studies demonstrated a clear and important role for caveolin in the trafficking of cholesterol from the ER to caveolae but they did not establish the mechanism of translocation.

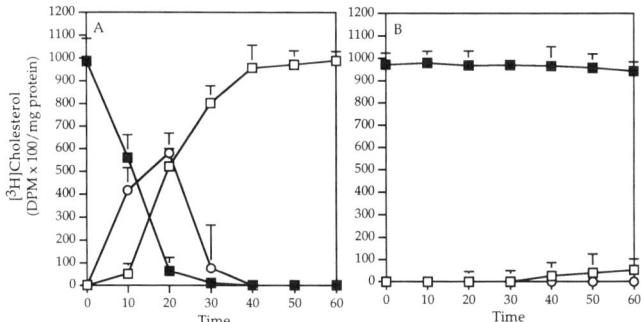

Fig. 4. **L1210-JF cells containing caveolin rapidly transport cholesterol to caveolae whereas L1210-JF cells lacking caveolin do not transport cholesterol to caveolae.** L1210-JF cells expressing caveolin (A) and L1210-JF cells not expressing caveolin (B) were chilled to 14°C and incubated with 30 μCi of [^{3}H]acetate and 10 μM cold acetate for 1 hr at 14°C. At the end of the labeling period the cells were washed and incubated at 37°C for the indicated times. The cells were then subfractionated to isolate caveolae (O) bulk plasma membranes (□), and intracellular membranes/endoplasmic reticulum (■). Mean ± S.E., n = 6.

Caveolin-Chaperone Complex

Two possible models can explain the mechanism of caveolin-dependent cholesterol transport: vesicle-dependent and vesicle-independent. Previous work by Simoni (107) and others (108, 109) have suggested that the direct transport of cholesterol to plasma membrane is dependent on a low-density transfer intermediate. However, the identity of the transfer intermediate is still unknown. Evidence for a caveolin-containing transport vesicle is also lacking. However, it is difficult to believe that caveolin, which has a relatively large hydrophobic domain and is acylated with three palmitoyl groups, can move through the cytosol as a soluble protein. One possibility to explain the movement of caveolin through the cytosol is that caveolin translocates in some sort of protein-lipid chaperone complex.

If caveolin traffics through the cytosol in a membrane-independent manner then a pool of caveolin must exist in a cytosol fraction. Uittenbogaard et al (12) used NIH 3T3 cells, subcellular fractionation, and immunoprecipitation to demonstrate that approximately 10% of the total cellular pool of caveolin is present in the cytosol. Immunoprecipitation with caveolin IgG under stringent conditions consistently co-precipitated three additional proteins. These proteins were identified as known chaperone proteins: HSP56, cyclophilin 40, and cyclophilin A. Immunoprecipitation with IgG raised against any of the four proteins precipitated the other three which further indicated that the proteins were truly in a cytosolic complex. Formation of the chaperone complex depended on the presence of caveolin because L1210-JF cells that did not express caveolin did not have the complex even though all the components (except caveolin) were present in the cells. Cyclosporin A and rapamycin, pharmacological reagents that disrupt the actions of cyclophilin 40, cyclophilin A, and HSP56, disrupted the complex and prevented co-immunoprecipitation. This study was the first to provide a possible mechanism for the transport of caveolin through the cytosol.

Does caveolin or rather the caveolin-chaperone complex transport cholesterol from the ER to the caveolae? Two pieces of evidence support the hypothesis that the caveolin-chaperone complex does indeed transport newly synthesized cholesterol directly from the ER to the caveolae. First, the same pharmacological reagents that disrupt the caveolin-chaperone complex, cyclosporin A and rapamycin, prevent the rapid transport of newly synthesized cholesterol to caveolae (12). Cyclosporin A and rapamycin do not effect cholesterol synthesis and both are completely reversible. Importantly, reformation of the caveolin-chaperone complex directly correlates with restoration of cholesterol trafficking. Second, the L1210-JF lymphocytes that do not express caveolin do not rapidly transport cholesterol to the caveolae (11, 12). However, transfection of caveolin into these cells permits cholesterol to traffic to the caveolae in about ten minutes. These studies strongly suggest that caveolin, HSP56, cyclophilin 40, and cyclophilin A form a cytosolic complex which translocates cholesterol from the ER to caveolae.

SUMMARY AND CONCLUSIONS

The bi-directional trafficking of cholesterol or cholesterol ester between HDL particles, and the ER is of unquestionable importance. The concept of "selective lipid uptake" was put forth to explain the phenomenon. Although many studies have examined this model, few molecular details were elucidated. In the past few years a convergence of research has filled in many of the potential details. One of the key components of the selective lipid uptake model was the hypothesized existence of a specific receptor for HDL. The groundbreaking work of Monty Krieger's laboratory (13, 14, 66, 110) clearly demonstrated that SR-BI mediates the selective uptake of HDL-derived cholesterol esters. The influx and efflux of sterol between the plasma membrane and HDL particles has been proposed to occur in a specific microdomain that can be alternatively enriched or de-enriched in cholesterol. Caveolae are a perfect match for the hypothesized microdomain. Caveolae can be highly enriched in cholesterol and invaginated or depleted of cholesterol and flat. Consequently, caveolae may function as acceptors or donors of sterol. The third element of selective lipid uptake is the transport of sterol from the plasma membrane (caveola) to the ER. The identification and characterization of a caveolin-chaperone-cholesterol complex provides a potential vesicle-independent mechanism for this uptake/efflux. The fact that SR-BI is localized to caveolae and that the caveolin-chaperone complex specifically traffics sterol to caveolae strongly suggest that the sites of cholesterol influx and efflux are caveolae.

REFERENCES

1. Fielding CJ, Fielding PE. Molecular physiology of reverse cholesterol transport. J. Lipid Res. 1995;36:211-228
2. Brown MS, Goldstein JL. A receptor-mediated pathway for cholesterol homeostasis. Science 1986;232:34-47
3. Rinninger F, Jaeckle S, Pittman RC. A pool of reversibly cell-associated cholesteryl esters involved in the selective uptake of cholesteryl esters from high-density lipoproteins by Hep G2 hepatoma cells. Biochim. Biophys. Acta 1993;1166:275-283
4. Mendez AJ, Oram JF, Bierman EL. Protein kinase C as a mediator of high density lipoprotein receptor-dependent efflux of intracellular cholesterol. J. Biol. Chem. 1991;266:10104-10111
5. Schroeder F, Jefferson JR, Kier AB, et al. Membrane cholesterol dynamics: cholesterol domains and kinetic pools. Proc. Soc. Exp. Biol. Med. 1991;196:235-252
6. Fielding CJ, Fielding PE. Intracellular cholesterol transport. J. Lipid Res. 1997;38:1503-1521
7. Fielding CJ, Bist A, Fielding PE. Caveolin mRNA levels are up-regulated by free cholesterol and down-regulated by oxysterols in fibroblast monolayers. Proc. Natl. Acad. Sci 1997;94:3753-3758

8.   Fielding PE, Fielding CJ. Intracellular transport of low density lipoprotein derived free cholesterol begins at clathrin-coated pits and terminates at cell surface caveolae. Biochemistry 1996;35:14932-14938
9.   Fielding PE, Fielding CJ. Plasma membrane caveolae mediate the efflux of cellular free cholesterol. Biochemistry 1995;34:14288-14292
10.   Smart EJ, Ying Y-Y, Conrad PA, Anderson RGW. Caveolin moves from caveolae to the Golgi apparatus in response to cholesterol oxidation. J. Cell Biol. 1994;127:1185-1197
11.   Smart EJ, Ying YS, Donzell WC, Anderson RGW. A role for caveolin in transport of cholesterol from ER to plasma membrane. J. Biol. Chem. 1996;271:29427-29435
12.   Uittenbogaard A, Ying Y-S, Smart EJ. Characterization of a cytosolic heat-shock protein-caveolin chaperone complex. J. Biol. Chem. 1998;273:6525-6532
13.   Babitt J, Trigatti B, Rigotti A, et al. Murine SR-B1, a high density lipoprotein receptor which mediates selective lipid uptake, is N-glycosylated, fatty acylated, and resides in plasma membrane caveolae. J. Biol. Chem. 1997;272:13242-13249
14.   Acton S, Rigotti A, Landschulz KT, Xu S, Hobbs HH, Krieger M. Identification of scavenger receptor SR-B1 as a high density lipoprotein receptor. Science 1996;271:518-520
15.   Palade GE. Fine structure of blood capillaries. J. Appl. Physics 1953;24:1424
16.   Kamen BA, Wang M, Streckfuss AJ, Peryea X, Anderson RGW. Delivery of folates to the cytoplasm of MA104 cells is mediated by a surface membrane receptor that recycles. J. Biol. Chem. 1988;263:13602-13609
17.   Kamen BA, Johnson CA, Wang M, Anderson RGW. Regulation of the cytoplasmic accumulation of 5-methylteterahydrofolate in MA104 cells is independent of folate receptor regulation. J. Clin. Invest. 1989;84:1379-1386
18.   Kamen BA, Smith AK, Anderson RGW. The folate receptor works in tandem with a probenecid-sensitive carrier in MA104 cells in vitro. J. Clin. Invest. 1991;87:1442-1449
19.   Rothberg KG, Ying Y, Kolhouse JF, Kamen BA, Anderson RGW. The glycophospholipid-linked folate receptor internalizes folate without entering the clathrin-coated pit endocytic pathway. J. Cell Biol. 1990;110:637-649
20.   Rothberg KG, Ying Y, Kamen BA, Anderson RGW. Cholesterol controls the clustering of the glycophospholipid-anchored membrane receptor for 5-methyltetrahydrofolate. J. Cell Biol. 1990;111:2931-2938
21.   Chang W-J, Rothberg KG, Kamen BA, Anderson RGW. Lowering the cholesterol content of MA104 cells inhibits receptor-mediated transport of folate. J. Cell Biol. 1992;118:63-69
22.   Anderson RGW, Kamen BA, Rothberg KG, Lacey SW. Potocytosis: sequestration and transport of small molecules by caveolae. Science 1991;255:410-411
23.   Chang W-J, Ying Y-S, Rothberg KG, et al. Purification and characterization of smooth muscle cell caveolae. J. Cell Biol. 1994;126:127-138
24.   Smart EJ, Ying Y-S, Mineo C, Anderson RGW. A detergent-free method for purifying caveolae membrane from tissue culture cells. Proc. Natl. Acad. Sci. USA 1995;92:10104-10108
25.   Rothberg KG, Heuser JE, Donzell WC, Ying Y, Glenney JR, Anderson RGW. Caveolin, a protein component of caveolae membrane coats. Cell 1992;68:673-682
26.   Schnitzer JE, Liu J, Oh P. Endothelial caveolae have the molecular transport machinery for vesicle budding, docking, and fusion including VAMP, NSF, SNAP, Annexins, and GTPases. J. Biol. Chem. 1995;270:14399-14404
27.   Fujimoto T. Calcium pump of the plasma membrane is localized in caveolae. J. Cell Biol. 1993;120:1147-1157
28.   Cameron PL, Cameron RS, Rasmussen H, Bollag R, Rubbin JW. Identification of caveolin and caveolin-related proteins in the brain. J. Neurosci. 1997;17:9520-9535
29.   Scherer PE, Lisanti MP, Baldini G, Sargiacomo M, Mastick CC, Lodish HF. Induction of caveolin during adipogenesis and association of GLUT4 with caveolin-rich vesicles. J. Cell Biol. 1994;127:1233-1243
30.   Fra AM, Williamson E, Simons K, Parton RG. De novo formation of caveolae in lymphocytes by expression of VIP21-caveolin. Proc. Natl. Acad. Sci. USA 1995;92:8655-8659
31.   Kiss AL, Geuze HJ. Caveolae can be alternative endocytotic structures in elicited macrophages. Eur. J. Cell Biol. 1997;73:19-27
32.   Baorto DM, Abraham SN, Lublin DM, van der Merwe A. Survival of FimH-expressing enterobacteria in macrophages relies on glycolipid traffic. Nature 1997;389:636-639
33.   Garver WS, Heidenreich RA, Kozloski MA, et al. Altered expression of caveolin-1 and increased cholesterol in detergent insoluble membrane fractions from liver in mice with Niemann-Pick disease type C. Biochim. Biophys. Acta. 1997;1361:272-280
34.   Fra AM, Williamson E, Simons K, Parton RG. Detergent-insoluble glycolipid microdomains in lymphocytes in the absence of caveolae. J. Biol. Chem. 1994;269:30745-30748
35.   Fujimoto T, Nakade S, Miyawaki A, Mikoshiba K, Ogawa K. Localization of inositol 1,4,5-trisphosphate receptor-like protein in plasmalemmal caveolae. J. Cell Biol. 1992;119:1507-1513

270

36. Smart EJ, Foster DC, Ying Y-S, Kamen BA, Anderson RGW. Protein kinase C activators inhibit receptor-mediated potocytosis by preventing internalization of caveolae. J. Cell Biol. 1994;124:307-313

37. Smart EJ, Ying Y-S, Anderson RGW. Hormonal regulation of caveolae internalization. J. Cell Biol. 1995;131:929-938

38. Schnitzer JE, Allard J, Oh P. NEM inhibits transcytosis, endocytosis, and capillary permeability: implication of caveolae fusion in endothelia. Am. J. Physiol. 1995;268:H48-H55

39. Brown DA, Rose JK. Sorting of GPI-anchored proteins to glycolipid-enriched membrane subdomains during transport to the apical cell surface. Cell 1992;68:533-544

40. Liu P, Anderson RGW. Compartmentalized production of ceramide at the cell surface. J. Biol. Chem. 1995;270:27179-27185

41. Pike LJ, Casey L. Localization and turnover of phosphatidylinositol 4,5-bisphosphate in caveolin-enriched membrane domains. J. Biol. Chem. 1996;271:26453-26456

42. Mineo C, James GL, Smart EJ, Anderson RGW. Localization of epidermal growth factor-stimulated ras/raf-1 interaction to caveolae membrane. J. Biol. Chem. 1996;271:11930-11935

43. Shaul PW, Smart EJ, Robinson LJ, et al. Acylation targets endothelial nitric-oxide synthase to plasmalemmal caveolae. J. Biol. Chem. 1996;271:6518-6522

44. Emeis JJ, Lupu F, Westmuckett A, de Priester W, van den Hoogen CM, van den Eijnden-Schrauwen Y. An endothelial storage granule for tissue-type plasminogen activator. J. Cell Biol. 1997;139:245-256

45. Dorahy DJ, Burns GF, Meldrum CJ, Lincz LF. Biochemical isolation of a membrane microdomain from resting platelets highly enriched in the plasma membrane glycoprotein CD36. Biochem J. 1996;319:67-72

46. Song KS, Lisanti MP, Parenti M, Galbiati F, Sargiacomo M. Targeting of a G alpha subunit (Gil alpha) and c-Src tyrosine kinase to caveolae membranes: clarifying the role of N-myristoylation. Cell Mol. Biol. 1997;43:293-303

47. Michel JB, Michel T, Sacks D, Feron O. Reciprocal regulation of endothelial nitric-oxide synthase by Ca^{2+}-calmodulin and caveolin. J. Biol. Chem. 1997;272:15583-15586

48. Stauffer TP, Meyer T. Compartmentalized IgE receptor-mediated signal transduction in living cells. J. Cell Biol. 1997;139:1447-1454

49. Shenoy-Scaria AM, Timson LK, Kwong J, Shaw AS, Lublin DM. Palmitylation of an amino-terminal cysteine motif of protein tyrosine kinases p56[lck] and p59[fyn] mediates interaction with glycosyl-phosphatidylinositol-anchored proteins. Mol. Cell. Biol. 1993;13:6385-6392

50. Lisanti MP, Scherer PE, Tang Z, Sargiacomo M. Caveolae, caveolin and caveolin-rich membrane domains: a signalling hypothesis. Trends in Cell Biology 1994;4:231-235

51. Anderson RGW. Caveolae: where incoming and outgoing messengers meet. Proc. Natl. Acad. Sci. USA 1993;90:10909-10913

52. van Deurs B, Holm PK, Sandvig K, Hansen SH. Are caveolae involved in clathrin-independent endocytosis. Trends in Cell Biology 1993;3

53. Maxfield FR, Mayor S. Cell surface dynamics of GPI-anchored proteins. Adv. Exp. Med. Biol. 1997;419:355-364

54. Smart EJ, Mineo C, Anderson RGW. Clustered folate receptors deliver 5-methyltetrahydrofolate to cytoplasm of MA104 cells. J. Cell Biol. 1996;134:1169-1177

55. Ritter TE, Fajardo O, Matsue H, Anderson RGW, Lacey SW. Folate receptors targeted to clathrin-coated pits cannot regulate vitamin uptake. Proc. Natl. Acad. Sci. USA 1995;92:3824-3828

56. Turek JJ, Leamon CP, Low PS. Endocytosis of folate-protein conjugates: ultrastructural localization in KB cells. J. Cell Sci. 1993;106:423-430

57. Rijnboutt S, Jansen G, Posthuma G, Hynes JB, Schornagel JH, Strous GJ. Endocytosis of GPI-linked membrane folate receptor-α. J. Cell Biol. 1996;132:35-47

58. Krieger M, Herz J. Structures and functions of multiligand lipoprotein receptors: macrophage scavenger receptors and LDL receptor-related protein (LRP). Annu. Rev. Biochem. 1994;63:601-637

59. Krieger M, Acton S, Ashkenas J, Pearson A, Penman M, Resnick D. Molecular flypaper, host defense, and atherosclerosis. Structure, binding properties, and functions of macrophage scavenger receptors. J. Biol. Chem. 1993;268:4569-4573

60. Rigotti A, Trigatti B, Babitt J, Penman M, Zu S, Krieger M. Scavenger receptor B1 - a cell surface receptor for high density lipoprotein. Cur. Opin. Lipidol. 1997;8:181-188

61. Rigotti A, Trigatti BL, Penman M, Rayburn H, Herz J, Krieger M. A targeted mutation in the murine gene enclding the high density lipoprotein (HDL) receptor scavenger receptor class B type I reveals its key role in HDL metabolism. Proc. Natl. Acad. Sci. 1997;94:12610-12615

62. Kozarsky KF, Donahee MH, Rigotti A, Iqbal SN, Edelman ER, Krieger M. Overexpression of the HDL receptor SR-B1 alters plasma HDL and bile cholesterol levels. Nature 1997;387:414-417

63. Jian B, de la Llera-Moya M, Ji Y, et al. Scavenger receptor class B type I as a mediator of cellular cholesterol efflux to lipoproteins and phospholipid acceptors. J. Biol. Chem. 1998;273:5599-5606

64. Ji Y, Jian B, Wang N, et al. Scavenger receptor BI promotes high density lipoprotein-mediated cellular cholesterol efflux. J. Biol. Chem. 1997;272:20982-20985

65. Calvo D, Vega MA. Identification, primary structure, and distribution of CLA-1, a novel member of the CD36/LIMPII gene family. J. Biol. Chem. 1993;268:18929-18935

66. Acton SL, Scherer PE, Lodish HF, Krieger M. Expression cloning of SR-B1, a CD36-related class B scavenger receptor. J. Biol. Chem. 1994;269:21003-21009

67. Tao N, Wagner SJ, Lublin DM. CD36 is palmitoylated on both N- and C-terminal cytoplasmic tails. J. Biol. Chem. 1996;271:22315-22320

68. Pearse SF, Wu J, Silverstein RL. A carboxyl terminal truncation mutant of CD36 is secreted and binds thrombospondin: evidence for a single transmembrane domain. Blood 1994;84:384-389

69. Shenoy-Scaria AM, Dietzen DJ, Kwong J, Link DC, Lublin DM. Cysteine[3] of src family protein tyrosine kinases determines palmitoylation and localization in caveolae. J. Cell Biol. 1994;126:353-363

70. Cao G, Garcia CK, Wyne KL, Schultz RA, Parker KL, Hobbs HH. Structure and localization of the human gene encoding SR-BI/CLA-1. J. Biol. Chem. 1997;272:33068-33076

71. Welch CL, Xia Y, Gu L, et al. *srb1* maps to mouse Chromosome 5 in a region harboring putative QTLs for plasma lipoprotein levels. Mamm. Genome 1997;8:942-943

72. Webb NR, de Villiers WJS, Connell PM, de Beer FC, van der Westhuyzen DR. Alternative forms of the scavenger receptor BI (SR-BI). J. Lipid Res. 1997;38:1490-1495

73. Webb NR, Connell PM, Graf GA, et al. SR-BII, an isoform of the scavenger receptor B1 containing an alternate cytoplasmic tail, mediates lipid transfer between high density lipoprotein and cells. J. Biol. Chem. 1998;In Press

74. Rothblat GH, Mahlberg FH, Johnson WJ, Phillips MC. Apolipoproteins, membrane cholesterol domains, and the regulation of cholesterol efflux. J. Lipid Res. 1992;33:1091-1097

75. Calvo D, Gomez-Coronado D, Lasuncion MA, Vega MA. CLA-1 is an 85-kD plasma membrane glycoprotein that acts as a high-affinity receptor for both native (HDL, LDL, and VLDL) and modified (OxLDL and AcLDL) lipoproteins. Arterioscler. Thromb. Vasc. Biol. 1997;17:2341-2349

76. Xu S, Laccotripe M, Huang X, Rigotti A, Zannis VI, Krieger M. Apolipoproteins of HDL can directly mediate binding to the scavenger receptor SR-B1, an HDL receptor that mediates selective lipid uptake. J. Lipid Res. 1997;38:1289-1298

77. Plump AS, Erickson SK, Weng W, Partin JS, Breslow JL, Williams DL. Apolipoprotein A-I is required for cholesteryl ester accumulation in steroidogenic cells and for normal adrenal steroid production. J. Clin. Invest. 1996;97:2660-2671

78. Landschulz K, Pathak RK, Rigotti A, Krieger M, Hobbs HH. Regulation of scavenger receptor, class B, type 1, a high density lipoprotein receptor, in liver and steroidogenic tissues of the rat. J. Clin. Invest. 1996;98:984-995

79. Rigotti A, Edelman ER, Seifert P, et al. Regulation by adrenocorticotropic hormone of the in vivo expression of scavenger receptor class B type I (SR-B1), a high density lipoprotein receptor, in steroidogenic cells of the murine adrenal gland. J. Biol. Chem. 1996;271:33545-33549

80. Wang N, Weng W, Breslow JL, Tall AR. Scavenger receptor B1 (SR-B1) is up-regulated in adrenal gland in apolipoprotein A-1 and hepatic lipase knock-out mice as a response to depletion of cholesterol stores. J. Biol. Chem. 1996;271:21001-21004

81. Ng DS, Francone OL, Forte TM, Zhang J, Haghpassand M, Rubin EM. Disruption of the murine lecithin:cholesterol acyltransferase gene causes impairment of adrenal lipid delivery and up-regulation of scavenger receptor class B type I. J. Biol. Chem. 1997;272:15777-15781

82. Rinninger F, Pittman RC. Regulation of selective uptake of high density lipoprotein-associated cholesteryl esters. J. Lipid Res. 1987;28:1313-1325

83. Fluiter K, van der Westhuyzen DR, van Berkel TJC. In vivo regulation of scavenger receptor BI and the selective uptake of high density lipoprotein cholesteryl esters in rat liver parenchymal and Kupffer cells. J. Biol. Chem. 1998;273:8434-8438

84. Murao K, Terpstra V, Green SR, Kondratenko N, Steinberg D, Quehenberger O. Characterization of CLA-1, a human homologue of rodent scavenger receptor BI, as a receptor for high density lipoprotein and adoptotic thymocytes. J. Biol. Chem. 1997;272:17551-17557

85. Knecht TP, Pittman RC. A plasma mambrane pool of cholesteryl esters that may mediate the selective uptake of cholesteryl esters from high density lipoproteins. Biochim. Biophys. Acta 1989;1002:365-375

86. Pittman RC, Knecht TP, Rosenbaum MS, Taylor CA. A non-endocylotic mechanism for the selective uptake of high density lipoprotein-associated cholesterol esters. J. Biol. Chem. 1987;262:2443-2450

87. Kurzchalia TV, Dupree P, Parton RG, et al. VIP21, a 21-Kd membrane protein is an integral component of trans-Golgi network-derived transport vesicles. J. Cell Biol. 1992;118:1003-1014

88. Glenney JR. The sequence of human caveolin reveals identity with VIP21, a component of transport vesicles. FEBS Lett. 1992;314:45-48

89. Lisanti MP, Tang ZL, Sargiacomo M. Caveolin forms a hetero-oligomeric protein complex that interacts with an apical GPI-linked protein: implications for the biogenesis of caveolae. J. Cell Biol. 1993;123:595-604

90. Monier S, Parton RG, Vogel F, Behlke J, Henske A, Kurzchalia TV. VIP21-caveolin, a membrane protein constituent of the caveolar coat, oligomerizes in vivo and in vitro. Mol. Biol. Cell 1995;6:911-927

91. Monier S, Dietzen DJ, Hastings WR, Lublin DM, Kurzchalia TV. Oligomerization of VIP21-caveolin in vitro is stabilized by long chain fatty acylation or cholesterol. FEBS Letters 1996;388:143-149

92. Dietzen DJ, Hastings WR, Lublin DM. Caveolin is palmitoylated on multiple cysteine residues. J. Biol. Chem. 1995;270:6838-6842

93. Tang Z, Scherer PE, Lisanti MP. The primary sequence of murine caveolin reveals a conserved consensus site for phosphorylation by protein kinase C. Gene 1994;147:299-300

94. Sargiacomo M, Scherer PE, Tang ZL, Casanova JE, Lisanti MP. In vitro phosphorylation of caveolin-rich membrane domains: identification of an associated serine kinase activity as a casein kinase II-like enzyme. Oncogene 1994;9:2589-2595

95. Mastick CC, Brady MJ, Saltiel AR. Insulin stimulates the tyrosine phosphorylation of caveolin. J. Cell Biol. 1995;129:1523-1531

96. Li S, Seitz R, Lisanti MP. Phosphorylation of caveolin by Src tyrosine kinases: the α-isoform of caveolin is selectively phosphorylated by v-Src *in vivo*. J. Biol. Chem. 1996;271:3863-3868

97. Tang Z, Scherer PE, Okamoto T, et al. Molecular cloning of caveolin-3, a novel member of the caveolin gene family expressed predominantly in muscle. J. Biol. Chem. 1996;271:2255-2261

98. Scherer PE, Okamoto T, Chun M, Nishimoto I, Lodish HF, Lisanti MP. Identification, sequence, and expression of caveolin-2 defines a caveolin gene family. Proc. Natl. Acad. Sci. USA 1996;93:131-135

99. Scherer PE, Tang Z, Chun M, Sargiacomo M, Lodish HF, Lisanti MP. Caveolin isoforms differ in their N-terminal protein sequence and subcellular distribution. J. Biol. Chem. 1995;270:16395-16401

100. Couet J, Lisanti MP, Ikezu T, Okamoto T, Li S. Identification of peptide and protein ligands for the caveolin-scaffolding domain. Implications for the interaction of caveolin with caveolae-associated proteins. J. Biol. Chem. 1997;272:6525-6533

101. Conrad PA, Smart EJ, Ying Y-S, Anderson RGW, Bloom GS. Caveolin cycles between plasma membrane caveolae and the Golgi complex by microtubule-dependent and microtubule-independent steps. J. Cell Biol. 1995;131:1421-1433

102. Murata M, Peranen J, Schreiner R, Wieland F, Kurzchalia TV, Simons K. VIP21/caveolin is a cholesterol-binding protein. Proc. Natl. Acad. Sci. USA 1995;92:10339-10343

103. Trigatti BL, Mangroo D, Gerbers GE. Photoaffinity labeling and fatty acid permeation in 3T3-L1 adipocytes. J. Biol. Chem. 1991;266:22621-22625

104. Lange Y. Tracking cell cholesterol with cholesterol oxidase. J. Lipid Res. 1992;33:315-321

105. Gronberg L, Slotte JP. Cholesterol oxidase catalyzed oxidation of cholesterol in mixed lipid monolayers: effects of surface pressure and phospholipid composition on catalytic activity. Biochemistry 1990;29:3173-3178

106. Joliot A, Trembleau A, Raposo G, Calvet S, Volovitch M, Prochiantz A. Association of engrailed homeoproteins with vesicles presenting caveolae-like properties. Development 1997;124:1865-1875

107. Kaplan MR, Simoni RD. Transport of cholesterol from the endoplasmic reticulum to the plasma membrane. J. Cell Biol. 1985;101:446-453

108. Lange Y, Matthies HJG. Transfer of cholesterol from its site of synthesis to the plasma membrane. J. Biol. Chem. 1984;259:14624-14630

109. Liscum L, Dahl NK. Intracellular cholesterol transport. J. Lipid Res. 1992;33:1239-1254

110. Rigotti A, Atton SL, Krieger M. The class B scavenger receptors SR-B1 and CD36 are receptors for anionic phospholipids. J. Biol. Chem. 1995;270:16221-16224

SELECTIVE UPTAKE OF LIPOPROTEIN FREE CHOLESTEROL AND ITS INTRACELLULAR TRANSPORT - ROLE OF CAVEOLIN

Christopher J. Fielding[1], Anita Bist and Phoebe E. Fielding[2]

Cardiovascular Research Institute and Departments of Physiology[1] and Medicine[2]. University of California, San Francisco, California 94143; e-mail: cfield@itsa.ucsf.edu

KEY WORDS: selective uptake, low density lipoprotein, trans-Golgi network, cholesterol synthesis, caveolin, caveolae

ABSTRACT

Free cholesterol (FC) from low density lipoprotein (LDL) enters peripheral cells via a N-ethylmaleimide-dependent selective uptake pathway. This pathway regulates cholesterogenesis, and promotes FC efflux via its effect on the transcriptional regulation of caveolin, the major structural protein of caveolae. Caveolae are the terminus at the cell surface of both newly synthesized FC and FC recycled from LDL. Caveolar FC is effectively transferred to extracellular high density lipoprotein (HDL). The activity of this new pathway may explain several paradoxes in current models of the regulation of intracellular FC transport.

CONTENTS

SELECTIVE UPTAKE OF LDL FREE CHOLESTEROL (FC)

Pathways contributing to FC homeostasis   In vivo, the great majority of mammalian cells are not actively dividing. They synthesize no steroid hormones or bile acids, and secrete no lipoproteins to the extracellular medium. Nevertheless, these quiescent cells continue to efflux FC at significant rates to lipoprotein acceptors, particularly high density lipoprotein (HDL) (1,2). To maintain cellular FC content, this efflux must be balanced by the sum of new FC synthesis and uptake of preformed lipoprotein cholesterol.

Multiple pathways are now recognized which could contribute to FC balance under these conditions. The endocytosis of intact LDL particles via the high affinity LDL receptor is well recognized, although the level of functional LDL receptors is very low in most peripheral tissues *in vivo* (3), and in cultured peripheral cells *in vitro* in normal (lipoprotein-containing) media (4). The uptake of intact high density lipoprotein (HDL) particles by peripheral cells also appears to be very low (5). Cholesteryl ester (CE) can be selectively internalized from LDL and HDL without the uptake of the corresponding lipoprotein protein moieties. This pathway, which was recently shown to be catalysed by the SR-B1 receptor protein, is most active in hepatocytes, adrenal cells and gonadal cells although low levels of SR-B1 receptors are expressed in many peripheral cells (6,7). Finally, FC can be selectively internalized from LDL (8,9). The selective internalization of LDL-FC was as active in LDL-receptor deficient as in normal cells, indicating that LDL endocytosis and LDL-FC selective uptake represent independent transport pathways.

FC homeostasis in quiescent fibroblasts   A widely used model to study intracellular FC transport and homeostasis in peripheral tissues is the human peripheral skin fibroblast. When fibroblast monlayers are grown to near confluence, >95% of   cells are quiescent (non-dividing) in terms of incorporation of ^{3}H-thymidine into DNA. The magnitude of most potential contributors to FC homeostasis has been measured in these cells, often by several laboratories, allowing a consensus to be reached on the contributions of different pathways. Additionally, several fibroblast lines are available which are genetically deficient in the expression of single well-characterized pathways of cholesterol transport. These include LDL-receptor deficient cells (10), Wolman Disease cells deficient in lysosomal acid cholesterol esterase (11) and Niemann-Pick C Disease cells, which lack NPC-1, a lysosomal protein required for the transport of FC from lysosomes to the endoplasmic reticulum and possibly other compartments (12).

Estimates of rates of individual FC homeostatic pathways in confluent normal human skin fibroblasts are given below.

Table 1. Contributions of different pathways to FC homeostasis

Pathway	Rates[a]
FC efflux	295 ± 6[b] (ref 9); 131 (ref 13); 155 (ref 14)
FC synthesis[e]	1.7[c] (ref 15);1.8[b] (unpublished)
LDL receptor mediated	2.7[c] (ref 4);0.5[b] (unpublished)
SR-B1 receptor mediated	2.1[c] (ref 16); 0.1 (ref 17)
LDL-FC selective uptake	318 ± 54[b] (ref 9); 85 (ref 8)[d,f]

[a]For comparison, rates have been converted to pmoles cholesterol synthesized or transported μg^{-1}cell FC h^{-1}. [b]In 10% fetal calf serum-DME medium. [c]Upregulated in lipoprotein-deficient serum or serum free medium. [d]LDL-receptor deficient cells [e]synthesis from ^{14}C-acetate. [f]From Figure 2, ref 8. The following conversion factors were used: 1 μg cell FC is associated with 50 μg cell protein. MW of FC 387.

These data show that rates of FC synthesis and receptor-mediated endocytosis, although determined in a number of different laboratories, are not adequate to explain fluxes of FC observed in normal human skin fibroblasts when these cells were cultured in normal media. On average, rates of FC efflux were about 50-fold greater than rates of either FC synthesis or LDL receptor-mediated endocytosis. Put another way, if FC efflux balanced only the inputs from new FC synthesis and receptoir-mediated endocytosis, its $t_{1/2}$ would be about 25 days, rather than the 10-20 h observed in a large range of experimental studies (review: 1). These differences appear to be much too large to be explained by technical differences between laboratories.

FC homeostasis in vivo It might be argued that the relatively low rates of FC synthesis and LDL receptor-mediated endocytosis, obtained with cultured cell monolayers, were unrepresentative of the same fluxes _in vivo_. Fortunately excellent data from several mammalian species (including human) has been recently reported (review: 18). FC synthesis rates were obtained using tritiated water, while rates of LDL turnover were obtained from the degradation of ^{125}I-labeled LDL. Using data summarized in ref 18, calculations were made of cholesterogenesis and LDL turnover, in terms of the same units as used in Table 1.

FC synthesis rates for the adult human were 15 mg day^{-1} kg^{-1} body weight while tissue FC content was 1.5-2 g kg^{-1}. For a 70 kg man, FC synthesis would be (15 x 70) = 1040 mg day^{-1} for a whole body FC pool of 105-140 g (average 122 g). FC synthesis expressed in pmoles μg^{-1} FC h^{-1} is given by:

$$(1040 \times 10^9)/(387) \times 1/(122 \times 10^6) \times (1/24) = 0.92 \text{ pmoles } \mu g^{-1} \text{ FC h}^{-1}.$$

This figure is similar to those obtained *in vitro* (Table 1). Similar calculations can be made for LDL total endocytosis (receptor mediated + nonspecific). From ref 18, clearance of LDL cholesterol was 13 mg day^{-1} kg^{-1} or 910 mg for a 70 kg human subject. Using the same whole body FC content as before, total LDL endocytosis would contribute:

$$(910 \times 10^9)/(387) \times 1/(122 \times 10^6) \times (1/24) = 0.7 \text{ pmoles FC } \mu g^{-1} \text{ FC } h^{-1}.$$

This value is also fairly consistent with the in vitro measurement of the same parameter shown in Table 1.

Because selective uptake of FC was proportional to LDL-FC concentrations below about 100 μg FC ml (9), its magnitude *in vivo* can be estimated if the extracellular LDL concentration surrounding the peripheral cells is known. Except where stated, rates in Table 1 were obtained in medium containing 10% fetal calf serum, where LDL makes up about ~50% of total cholesterol and 85% of total FC (19,20). In 10% fetal calf serum, LDL-FC is 4.5-6 μg ml^{-1}. In human lymph, LDL protein concentration was ~6% that of plasma (21) equivalent to 10-15 μg LDL-FC ml^{-1}. Consequently, the FC fluxes given in Table 1 are probably representative of those in peripheral cells *in vivo*. They suggest major roles for LDL-FC selective uptake, and FC efflux, in the homeostasis of quiescent peripheral cells in normal (lipoprotein-containing) media, and only minor roles for LDL receptor mediated endocytosis and FC synthesis.

Properties of selective uptake of FC   If FC uptake and efflux (Table 1) reflected mainly simple exchange at the cell surface, then the kinetic properties of both fluxes should be the same. In contrast, the experimental data indicated that selective uptake of FC (mainly from LDL) and FC efflux (mainly to HDL) had quite different properties (Table 2).

These data strongly suggest that the uptake and efflux of FC by peripheral cells reflect different biochemical pathways, rather than two limbs of the same nonspecific diffusional exchange.

The selective uptake of FC from LDL has kinetic properties characteristic of NEM-sensitive ATPases resistant to azide and vanadate. The best characterized of these proteins, NEM-sensitive factor (NSF) plays a key role in vesicular transport. Other ATPases with these properties are involved in endocytosis via the coated pit mechanism. The selective uptake of FC from LDL is membrane-dependent, but it is not yet clear whether this FC binds to a specific sterol receptor (distinct from the LDL receptor) at the cell surface; or if it is adsorbed into the membrane lipids of the clathrin-coated vesicle, prior to endocytosis.

Table 2. Kinetic and biochemical properties of FC influx and efflux

Parameter	FC Influx	FC efflux
Apparent Km (μg FC ml^{-1})	250 ± 20	18 ± 2
Specificity	LDL > HDL	HDL > LDL
No inhibitor	100%	100%
2.0mM NEM	30%	70%
40 μM NC	100%	40%
0.35M NaCl	40%	100%
7-KC	100%	45%

NEM, N-ethyl maleimide; NC, nocodazole; 7KC, 7-ketocholesterol. Data are from ref. 9, ref. 22 and unpublished.

CELLULAR TRANSPORT OF LDL-DERIVED FC

Further information on the intracellular transport of FC internalized from LDL was obtained using human fibroblast monolayers pulsed with ^{3}H-FC labeled LDL, under conditions shown to trap ligands internalized from coated pits as dense clathrin-coated vesicles. LDL labeled with ^{3}H-FC was briefly (0.5 min, 31 °C) incubated with human skin fibroblasts (22). After a chase with unlabeled LDL, the cells were homogenized at 0-2 °C and fractionated by density gradient ultracentrifugation. A major part of label was associated with clathrin in a dense vesicle fraction (d 1.12 g ml^{-1}) which also co-migrated with transferrin. A similar proportion of both transferrin and ^{3}H-FC remained associated with the plasma membrane fraction (d 1.03 g ml^{-1}). This probably represents extracellular ligand not yet transferred into endocytic vesicles. In human fibroblasts, transferrin is internalized exclusively via clathrin-coated pits. It is very likely that FC from LDL in the d 1.12 g ml^{-1} fraction remains associated with these vesicles during the subsequent dissociation of clathrin, as ^{3}H-FC and transferrin were subsequently recovered without clathrin in d 1.06 g ml^{-1} vesicles (Figure 1).

There was no effect of inhibitors of lysosomal function on the intracellular transport of LDL-derived FC. It appears likely that the FC label is fractionated into a recycling compartment prior to complete acidification of the endocytic vesicle.

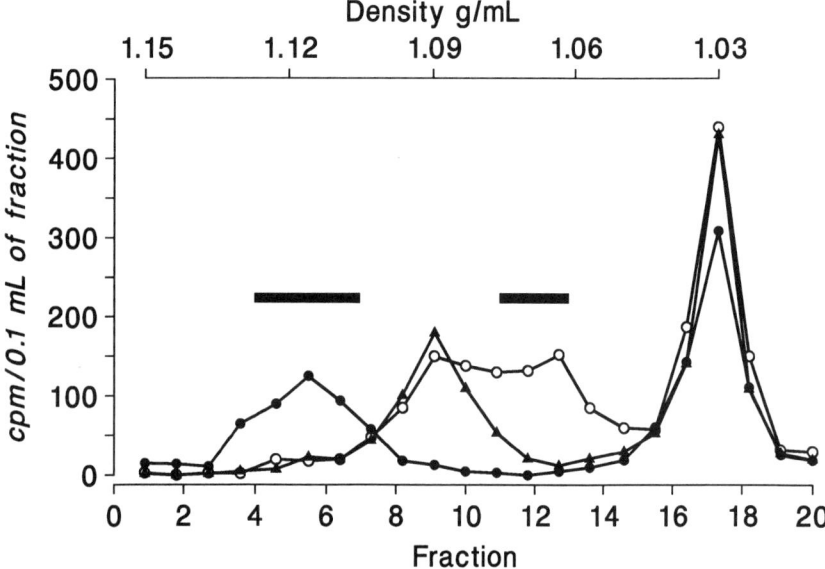

Figure 1. **Density gradient fractionation of human skin fibroblasts pulse-labeled with ³H-FC labeled LDL.** Closed circles: 0.5 min pulse with ³H-FC labeled LDL at 31 °C. Open circles, the same, followed by 2 min without LDL. Closed triangles, the same, followed by 15 min incubation at 31 C without LDL. Bars, fractions containing transferrin. From ref.23.

In contrast, intact LDL particles endocytosed into clathrin-coated pits follow a quite different pathway. The CE component of LDL internalized via the LDL receptor requires hydrolysis by lysosomal acidic cholesterol esterase, before the FC generated in this way can enter other cellular pools. It had not been determined if the FC of LDL internalized by the LDL receptor pathway was transported by the lysosomal or the non-lysosomal pathway. To study this point further, ³H-FC labeled LDL was equilibrated in lipoprotein-deficient medium. LDL-FC rapidly dissociated from LDL (Figure 2) and did not enter the cell. Under lipoprotein-deficient conditions, it is likely that the cell becomes totally dependent on the lysosomal pathway to hydrolyse CE to provide preformed FC to intracellular pools. In contrast in normal (lipoprotein-containing) media, most cholesterol enters the cell as FC (Table 1)

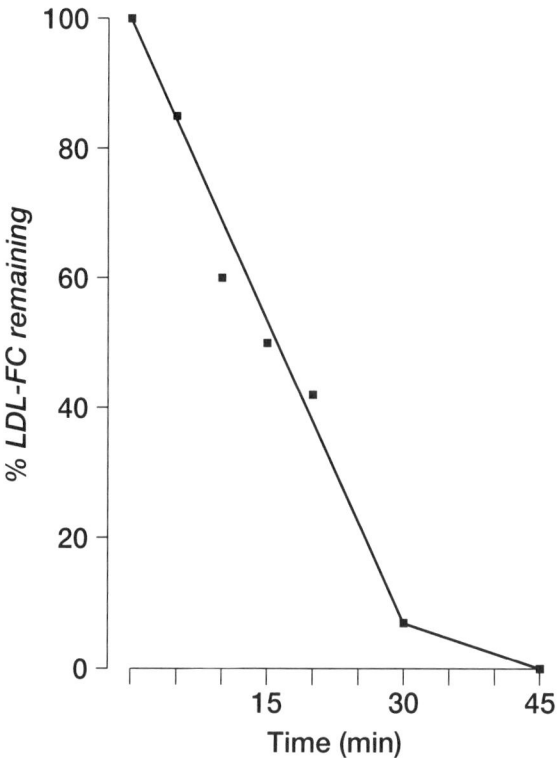

Figure 2. **Dissociation of FC from LDL in lipoprotein-deficient serum (LPDS).** LDL-FC was labeled from ^{3}H-albumin-agarose covalent complex (ref. 9) then incubated with LPDS (4 mg ml^{-1}) for the indicated period. At intervals, MgCl$_2$ and dextran sulfate were added to precipitate LDL. ^{3}H-label was determined in the supernatant fraction. LDL-FC was rapidly transferred away from LDL. There was no significant loss of CE under the same conditions (data not shown).

When the pulse-labeled cells described above were incubated at 15 °C, FC label was not returned to the cell surface but accumulated in a vesicle fraction that co-migrated on density gradients with protein markers of the *trans*-Golgi network (TGN) including mannose 6-phosphate receptor protein and caveolin, which cycles between the TGN and cell-surface caveolae. The TGN fraction was purified but not isolated in these experiments, so it remains possible that these FC-containing complexes are associated with but not integral to the TGN. When the temperature was raised to 31 °C, label associated with this fraction appeared in cell surface caveolae, and was rapidly released to HDL (22). It is possible that this pathway involves the cholesterol trafficking complex described elsewhere (23).

REGULATION OF CELLULAR FC CONTENT

Equilibration of cells in lipoprotein-deficient medium in the presence of purified LDL makes these cells dependent for FC upon a single (lysosomal) pathway, specifically, upon the hydrolysis of LDL-CE by lysosomal acidic cholesterol esterase (Figure 2). In normal plasma media, can the selective uptake of LDL-FC described above substitute for LDL receptor-mediated endocytosis? Key information as to this point can be obtained from the FC and CE content and from the rates of cholesterogenesis, from mutant cell lines equilibrated in normal (lipoprotein-containing) medium. FC balance was studied in LDL-receptor deficient cells, in Wolman disease cells deficient in acidic cholesterol esterase, and in Niemann-Pick C cells deficient in NPC-1, a protein catalyzing the transport of FC from lysosomes to the endoplasmic reticulum.

Table 3. FC and CE content and FC synthesis in normal and mutant fibroblasts

Cell line	FC $nmoles\ mg^{-1}\ cell\ protein$	CE	Sterol synthesis $pmoles\ h^{-1}\mu g^{-1}\ FC$
Wild type	97.7 ± 2.8	0.0	2.7 ± 0.3
LDL-R deficient	92.0 ± 4.7	0.0	5.8 ± 0.6
Wolman	113.2 ± 0.3 *	95.4 ± 4.3**	4.1 ± 0.2*
Niemann-Pick C	164.3 ± 5.2**	0.0	1.9 ± 0.2*

LDL-R deficient cells were the GM2000 line; Wolman disease cells were the GM1606 line; Niemann-Pick C cells were the GM3123 line. All cells were equilibrated in 10% fetal calf serum-DME medium. FC and CE mass were determined fluorimetrically with cholesterol oxidase. Sterol synthesis from 2-^{14}C-acetate was determined in the presence of 2 mM or 20 mM cold acetate. Equivalent rates were obtained, indicating complete equilibration of intracellular acetate pools. Values shown are means $\pm$ one SD of five independent determinations. *, p < 0.05 relative to normal cells; **, p< 0.01 relative to normal cells.

Selective uptake of FC from LDL, FC efflux, and growth rate were within normal limits. In each of the three mutant cell lines, transport of FC from lysosomes is markedly inhibited (10-12). In LDL-R cells, lysosomally-directed LDL cholesterol does not enter the cell. In Wolman cells, LDL-CE is delivered to the lysosomes but not hydrolyzed. In Niemann-Pick C cells, LDL-CE is hydrolysed in the lysosomes but not delivered to the regulatory pool in the endoplasmic reticulum. Nevertheless, sterol synthesis rates, even if slightly raised (LDL-R deficient and Wolman cells) remained much too small to maintain FC homeostasis (see Table 1) while in Niemann-Pick C cells, sterol synthesis rates were actually lower than in normal cells. These results contrast with data obtained with cells equilibrated with lipoprotein-deficient serum, and incubated with purified LDL. Under these conditions, sterol synthesis was significantly increased (11,12,15).

The most likely explanation of this difference is that in normal (lipoprotein-containing) medium, selective uptake of LDL-FC provides most of the FC which regulates cholesterogenesis in the endoplasmic reticulum. Consistent with this interpretation, FC synthesis in skin slices from LDL-R deficient patients was almost normal, while FC synthesis in skin slices from abetalipoproteinemic patients, who had no circulating LDL, was significantly increased (24).The increase of CE within Wolman disease fibroblasts probably represents unmetabolized LDL-CE. The increased FC content of Niemann-Pick C cells, previously found in these cells, it likely to represent mainly or exclusively a lysosomal pool, although increased FC was also observed by electron microscopy within the region of the trans-Golgi network (25). On the basis of pulse-chase experiments with normal cells, the trans-Golgi network was proposed as the 'cholesterostat' for the LDL-FC recycling pathway (22). 'Spilling over' of some lysosomal FC to the trans-Golgi network in Niemann-Pick C cells may account for the slightly lower rate of sterol synthesis (compared to normal fibroblasts) which was observed in normal medium in these cells (Table 3).

These data suggest that the selective uptake of FC from LDL, and the endocytosis of LDL via the LDL receptor, represent alternative mechanisms to supply cells with preformed FC and downregulate cholesterogenesis. In peripheral cells in normal medium, as shown in Table 1, and probably *in vivo*, selective uptake of FC from LDL predominates. In the liver, the situation may be reversed, owing to the expression of functional LDL receptors in this tissue (3). Equilibration of peripheral cells with lipoprotein-deficient serum *in vitro* 'hepaticizes' them, by removing FC from the extracellular medium (Figure 2) and as a result making these cells totally dependent on the LDL receptor to supply preformed cholesterol.

In Figure 3 below, a general model of both selective and receptor-mediated pathways is presented. It is not clear whether these pathways interact at the level of the *trans*-Golgi network. The low levels of FC synthesis (relative to FC

282

influx) make it likely that both pathways interact with similar efficiency with the FC pool in the endoplasmic reticulum which regulates FC synthesis.

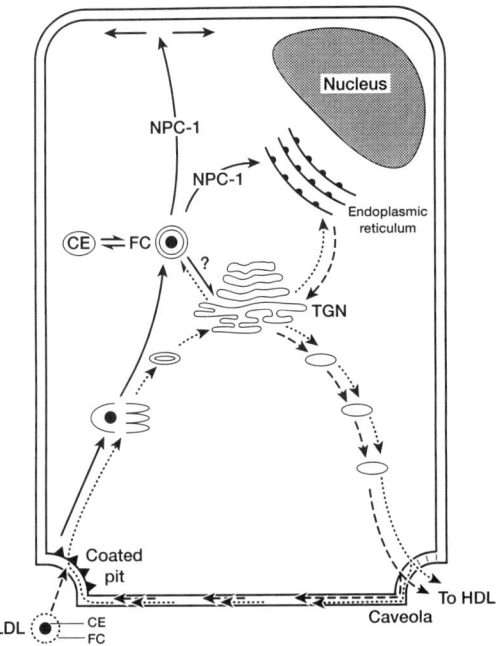

Figure 3. **Selective and endocytic pathways of intracellular cholesterol transport.**Continuous lines, cholesterol from LDL-CE; dotted arrows, FC from LDL-FC; dashed arrows, newly synthesized FC.

CAVEOLIN AND CAVEOLAE IN FREE CHOLESTEROL TRANSPORT

Caveolae are clathrin-free cell surface invaginations (60 - 80 nm) present in most peripheral cells. Caveolae are most abundant in quiescent cells including those of endothelium and vascular smooth muscle, and are reduced or absent in transformed and other rapidly dividing cell lines, and in cells utilizing FC for steroid hormone synthesis or bile acid production. Functions proposed for caveolae now include facilitation of transcytosis (particularly in endothelium)(26), the uptake of small solutes (such as folate) by receptor proteins within the caveola (27), sequestration of inactive forms of signalling molecules such as Ras (28) and the promotion of intracellular FC transport and efflux (29,30). Whatever the functions of caveolae in individual tissues, there is now strong cumulative evidence of a close relationship linking caveolae and FC. The expression of caveolae is regulated by cell FC content (31,32). The caveolae are the terminus of both newly-synthesized and LDL-derived recycling

FC (22,29-30). The expression of caveolin, the major protein of caveolae, is regulated transcriptionally by FC (31,32) via two G/C-rich promoter elements one of which binds sterol regulatory element binding protein-1 a regulator of numerous FC-dependent genes (33). When human skin fibroblasts equilibrated in 10% fetal calf serum were transferred to increased concentrations of plasma or purified LDL, caveolin gene transcription rates were upregulated, as were equilibrium caveolin mRNA levels (31,32). LDL receptors were essentially absent under these conditions. The upregulation of caveolin mRNA was blocked by inhibitors of LDL-FC selective uptake. Based on the transcriptional regulation of other FC-dependent genes, the mechanism shown in Figure 4 was proposed (34), with the difference from a previous model (33) that FC *up*regulates caveolin expression, while it *down*regulates the expression of LDL receptors and enzymes of FC synthesis. Up- and downregulation of caveolin mRNA and protein was associated with a major reduction in the expression of caveolae at the cell surface, as determined by electron microscopy (32).

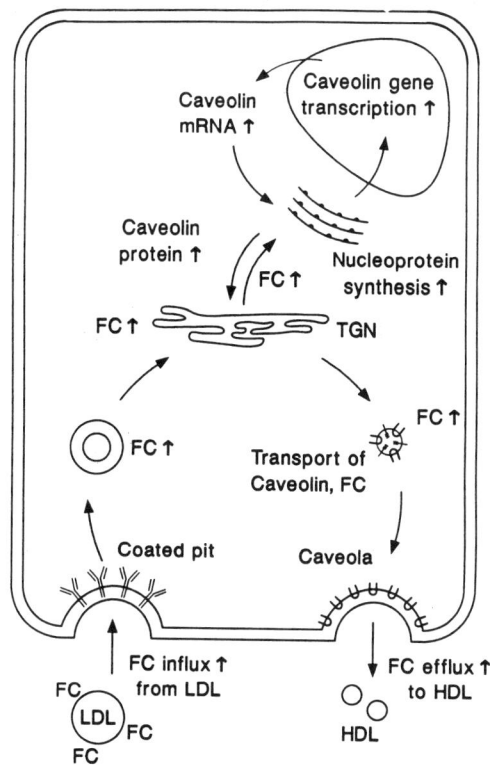

Figure 4. **Regulation of caveolin expression in response to cell FC content**. From ref. 34.

The observation that FC regulates caveolin expression represents strong additional evidence that the selective uptake of FC from LDL reflects FC mass transport, not exchange, even though at equilibrium, in normal serum medium, FC influx is balanced by an equal but opposite mass transport of FC to HDL.

The mechanism by which caveolae regulate cell FC content has been the subject of recent intense investigation. In addition to representing the major structural protein of caveolae, it was recently shown that caveolin was also a component of a transport complex including cyclophilins and heat shock proteins, which carry FC from the endoplasmic reticulum to cell surface caveolae (23). Both of these functions are likely to depend on the unusual lipid-binding properties of caveolin. Unlike plasma apolipoproteins and most membrane proteins, caveolin will only bind to phospholipid bilayers in the presence of FC (35,36). While the transport complex has not yet been fully characterized it seems likely that its FC binds to caveolin, while the additional proteins in this complex stabilize the association, possibly with the assistance of sphingolipids. At the cell surface, the FC-binding site of caveolin may be responsible for the unique kinetic properties of caveolar FC. In mammalian fibroblasts, the major part (80-85%) of total cell FC is associated with the plasma membrane, but most of this appears to be associated with the cytofacial leaflet of the bilayer (37). Filipin binds selectively to FC-rich targets in the plasma membrane. Extracellular filipin localized to the caveolae. Additionally, only caveolar FC could be modified by cholesterol oxidase in unfixed cells (38). These data, taken together, make it likely that in caveolae, unlike other microdomains of the plasma membrane, FC is localized to the exofacial leaflet of the membrane bilayer.

It has been argued on thermodynamic grounds that FC would be more effectively desorbed from FC-poor than from FC-rich domains of the plasma membrane, because of the condensing effect of FC on the acyl chains of membrane phospholipids (1). However the balance of evidence no longer supports the hypothesis that FC efflux from living cells is a consequence of simple diffusion. Rather, it appears to involve direct binding of acceptor lipoproteins (particularly HDL) to components of the plasma membrane (7,34).

Recent data from several laboratories provides the first indication of the nature of this interaction. It has been recognized for a number of years that caveolae were enriched not only in FC, but also in sphingolipids (particularly ganglioside) and in GPI-anchored proteins. One recent model assigns a 'rim' of GPI-anchored proteins around the circumference of a 'bowl' containing FC and sphingolipids (39). HDL were recently identified by electron microscopy concentrated within caveolae (40). Purification of the caveolar fraction under detergent-free conditions showed association of HDL with membrane proteins of ~80 and ~120 kDa. Pretreatment of the cells with phosphatidylinositol-specific phospholipase C, which cleaves the lipid links between GPI-anchored proteins and the cell surface, markedly inhibited HDL binding (41). These properties appear most consistent with direct binding of FC-acceptor HDL to cell surface caveolae as the terminus of the intracellular recycling pathway for LDL-derived FC. SR-B1 receptor protein (82 kDa) was identified in the caveolae of fibroblastic CHO cells transfected with the corresponding cDNA (42). It remains to be determined if SR-B1, or a related protein, represents the HDL-binding membrane protein reported above.

CONCLUSIONS

Peripheral cells normally require little FC for growth and none for metabolism, yet maintain a rapid flux of FC through the cell. Data now available suggest that most of this flux may be generated by the selective uptake of FC from LDL, and its recycling through the cell prior to efflux.

Data from LDL-receptor deficient, Wolman's Disease, and Niemann-Pick C Disease fibroblasts are consistent with the hypothesis that the selective uptake of FC from LDL, which is unaffected in these mutant cells, is the major determinant of FC homeostasis in normal (lipoprotein-containing) medium.

Caveolae are now indicated as a significant determinant of intracellular FC transport. Caveolae are the terminus for FC delivered from new synthesis and from LDL. The number of caveolae at the cell surface is determined by the level of caveolin, whose expression in turn is transcriptionally regulated by the level of FC within the cell. FC efflux is mediated by HDL binding to the exofacial surface of the caveolae, via one or more GPI-anchored proteins.

ACKNOWLEDGMENT

The authors' research cited in this chapter was supported by the National Institutes of Health through HL 57976.

REFERENCES

1. Rothblat GH., Mahlberg FH, Johnson WJ, Phillips MC: Apolipoproteins, membrane cholesterol domains, and the regulation of cholesterol efflux. J. Lipid Res. 1992; 33:1091-1097.
2. Fielding CJ, Fielding PE: Molecular physiology of reverse cholesterol transport. J. Lipid Res.1995: 36: 211-228.
3. Spady DK, Bilheimer DW, Dietschy JM: Rates of receptor -dependent and -independent low density lipoprotein uptake in the hamster. Proc. Natl. Acad. Sci. USA 1983: 80:3499-3503
4. Goldstein JL, Brown MS: Binding and degradation of low density lipoproteins by cultured human fibroblasts. J. Biol. Chem.1974; 249:5153-5162.

5. Miller NE, Weinstein DB, Steinberg D: Binding, internalization and degradation of high density lipoprotein by cultured normal human fibroblasts. J. Lipid Res. 1977; 18:438-450.

6. Acton S, Rigotti A, Landschulz KT, Xu S, Hobbs HH, Krieger M: Identification of scavenger receptor SR-B1 as a high density lipoprotein receptor. Science1996; 271:518-520.

7. Ji Y, Jian B, Wang N, Sun Y, de la Lhera Moya M, Phillips MC, Rothblat GH, Swaney JB, Tall AR: Scavenger receptor B1 promotes high density lipoprotein mediated cellular cholesterol efflux. J. Biol. Chem.1997; 272:20982-20985.

8. Slotte JP, Ekman S, Bjorkerud S: Uptake and esterification of exogenous cholesterol by low density lipoprotein receptor negative human fibroblasts in culture. Biochem. J. 1984; 222:821-824.

9. Fielding CJ, Fielding PE: Role of an N-ethylmaleimide sensitive factor in the selective cellular uptake of low density lipoprotein free cholesterol. Biochemistry 1995; 34:14237-14244.

10. Goldstein JL, Hobbs HH, Brown MS: Familial Hypercholesterolemia. In *Metabolic and Molecular Bases of Inherited Disease* (C.R. Schriver, A. Beaudet, W.S. Sly and D. Valle, eds), 1995. McGraw-Hill, NY. Pp 1981-2030.

11. Assmann G, Seedorf U: Acid Lipase Deficiency: Wolman Disease and Cholesteryl Ester Storage Disease. In *Metabolic and Molecular Bases of Inherited Disease* (C.R.

12. Carstea ED, Morris JA, Coleman KG, Loftus SK, Zhang D, Cummins C, Gu J, Rosenfeld MA, Pavan WJ, Krizman DB, Nagle J, Polymeropoulos MH, Sturley SL, Ioannou YA, Higgis ME, Comly M, Cooney A, Brown A *et al*: Niemann-Pick C disease: homology to mediators of cholesterol homeostasis. Science 1997; 277:228-231.

13. Fielding CJ, Fielding PE: Evidence for a lipoprotein carrier in human plasma catalyzing sterol efflux from cultured fibroblasts and its relationship to lecithin:cholesterol acyltransferase. Proc. Natl. Acad. Sci.USA 1981; 78:3911-3914.

14. Wu J-D, Bailey JM: 1980. Lipid metabolism in cultured cells. Studies on lipoprotein-catalyzed reverse cholesterol transport in normal and homozygous familial hypercholesterolemic skin fibroblasts. Arch. Biochem. Biophys. 1980; 202:467-473.

15. Brown MS, Goldstein JL: Suppression of 3-hydroxy-3-methylglutaryl CoA Schriver, A. Beaudet, W.S. sly and D. Valle, eds), 1995. McGraw-Hill, NY. Pp 2562-2587. reductase activity and inhibition of growth by 7-ketocholesterol. J. Biol. Chem. 1974; 249:7306-7314.

16. Pittman RC, Knecht TP, Rosenbaum MS, Taylor CA: A nonendocytotic mechanism for the selective uptake of high density lipoprotein-associated cholesteryl esters. J. Biol. Chem. 1987; 262:2443-2450.

17. Rinninger F, Pittman RC: Regulation of the selective uptake of high density lipoprotein-associated cholesteryl esters. J. Lipid Res. 1987; 28:1313-1325.

18. Dietschy JM, Turley SD, Spady DK: Role of the liver in the maintenance of cholesterol and low density lipoprotein homeostasis in different animal species, including humans. J. Lipid Res. 1993; 34:1637-1659.

19. Forte TM, Bell-Quint JJ, Cheng F: Lipoproteins of fetal and newborn calves and adult steer: a study of developmental changes. Lipids 1981; 16:240-245.

20. Bauchart D, Durand D, Laplaud PM, Forgez P, Goulinet S, Chapman MJ: Plasma lipoproteins and apolipoproteins in the preruminant calf Bos spp. Density distribution, physicochemical properties and the in vivo evaluation of the contribution of the liver to lipoprotein homeostasis. J. Lipid Res. 1989; 30:1499-1514.

21. Reichl D, Myant NB, Pflug JJ: Concentration of lipoproteins containing apolipoprotein B in human peripheral lymph. Biochim. Biophys. Acta 1977; 489: 98-105.

22. Fielding PE, Fielding CJ: Intracellular transport of low density lipoprotein derived free cholesterol begins at clathrin coated pits and terminates at cell surface caveolae. *Biochemistry 1996; 35*:14932-14938.

23. Uittenbogaard A, Ying YS, Smart EJ: Characterization of a cytosolic heat shock protein-caveolin chaperone complex. Involvement in cholesterol trafficking. J. Biol. Chem.1998; 273:6525-6532.

24. Brown MS, Brannon PG, Bohmfalk HA, Brunschede GY, Dana SE, Hegelson J, Goldstein JL: Use of mutant fibroblasts in the analysis of the regulation of cholesterol metabolism in human cells. J. Cell Physiol. 1975; 85:425-436.

25. Coxey RA, Pentchev PG, Campbell G, Blanchette-Mackie EJ: Differential accumulation of cholesterol in Golgi compartments of normal and Niemann-Pick type C fibroblasts incubated with LDL: a cytochemical freeze-fracture study. J. Lipid Res. 1993; 34:1165-1176.

26. Milici AJ, Watrous NE, Stukenbrok H, Palade GE: Transcytosis of albumin in capillary endothelium. J. Cell Biol. 1987; 105:2603-2612.

27. Anderson RGW, Kamen BA, Rothberg KG, Lacey SW: Potocytosis: sequestration and transport of small molecules by caveolae. Science 1992; 255:410-411.

28. Okamoto T, Schlegel A, Scherer PE, Lisanti MP: Caveolins, a family of scaffolding proteins for organizing "preassembled signaling complexes" at the plasma membrane. J. Biol. Chem. 1998; 273:5419-5422.

29. Fielding PE, Fielding CJ: Plasma membrane caveolae mediate the efflux of cellular free cholesterol. Biochemistry 1995; 34:14288-14292.

30. Smart EJ, Ying YS, Donzell WC, Anderson RGW: A role for caveolin in transport of cholesterol from endoplasmic reticulum to plasma membrane. J. Biol. Chem. 1996; 271:29427-29435.

31. Fielding CJ, Bist A, Fielding PE: Caveolin mRNA levels are upregulated by free cholesterol and down-regulated by oxysterols in fibroblast monolayers. Proc. Natl. Acad. Sci. USA 1997; 94:3753-3758.

32. Hailstones D, Sleer LS, Parton RG, Stanley KK: Regulation of caveolin and caveolae by cholesterol in MDCK cells. J. Lipid Res. 1998; 39:369-379.

33. Goldstein JL, Brown MS: Regulation of the mevalonate pathway. Nature 1990; 343:425-430.

34. Fielding CJ, Fielding PE: 1997. Intracellular cholesterol transport. J. Lipid Res. 1997; 38:1503-1521.

35. Li S, Song KS, Lisanti MP: Expression and characterization of recombinant caveolin. Purification by polyhistidine tagging and cholesterol-dependent incorporation into defined lipid membranes. J. Biol. Chem. 1995; 271:568-573.

36. Murata K, Peranen J, Schreiner R, Wieland F, Kurzchalia TV, Simons K: 1995. VIP21/caveolin is a cholesterol binding protein. Proc. Natl. Acad. Sci. USA 1995; 92:10339-10343.

37. Schroeder F, Nemecz G: Transmembrane cholesterol distribution. In *Advances in Cholesterol Research* (Esfahani M, Swaney J, eds). Telford Press, Caldwell N.J., 1990. Pp 47-88.

38. Smart EJ, Ying YS, Conrad PA, Anderson RGW: Caveolin moves from caveolae to the Golgi apparatus in response to cholesterol oxidation. J. Cell Biol. 1994; 133:1265-1276.

39. Schnitzer JE, McIntosh DP, Dvorak AM, Liu J, Oh P: Separation of caveolae from associated microdomains of GPI-anchored proteins. Science 1995; 269:1435-1439.

40. Lestavel S, Briand O, Nion S, Torpier G, Copin C, Fruchart JC, Clavey V: Caveolae and GPI-anchored proteins - a specific binding membrane domain for high density lipoproteins. Atherosclerosis 1997;134(S1):369.

41. Nion S, Briand O, Lestavel S, Torpier G, Nazih F, Delbart C, Fruchart JC, Clavey V: High density lipoprotein subfraction-3 interacts with GPI-anchored proteins. Biochem. J.1997; 328:415-423.

42. Babitt J,Trigatti B, Rigotti A, Smart EJ, Anderson RGW, Xu S, Krieger M: 1997. Murine SR-B1, a high density lipoprotein receptor that mediates selective lipid uptake, is N-glycosylated and fatty acylated and colocalizes with plasma membrane caveolae. J. Biol. Chem. 1997; 272:13242-13249.

SUMMARY AND FUTURE PERSPECTIVES

Ta-Yuan Chang

Department of Biochemistry, Dartmouth Medical School, Hanover, NH 03755;
e-mail: Ta.Yuan.Chang@Dartmouth.Edu

The process of intracellular cholesterol trafficking is involved in many physiological events, including lipoprotein synthesis and secretion in hepatocytes and intestinal enterocytes, steroidogenesis in steroid hormone-producing cells, cholesterol accumulation in macrophages, synthesis and maintenance of neuronal cell membranes. It is also involved in embryogenesis and development. Abnormalities in these events often lead to diseases in animals and humans. As editors of this book, Dale Freeman and I are grateful to have had many leading scientists contributing chapters and furthering the advancement of this field. It is clear that this area of research will continue to grow during the next decade, and I wish to recommend the following topics as key areas for future investigations:

A. The cellular and molecular mechanisms that govern the fates of free cholesterol and cholesteryl esters derived from various lipoproteins (LDL, HDL, atherogenic lipoproteins, etc.) in various cell types.

This is one of the most active areas of research in this field, and it has been covered from various angles by almost every chapter in this book. Cholesterol derived from various lipoproteins eventually becomes a major lipid component in the plasma membrane, or moves to the ER for storage as cholesteryl esters. Several trafficking pathways have been identified, and several protein molecules have been implicated in these pathways. To gain further knowledge in this area, a challenging but potentially fruitful approach may be to continue to develop cell mutants that are defective in various discrete steps in these processes; phenotypic characterization of these mutants can demonstrate the biological relevance of intermediate steps in the overall pathway. In addition, a well-characterized mutant can then serve as an appropriate host for cloning and indentifying the gene encoding the factor involved in each step. Another challenging but valuable approach may be to develop biochemical assays to demonstrate each discrete step *in vitro*, such that the trafficking process can be studied in cell-free systems.

B. The cellular and molecular mechanisms that direct the roles of newly synthesized cholesterol or its precursors for plasma membrane synthesis, for efflux, for storage as cytoplasmic cholesteryl esters, and for lipoprotein synthesis.

The rapid movement of newly synthesized cholesterol from the ER to the plasma membrane was first demonstrated by Simoni and colleagues many years ago (1). This is an important area that deserves much attention. The status of this field, which focuses on describing the mechanistic and regulatory aspects, has been reviewed in several chapters (by Smart and van der Westhuyzen, by Field, by Lange, by Phillips and Johnson, and by Chang, Chang, and Lee). In addition, the chapter by Phillips and Johnson describes the fates of certain biosynthetic precursor sterols in extrahepatic cells. These authors indicate that the undesirable accumulation of certain precursor sterols in various tissues can cause abnormalities in embryogenesis and development in animals and humans.

C. The cellular and molecular mechanisms that regulate the cholesterol-cholesteryl ester cycle.

Brown, Goldstein, and colleagues in 1980 demonstrated the existence of this cycle in macrophages (2). In addition, an important paper by Glick and colleagues describes certain unique aspects of the cycle (3). The status of this area is briefly reviewed in the chapter by Chang, Chang, and Lee. The molecular nature of this cycle still needs to be identified.

D. The role of plasma membranes in participating and controlling various cholesterol trafficking events, including cholesterol internalization, cholesterol efflux from caveolae, etc.

This is also a burgeoning area of research, and it has been covered from various viewpoints in several chapters (by Lange, by Tabas, by Liscum, by Fielding, Bist, and Fielding, by Choi and Freeman, by Field, by Phillips and Johnson, and by Schroeder and colleagues). Several investigators have implied that a certain cholesterol sensing mechanism must exist in the plasma membrane to mediate (and perhaps to control) the cholesterol trafficking events. Could this sensing mechanism involve the multiple drug resistance protein (MDR), or caveolae, or a receptor that recognizes the NPC-1 protein, or a specific lipid microdomain within the plasma membrane? To identify the proposed sensing mechanism, as well as the molecular nature of the putative "cholesterol sensor" in the plasma membrane, further work is needed.

E. The roles of NPC-1 protein in intracellular cholesterol trafficking.

The gene encoding the NPC-1 protein was discovered by Peter Pentchev and his colleagues in 1997. This is one of the most important genes involved in intracellular cholesterol trafficking. The research in this area is progressing at a rapid rate. Three chapters (by Blanchette-Mackie and Pentchev, by Liscum, and by Neufeld) describe potential roles of the NPC-1 protein in the trafficking process(es).

It is worthwhile to note that the chapter by Blanchette-Mackie and Pentchev contains several original figures and tables that have never before been published and are very valuable. Also, as briefly mentioned in these chapters, it has been speculated that the major cause of death in mice (and probably humans) with a defective NPC-1 gene is neuronal cell death in the brain, implying a close yet unknown relationship between intracellular cholesterol/lipid trafficking and various neuronal functions in the brain. More than half of the dry weight in the brain is composed of lipids. At present, the process of lipid trafficking in the central nervous system remains largely unknown and needs to be explored.

F. The cholesterol trafficking roles of various cholesterol binding proteins, including caveolin, multiple drug resistant protein (MDR), sterol carrier protein-2 (SCP-2), and steroidogenic acute regulatory protein (StAR).

There is growing evidence that each of these proteins is involved in various stages of intracellular cholesterol trafficking. Further evidence, particularly at the genetic level, is needed to sort out the functional role for each of these proteins in a tissue-specific manner. In this regard, an elegant example is presented in the chapter by Stocco and Strauss, who describe the identification and characterization of the novel protein StAR. This is a cycloheximide-sensitive, labile protein that plays a critical role in intramitochondrial cholesterol transfer in steroidogenic cells; mutations in the StAR gene in humans have been shown to cause a potentially lethal disease, congenital lipoid adrenal hyperplasia. Two different views on the roles of SCP-2 are represented in this book: the chapters by Scallen and colleagues and by Schroeder and colleagues argue strongly for a prominent role of SCP-2 in cholesterol trafficking, while the chapter by Seedorf describes an unexpected finding implying that the main physiological function of SCP-2 may involve lipid catabolism within the peroxisomes. It is noteworthy that the chapter by Scallen and colleagues contains many valuable results that have not previously been published.

G. The application of mammalian cell genetics, yeast genetics, and mouse genetics to dissect intracellular sterol trafficking processes.

Somatic cell genetics has played an important role in helping to understand intracellular cholesterol metabolism in mammalian cells. The chapter by Liscum provides an excellent example. For additional information, the readers are referred to an earlier review that summarizes the characteristics of all available CHO cell mutants defective in intracellular cholesterol metabolism (4). Yeast genetics has played a powerful role in elucidating the mechanisms involved in intracellular protein trafficking. It is predicted that this approach, in combination with other approaches, will contribute significantly to the dissection of various intracellular sterol trafficking events. The chapter by Sturley and colleagues describes the status of this approach. The power of mouse genetics in determining the physiological function of a certain specific gene product in intact animals is demonstrated by the chapter by Farese and colleagues (on ACAT-1 gene knockout mice) and the chapter by Seedorf (on SCP-2 knockout mice).

H. The functional roles of ACAT-1 and ACAT-2 in mammals.

The ACAT-1 gene was discovered in 1993. The chapters by Farese and colleagues and by Sturley and colleagues describe a series of events that led to the discovery of ACAT-2 in their own laboratories. In addition, the chapter by Tabas highlighted many interesting features of ACAT-1 protein in the macrophage cells. Based on results available from different laboratories, it is tempting to speculate that the physiological roles of ACAT-1 and ACAT-2 in different tissues may be different in mice and humans. In the future, the functional significance of ACAT-1 and ACAT-2, as well as their mode(s) of regulation by sterol and/or by other regulatory signals in various tissues and cells, needs to be clarified.

I. Molecular actions of cholesterol on various cholesterol-sensing proteins including ACAT-1, ACAT-2, sterol regulatory element cleavage activating protein (SCAP), HMG-CoA reductase, NPC-1, etc.

These cholesterol-sensing proteins are known to be integral membrane proteins located in the ER or in other internal membranes. This topic has been discussed in several chapters (by Chang, Chang, and Lee, by Blanchette-Mackie and Pentchev, by Liscum, and by Neufeld). In the future, the mode(s) of interaction between cholesterol and each of these putative sterol sensing proteins need to be demonstrated *in vitro*, and should be analyzed by various biochemical and biophysical techniques. The chapter by Schroeder and colleagues provides an example of using various physicochemical methods for studying the interactions between soluble proteins that bind sterol or fatty acids *in vitro*. This chapter also describes the use of a naturally occuring fluorescent sterol (dehydroergosterol) for fluorescence imaging in intact mammalian cells.

The process of intracellular cholesterol trafficking is indeed complex. To dissect these problems, collective efforts from cell biologists, biochemists, geneticists, biophysicists, structural biologists, and neurobiologists are needed. Hopefully, ten to fifteen years from now, certain basic principles may emerge that can explain the bulk of the intracellular cholesterol trafficking processes at the cellular and molecular levels. The basic knowledge gained in this process will facilitate and improve the treatment of major diseases that include atherosclerosis, endocrine abnormalities, and neurodegenerative diseases.

REFERENCES

1. DeGrella RF and Simoni RD. 1982. *J. Biol. Chem.* 257: 14256-14262
2. Brown MS, Ho YK and Goldstein JL. 1980. *J. Biol. Chem.* 255: 9344-9352
3. Klansek JJ, Warner GJ, Johnson WJ and Glick JM. 1996. *J. Biol. Chem.* 271: 4923-4929
4. Chang TY, Hasan MT, Chin J, Chang CCY, Spillane DM and Chen J. 1997. *Curr. Opin. Lipidol.* 8:65-71

INDEX

Acyl CoA: Cholesterol acyl transferase (ACAT)
    ACAT inhibitors, 38
    ACAT 1, 3,29
    ACAT 2, 3,29
    ACAT related yeast proteins ARE_1, ARE_2, 44
    allosteric regulation, 1,188
Amphotericin B selection of mutants, 79
Atherosclerosis,183

bile acids in SCP_2 knockout mice, 247
Brefeldin A, 96, 119,134

CaCo-2 cells, 123
Caveolae, 81,138,156,253,282
    associated proteins,258
Caveolin, 81,138,156,220,253,282
Chaperone protein, 255
Cholesterol
    distribution in cells, 16,111
    precursors, 147
    sensors, 15
    transport defective mutants
    Ced-1 Ced-2, CT-60, 60,77,78
Cholesterol Oxidase, 16,83, 111, 135
Cholesteryl ester cycle, 1,183
Cyclodextrin, 18

DAMP as an acidic vesicle marker, 119
Dehydroergosterol, 227

Endoplasmic reticulum in cholesterol transport, 66, 93

Fatty-acid binding protein (L-FABP), 208, 214
Filipin as a cholesterol probe, 16,55
Foam cells

Golgi apparatus
    disruption and effect on cholesterol esterification and steroidogenesis, 54,85,96,119,154
    Niemann-Pick Type C disease, 57,93
    trans-Golgi network, 55, 93
Glyerosphingolipids (GSL's), 103

High density lipoproteins
    binding and cholesterol efflux, 147,253,273
    receptor B1 (SR-B1), 255,273
    selective cellular uptake of sterol, 100,273
Hydroxymethylglutaryl Coenzyme A Reductase HMG -CoA reductase, 9,95,133

Inhibitors of cholesterol transport
    class 1 and class 2 agents, 23
    drugs that mimic NPC-1 lesion, 24,60, 93
  drugs that inhibit steroidogenesis, 112, 117,171
  hydrophobic amines, 63,84,117

Knockout mice
    $ACAT_1$, 29
    SCP_2 , 235

Lipoid adrenal hyperplasia, 173
Lipoproteins
    source of lipoprotein sterols, 123
    transport within cells, 55,77,93,112
Low Density Lipoprotein
    modification, 185,172
LDL, selective uptake of lipid, 274
Lysosomal cholesterol transport
  inhibitors, 55,117
Lysosomotrophic agents, 63,117

MA-10 cells, 110,
Multiple drug-resistance protein (MDR),127,157
Mitochondria, 170
MLN 64,175

NBD-cholesterol, 120,226
Niemann-Pick Type C disease (NPC)
    NPC-1, 10,47,88,93
    NPC-2,93
    NCR-1 ,NPC analogue in yeast, 47